THE STUDENT CITY

The European Institute for Comparative Urban Research, EURICUR, was founded in 1988 and has its seat with Erasmus University Rotterdam. EURICUR is the heart and pulse of an extensive network of European cities and universities. EURICUR's principal objective is to stimulate fundamental international comparative research into matters that are of interest to cities. To that end, EURICUR coordinates, initiates and carries out studies of subjects of strategic value for urban management today and in the future. Through its network EURICUR has privileged access to crucial information regarding urban development in Europe and North America and to key persons at all levels, working in different public and private organizations active in metropolitan areas. EURICUR closely cooperates with the Eurocities Association, representing more than 100 large European cities.

As a scientific institution, one of EURICUR's core activities is to respond to the increasing need for information that broadens and deepens the insight into the complex process of urban development, among others by disseminating the results of its investigations by international book publications. These publications are especially valuable for city governments, supranational, national and regional authorities, chambers of commerce, real estate developers and investors, academics and students, and others with an interest in urban affairs.

EURICUR website: http://www.euricur.nl

This book is one of a series to be published by Ashgate under the auspices of EURICUR, the European Institute for Comparative Urban Research, Erasmus University Rotterdam. Titles in the series are:

Metropolitan Organising Capacity
Leo van den Berg, Erik Braun and Jan van der Meer

National Urban Policies in the European Union
Leo van den Berg, Erik Braun and Jan van der Meer

The European High-Speed Train and Urban Development
Leo van den Berg and Peter Pol

Growth Clusters in European Metropolitan Cities
Leo van den Berg, Erik Braun and Willem van Winden

Information and Communications Technology as Potential Catalyst for Sustainable Urban Development
Leo van den Berg and Willem van Winden

Sports and City Marketing in European Cities
Leo van den Berg, Erik Braun and Alexander H.J. Otgaar

Social Challenges and Organising Capacity in Cities
Leo van den Berg, Jan van der Meer and Peter M.J. Pol

City and Enterprise
Leo van den Berg, Erik Braun and Alexander H.J. Otgaar

The Student City

Strategic Planning for Student Communities in EU Cities

LEO VAN DEN BERG
and
ANTONIO RUSSO

Euricur
EUROPEAN INSTITUTE FOR COMPARATIVE URBAN RESEARCH

European Institute for Comparative Urban Research
Erasmus University Rotterdam
The Netherlands
www.euricur.nl

ASHGATE

Published by
Ashgate Publishing Limited
Gower House
Croft Road
Aldershot
Hants GU11 3HR
England

Ashgate Publishing Company
Suite 420
101 Cherry Street
Burlington VT 05401-4405
USA

Ashgate website: http://www.ashgate.com

British Library Cataloguing in Publication Data
Berg, Leo van den, 1948-
The student city : strategic planning for student communities in EU cities. - (EURICUR series)
1. Campus planning - European Union countries - Case studies
2. College facilities - European Union countries - Case studies 3. City planning - European Union countries - Case studies
I. Title II. Russo, Antonio
711.5'7'094

Library of Congress Cataloging-in-Publication Data
Berg, Leo van den.
The student city : strategic planning for student communities in EU cities / by Leo van den Berg and Antonio Russo.
p. cm. -- (EURICUR series)
Includes bibliographical references.
ISBN 0-7546-4140-6
1. Campus planning--European Union countries--Case studies. 2. College facilities--European Union countries--Planning--Case studies. 3. City planning--European Union countries--Case studies. I. Russo, Antonio (Antonio Paolo) II. Title. III. Series.

LB3223.5.E85B47 2004
711.57'094--dc22 2004010341

ISBN 0 7546 4140 6

Printed and bound in Great Britain by Antony Rowe Ltd, Chippenham, Wiltshire.

Contents

List of Figures

List of Tables

Preface

Student communities are without doubt a strategic resource for urban development. Students are the citizens and the highly-skilled working class of tomorrow. They keep cities lively and diverse. They are consumers of cultural and recreational facilities. They have a distinct expenditure pattern that in some cases is crucial to support the economy of specific areas or neighbourhoods. Increased student mobility at the international level is a major vector of socioeconomic cohesion between regions of Europe.

However, European cities are seldom ready to recognise and manage this potential – information on student communities is generally scarce: students are an 'invisible population', being given little space in local policy. Cooperation between education institutions and city planners is often lacking. Cities are passive towards their universities' foreign relations. Attention is paid to student settlements only on those occasions when they become a source of problems or costs for the local community.

Nevertheless, the importance of human capital as a determinant of cities' competitiveness demands *proactive city policies* targeting this community of city users. National or regional education programmes often neglect the 'urban' dimension of the issue, forgetting that human capital is highly mobile, and that it needs to be attracted and managed locally.

The present study contributes to the elaboration of a framework for strategic action aiming at the integration and valorisation of student communities for urban development. This final report is organised as follows.

Chapter 1 presents the main issues under investigation and their relevance for urban sustainability, and proceeds by setting out the methodological and practical lines of EURICUR's investigation.

Chapters 2 to 10 are dedicated to the nine cities that participated in this project as case studies: in alphabetical order, Birmingham, Eindhoven, Helsinki, Lille, Lyon, Munich, Rotterdam, Utrecht and Venice. Each city is presented through its main features, highlighting the importance and role of education and research as catalysts of urban development. There follows a description of the role and strategies of the main actors in the field of higher education, such as the institutes, the private sector, the community organisations and the students themselves, through the elaboration of available data and other information collected during field trips by the EURICUR research team. The links and the consistency between the strategies of these actors is assessed next, and on the basis of this, we highlight the main problems existing in the city-academy relationship, suggesting a number of recommendations to improve it and progress towards a genuine, sustainable 'student-friendly' city.

Chapter 11 concludes. Through a comparative analysis of the cities investigated, we identify typical profiles of student cities, as well as key elements determining the sustainability of the city-university relationship and cases of best practice in this relationship that can be transferred among the cities in this study.

The project has been carried out by a team of EURICUR researchers based at Erasmus University Rotterdam, Department of Regional, Port and Transport Economics. The team includes the authors of the present book: Professor van den Berg, director of EURICUR, and Dr. Russo, research assistant at the Erasmus University, as well as master graduate Mariangela Lavanga, who contributed to the elaboration and writing of the case studies of Helsinki, Utrecht, Birmigham, Lyon, Lille and Eindhoven. The authors also acknowledge Paulus Woets for contributing to the case studies of Rotterdam and Munich.

EURICUR takes this opportunity to acknowledge the members of the steering committee in this project, representatives of the partner cities who have helped us to identify the most relevant issues and to organise the field visits in each city. This report also reflects the challenging discussion of the results at different stages of the research with Mr Taylor of the City of Birmingham; Mr Pullen, Mr Tindemans and Mr Czech of the City of Eindhoven; Ms Manninen, Mr Cantell and Ms Vaattovaara of Helsinki Urban Facts; Mr Baert of the Lille Metropole Urban Planning Agency; Mr Tesse of the Chamber of Commerce of Lyon; Mr Schussman, Mr Heidebach and Ms Schweigard of the City of Munich; Mr van't Verlaat of the Rotterdam Development Corporation; Mr Kulik and Mr de Viet of the City of Utrecht; and Mr Del Mercato of the City of Venice. We also thank the several discussion partners in each city, a complete list of whom is the Annex.

Leo van den Berg
Antonio Russo
Rotterdam, October 2003

Chapter 1

Background Issues and Research Methodology

1.1 Student Cities in Europe: Models and Challenges

Universities and higher research institutions have relevant regional impacts. These have been the object of thorough academic research (see Florax 1992 and Phelps 1998 for extensive literature reviews). Today it is accepted that the formation of human capital and the production of scientific knowledge are strong determinants of regional growth. Moreover, it is typical of European cities that universities are strongly associated with urban communities (Hardy, in Elliott et al., 1996: 11; Hall, 1997) and contribute to determining their cultural status (Chatterton, 2000). However, seldom do such economic and cultural links between universities and cities translate into strategic common management of all the areas of contact between them.

New challenges for a serious effort in this sense come from the loss of relevance of nation-state identities and the increasing importance of cities as nodes of the global economy. In the emerging socioeconomic environment, the opportunities for an increased synergy between centres of creation of knowledge and local communities increase. Local universities and other higher education and research centres become fundamental elements of the competitiveness of cities and regions. However, this dependence is not monodirectional. Universities become increasingly embedded in highly specialised regional contexts, and – in an era of cutbacks to central government transfers – research comes to depend on private sponsors, donors and commercial partners, which are typically local actors (van den Berg, Braun and Otgaar, 2003). Moreover, the city – as the crucial node of contemporary economic and social development – is increasingly a source of 'problems', a constant stimulus to intellectual production, that is one of the university's main missions.

In any case, the relationship between universities and host communities is not limited to the institutional sphere. Universities are made up of people: students, academic staff and workers, a mass of city users who add (only partially overlapping) to local residents, and to other categories such as businesspeople, foreign workers and tourists. Students may be described as a *population*, with distinct patterns of organisation. They make use of the city's resources, and interact with other urban dwellers, in particular with the residents. The coexistence and interaction of such different populations, which according to Martinotti (1997) is one of the driving forces of contemporary urbanisation, is not without problems. At the same time, it is a huge opportunity that cities can turn into a competitive advantage, to the extent to which the human capital formed contributes *persistently* to the city's performance and quality of life.

Almost all major cities, regional capitals and metropolises of Europe host one or more large universities and other higher education institutions (HEI), which attract a number of students from regional hinterlands and from the whole country. In some cases, the 'catchment area' of large universities extends to foreign nations. Apart from students only temporarily staying in the city (e.g. exchange students), this is the case for communities from neighbouring countries or from specific areas linked to the host countries for historical and political reasons.

However, some cities are perceived or promote themselves as more like 'university cities' than others. There are a number of reasons. First, a large student population with respect to the resident mass (e.g. Heidelberg, Tilburg, Lund, Padova) clearly forges the city's identity. Secondly, it increases the prestige of their education institutions, which are able to attract students from all over the country and from abroad (e.g. London, Köln, Coimbra, Pisa). Finally, there is the specific sociocultural profile assumed by students' communities in the urban environment (e.g. Berlin, Bologna, Manchester).

The above-mentioned cities, among others, have learnt with time and experience to accommodate and and profit from this segment of city users, even though some issues and problems remain unsolved. For instance, in these cities the presence of students is often physically separated from the host communities through the organisation of dedicated facilities like one or several campus, dorms, cultural facilities and meeting places, etc. At the same time, opportunities for encounters with the host community remain numerous and are to some extent 'institutionalised' and publicly supported. This might explain the success of small universities and the high performance of small university cities. The university becomes an integral part, and often a leading actor, in networks where small and medium enterprises are able to capture the knowledge spillover generated by research and education activities. However, even in these 'success stories' there are problems as far as the long-term sustainability of such interrelations is concerned; the more so in big European cities where the economic impact of the academic settlements is more tenuous and the socioeconomic links between community and students is far more complex.

Though their student population is by no means irrelevant, many European cities are hardly aware of the potential of these communities for urban development, and of the necessity of strategic action to take full advantage of it. Most of the cities participating to this research belong to this latter category. Students are often perceived by the resident population as a nuisance with respect to the ordinary 'way of life', and because these settlements influence the market price of commodities such as housing, recreational facilities and consumption goods. The almost spontaneous development (in some cases a real explosion) of student communities, a consequence of the contemporary shift from a 'élite' model of higher education towards the 'mass university', has largely not been accompanied by the provision of adequate infrastructure and facilities for a smooth encounter of the two communities. In the absence of dedicated physical facilities, appropriate meeting places and cultural mediators, the occasions for confrontation and possibly conflict between the two communities are numerous (Parsons and Davis, 1971). Such conflicts, it is argued, hamper the possibility of establishing the appropriate synergies and diminish the capacity of cities to compete with other, better equipped and better organised regions.

Florax (1992: 3) notes that proactive behaviour of cities towards students' communities might have ancient origins: some Dutch provinces in the sixteenth century already granted fiscal benefits to members of students' communities as a means of attracting them to these areas. More recently, national policies for the decentralisation of higher education, implemented in the 1960s and 1970s with the aim of supporting weak regional economies, have had mixed results and have to face the recent trends towards university restructuring and government cutbacks. The 'regionalisation' of higher education through an array of strategies (the birth of new 'decentralised' universities; the decentralisation of special study curricula by existing universities; the diversification of university locations; etc.) has been pursued in other contexts (see Savino 1998 for an extensive account of Italy's case). However, this appears more a centralised effort by states to deal with regional disparities, matched by local institutions in search of legitimacy, than an integral and well-structured plan to favour community development.[1] Thus we assist in the failure of 'campus planning' both in Italy and France, while university location has had only a limited impact on local socioeconomic problems. *The challenge for large urban areas in Europe, like the ones included in this research, is to find the right balance between universities' needs and policies targeting the resident population.* The importance of improving the university-city relationship goes far beyond regional impacts and invests the wide field of urban management. It is evident that the dialogue between universities and cities must become truly bi-directional. By understanding the areas of conflict and organising the appropriate policy instruments and processes to solve them, European cities may finally fully exploit their precious education assets and define their identity as preferred locations for highly-skilled and sophisticated young citizens.

1.2 The Impacts of Student Communities in Cities

1.2.1 Students and Economic Development

Education and research activities produce direct economic effects (additional jobs, revenues, services). These can be quantified rather easily; one important caveat is that the scale of observation influences the results. In practice, if the university functions like an 'export base' the economic benefits are tangible; less so if the income generated by the university tends to 'leak away'. At the same time, local effects are more pronounced when the university 'dominates' competing sectors, representing a real 'additional' source of development. This is why economic impacts tend to be larger when universities are located in *very big* (the metropolitan university, e.g. the Paris University system) and *very small* urban areas lacking other important economic specialisation (the *campus* city, e.g. Coimbra, Urbino, Durham), whereas they tend to be more ambiguous in middle-size cities.

However, such effects are not central to the present research, which instead focuses on intangible, less easy to quantify effects. To this respect, the impact on human capital availability, generally referred to as *knowledge spillover*, has primary importance. According to district theorists such as Jane Jacobs (and *contra* to, among others, Porter), the generalistic and wide-ranging knowledge generated in publicly-subsidised HEIs has a wider impact on local growth than specialised knowledge produced in the private-sector R&D departments. In fact, the former is clearly linked to industrial diversification, which is the winning strategy in the 'new economy' environment. To facilitate knowledge dissemination, initiatives such as science parks, knowledge transfer organisations, and centres of excellence are organised.[2] The direct encounter between firms and knowledge centres occurs also via training programmes such as lifelong education and in-company courses. In turn, an innovative and stimulating working environment is supposed to bring forward those urban amenities that create further attractiveness, in a 'virtuous cycle' of knowledge-driven development. Cities like Cambridge, Bologna, Dortmund and Lund are good examples of places that benefit from the intangible elements linked with the presence of important education and research centres. Universities can also be considered counter-cyclical organisations: whenever there is a recession, a higher proportion of people goes to the university in search of higher skills to make them more employable in a more difficult job market. Instead, when the economy is booming, people tend to access the job market as soon as possible. This makes universities important elements of economic regeneration strategies for regions in transition (Noyelle and Stanback, 1983).

1.2.2 Students and Urban Infrastructure

Students consume cultural and recreational products and, in many instances, they are producers themselves. Facilities like sport halls, university theatres, music ensembles, multimedia workshops and cinemas, depend crucially on the students' demand. It is nowadays widely recognised that the provision of such services contributes to the improvement of the quality of life, and hence to the attractiveness of the city as a working and living location for businesses and workers.

City managers may base regeneration strategies on the demand for peculiar goods and services (especially food, sport and culture) when they decide to 'open up' specific neighbourhoods or areas of the city to student communities. In this way they wish to create 'cultural quarters' that are an increasingly important element of the post-modern urban landscape. However, they should be aware of the risks of gentrification and they should take into account that most of the time, social preferences are highly dynamic, that is, convenient accommodation for a student in one year might be less so in the following year. Moreover, students pay little towards such public infrastructure, as they are not citizens and they are only asked to pay local taxes in a few contexts. Therefore, the burden of this infrastructure is borne by citizens. It is then desirable that the latter perceive the benefits of their expenses, for instance, being able to enjoy the new facilities at reduced prices on special occasions.

Such infrastructure becomes a central asset in the cities' tourism policy, as it consists of high-class facilities and cultural services clearly targeting an international and educated audience. Often, student-managed facilities provide shows and events that are of strategic importance to position the city as a 'niche' destination for avant-garde culture (e.g. Berlin, Bologna, Amsterdam). In general, the informal, *bohemien* and extroverted climate of university cities and neighbourhoods adds elements of attractiveness to the urban environment. However, it is important that such assets are not located too far from tourist areas and that information and access are guaranteed to people who may not be familiar with the city's daily life and language.

At the same time, universities and cities compete for land use; in various cases this has led to the relocation of historically 'urban' universities to suburban locations (Hall, 1999: 301 quotes several examples), with mixed results (Vassal, 1987). More recently, though, the orientation of cities towards university settlements has changed as policy makers acknowledge that campuses and other university facilities represent an enormous stock of capital investment, most of which is used for less than half of the year (Hardy, op. cit.). Universities are therefore looking for ways to increase the outward use of these assets, generating extra revenue and benefiting the community. In the European urban context, where universities have expanded in already congested city centres, subtracting resources and functionality from the resident population and creating objective occasions for political confrontation.

1.2.3 Students and the Housing Market

Students have a crucial impact on the housing market. Their spending patterns and standards may differ greatly from those of the host community, with lower protection and a faster life cycle in respect to those of resident tenants. In the end, these characteristics are likely to push up the price levels and increase the rigidity of the market. However, the housing market is highly segmented and it is also possible that the residents' and the students' housing markets hardly overlap. This is shown, for instance, in a research conducted in Venice (Bonifacio, Vassallo and Zanus Fortes, 1995). Such a study allowed the identification of a 'false problem': for years, the university had been wrongly blamed as the culprit for the exodus of inhabitants from the most central areas of the city. Consequently, local institutions were under continuous political pressure to organise a settlement policy for the university's facilities. On the other hand, similar studies also show that the housing preferences of the transient students are not satisfied in the 'normal' market and require dedicated structures, like low-price units close to the university areas, of which there might be a shortage. Moreover, the opposite effect is also true. When an area is regenerated through a university settlement policy, the value of housing in that area increases to the benefit of house owners. This is a tool that may be used when costs are originally too high for restoration to be started, and neighbourhoods are left to decay. Costs and benefits of gentrification policies have to be exactly assessed in advance so as not to create social tensions and division.

In this respect, the 'campus formula', popular in Anglo-Saxon countries and in France might be a proper solution, even if in this case the occasions of fruitful encounter between students and host communities are limited, and the direct economic benefits from students' settlements are reduced (Vassal 1987). In some cases, as in the Bordeaux campus experiment quoted in Dubet and Sembel (1994), the creation of 'cities of studies' has produced unattractive peripheral enclosures rather than replicating the integral, comprehensive 'academic community' that is found in American campuses. The plan Université 2000 launched by culture minister Jack Lang (VVAA, 2001: 49–50) calls for a 'new university urbanism' in France, to be achieved either through a physical return of university in city centres, or through planning solutions which break the isolation of peripheral locations (recreation corridors, new, open architecture, better transport infrastructure, etc.). This apparent contradiction is solved through planning solutions that favour mixed functions and multiple users. However, as suggested by Pouyet (VVAA, 2001: 55) universities should not be asked to provide solutions to areas in crisis just because they are there. It will be necessary to establish case by case what the best strategy is, depending on the local socioeconomic context.

1.2.4 Students and Globalisation

Students, especially international ones, represent a *wedge* with respect to local cultures. They bring about social innovation and cultural change. For instance, as their life follows a different time schedule (more flexible and 'flowing' than that of the resident working class or other social groups), they demand a different organisation of the city's activity. As a consequence, this comes to resemble the 24-hour economy of the global mega-cities, even at the typical small scale of university cities, increasing the opportunities for residents but provoking a rupture in their social routines. Despite such possible occasions for friction, it must be recognised that students represent a huge resource for growth and innovation. In some contexts, such benefits are hardly emerging. This is the case in cities that host technical schools, whose students have a more 'introverted' behaviour. In other cases, a badly managed mixed use of the public space may create a 'cultural shock' which gives the university a bad reputation and hinders its role as an engine of local development.

1.3 Strategic Planning for Student Communities: Critical Issues, Actors and Priorities

1.3.1 Goals of City Marketing and the Need for Strategic Planning

The increasing dynamism, complexity and uncertainty of the socioeconomic environment, even at the very local scale, make it increasingly difficult for 'traditional' government styles to maintain a hold on development. In the information age, business, but also institutions and citizens, are less and less 'place bound'; when conditions change, these actors very easily switch location, and precious resources for development – physical and human – migrate toward more competitive regions. Location factors such as access to primary resources and plenty of space are not the same as in the past. Today, both individuals and organisations appreciate amenities such as a lively and culturally stimulating environment, a functional and accessible city, a clean and aesthetically interesting landscape. Urban resources do not have to be necessarily 'cheap', but they have to be convenient: quality and knowledge, rather than quantities and physical goods, are the key words. Ownership of knowledge is hardly possible in the Internet era. As quality is a social and volatile construct, what is convenient today may easily no longer be so tomorrow, if someone 'can do it better'. This means that whereas old-style industrial cities could count on sunk capital stocks for very long periods, new global places have to be continuously tuned to the marketplace to maintain their competitive position, interpreting and possibly anticipating the emerging trends in society and the economy.

The local scale is crucial, as cities and metropolitan areas are the engines of European economic development. If cities grow in an unbalanced way, or they do not make the best of the resources that are there, entire regional or national economies may suffer. European cities face stricter budgets and harsher competition; at the same time, they are given greater responsibilities (and pressure) by national urban policies as regards the strategies they use to achieve development goals. Therefore, it is natural that city authorities have placed greater attention on strategic planning. Development efforts are accordingly redirected towards stimulating those activities for which they offer a competitive location environment. At the same time, they seek to improve the conditions – infrastructure, workforce availability, communication and knowledge – which are needed to attract new important functions. The attraction of financial, economic and human resources becomes a stringent condition for local development. At all times, they have to know what opportunities are there for development, and how to avoid the pitfalls of fast and easy but unbalanced or inequitable growth. City marketing, described by Kotler (1993) as the fine-tuning of demand and supply at the city level, is a fundamental element to cope with these new demands, since it also allows the cities to become more customer-oriented, ready to 'give service to, and mind the interests of, the town's citizens' (van den Berg and Braun, 1999).

The success in the effort to steer and control the development process depends crucially on the 'organising capacity' of the urban government. This conceptual scheme – introduced by van den Berg et al. (1997) – addresses the capacity of the governments to 'adapt' their governance styles to a the impulses coming from a dynamic and diverse environment. This is mainly achieved by formulating a vision of the city which is acceptable for any of the actors involved, transforming it into a strategy when favourable conditions are present, and implementing it by stimulating innovative ways of sharing interests and responsibility among them. For this process to be effective, it is required that some of these actors – not necessarily within the public sector – take a strong leadership in the process, ensuring to it a broad political and social support.

The creation of strategic networks at intra- and inter-city levels, as well as the formation of public-private partnerships in relation to development programmes and projects – as fundamental means to govern the complexity of goals attached to physical and social growth – form the bulk of this approach. Also, attention to spatial and social economic conditions is a crucial issue. A development that neglects spatial balance, the harmonious mix of functions, accessibility and aesthetic landscape quality is at best short-lived and at worst destructive. The example of tourism is illuminating. Cities that develop tourist functions without caring to manage the conflicts that may emerge with other sectors of the city's economy and society – the use of central location, the wear and tear on urban resources, the pressure on price and quality levels – ultimately damages those very factors that make the city attractive in the long term (van den Berg et al., 1995; Russo, 2001). Students can indeed be considered a special kind of tourist.

1.3.2 Student Communities as a Strategic Resource for Development

From the above, it is apparent that universities and higher education institutions are fundamental elements of urban development strategies, from many points of view. First, they represent infrastructure needed to upgrade the human capital and improve the city's and region's position in the new socioeconomic environment. Secondly, universities are firms themselves, mostly growing ones, with strong relations with the territory. Finally, university communities (student, workers and lecturers) are non-negligible shares of the local consumers and city users. Universities produce revenues and jobs, and – though non-profit-making firms – are often incubators to for-profit institutions and research centres. Increasingly, research activities are attracted in specific locations by favourable environmental conditions of the 'intangible' kind. They generate knowledge and a variety of positive spillovers. Moreover, they are functional to the development of synergies with a wide range of urban sectors, ranging from the public (e.g. administration, planning, real estate) to the private (e.g. 'growth sectors', financial services, etc.). More generally, it could be argued that universities are firmly embedded in the local socioeconomic environment.[3] This is shown, for instance, by the increasing trend of 'lending' university personnel to local and regional administration.

On the other hand, students are a population of city users for which the city has to cater. If a city is not attractive or convenient for students, the university located in that city would not grow, no matter how good its reputation is. The relation of students to the education institution is neither direct nor objective, as Dubet suggests (VVAA, 2001: 62), but it is highly mediated by the environmental context. With an increasing mobility of lecturers and researchers, reputation advantages tend to get lost more rapidly than in the past.[4] At the end of their study period, the newly-skilled personnel go where the job market offers better opportunities. Thus we have examples of 'booming' university cities and of cities that experience a sudden decline as conditions for their prosperity (a safe and intimate location, state subsidies, an appealing workforce) are no longer available. Local firms that want to retain this precious resource improving their knowledge capital have to act strategically to convince graduate students to settle there after the completion of their studies. The quality of life in the period of permanence, as well as the sense of integration in the community, may be crucial elements for this choice, and they are clearly out of the control of the business sector.

Two extremes can be envisaged to this respect: one in which students, just like any other category of city-users, 'use' or 'consume' the city without ever becoming part of the community: in this case the university-city relationship is more likely to be one of conflict. At the other end, if the favourable conditions are present, the students find their own 'place' in the city, or better create their own spaces, which become assets for the community at large. In this way, they become integrated in the community, acting as a driving force of urban development and change (Pallares and Feixa, 2000). To Felsenstein

(1995), the outflow of students from the area after the completion of studies can be conceived as an 'opportunity cost' sustained by the community, which produced no localised return in terms of human capital.

It is interestingly suggested by the architect Lipsky (VVAA, 2001: 56) that the micro-scale, rather than big infrastructure projects, has a decisive impact on the way in which students feel 'comfortable' in a place. One can think (quoting Lipsky) of 'easy access to copying machines, the quality of mattresses in student residences, the flexibility of libraries opening times', etc. In the end, the opportunities for synergy between city and university are increased if small and banal details such as these are catered for. It is implied that the city has an interest for a real 'student orientation' of many services provided by the university, and the attitude of the city is critical to mobilising other actors around this goal.

Moreover, it should not be forgotten that students contribute directly and indirectly to the city's welfare as consumers and taxpayers. However, the opportunities of democratic participation to the city life are rather limited. In most cases, students do not vote in the city, and neither are they consulted in issues that do not directly concern university affairs. Of course, this is a general problem that regards 'new citizenships' and that is barely solved at local level only. However, the cities can do much to 'audit' this community and orient policy towards their needs.

As Indovina (1998) puts it, universities are 'good' for cities, and cities are 'good' for universities, but this virtuous relationship is far from automatically achieved. Local governments play a crucial role in bringing the favourable conditions about. Moreover, the city can actively support graduate students who wish to stay through a 'settlement policy'; for instance, facilitating the access to the job market and the real estate for young couples, or organising local alumni networks. Finally, even if the host community clearly benefits from a wider knowledge base, such benefits are not immediately obvious to most citizens and need to be communicated effectively to grant the necessary social and political support to such initiatives. In synthesis, a comprehensive urban strategy for a student-friendly city should include the following issues:

- *attracting the students*: marketing the city as a student-friendly community;
- *assisting the students*: offering high quality services for welcoming and assisting the students, in order to achieve a better integration with the host community;

- *housing the students*: the optimal pattern of student settlement according to local contexts must be identified in order to minimise the impact of students' communities on the local housing market;
- *empowering the students*: it is necessary to recognise a 'right of citizenship' of students, granting them a role in local decisions, which may vary from informal consultation arrangements to direct participation to democratic life;
- *increasing the opportunities of contact* between students and the other local stakeholders, to integrate them in the web of relations that characterises the local economic environment and enhance the 'embeddedness' of the university in the city fabric;
- *keeping the students linked to the city* so that the human capital is not dispersed after the completion of studies and the benefits from knowledge and cultural spillovers are maximised.

1.3.3 Negotiating and Governing a Sustainable University-city Relationship

More generally, different stakeholders are interested in university development. Among these, urban players who are affected by the 'location choice' of the university, such as the government – which provides the necessary infrastructure – and sectors of the society such as landowners, speculators, commerce producers, consumer services producers, and the community at large. Other actors are concerned with the quality of the scientific and human output from universities, such as the private sector, governmental agencies promoting regional growth, the trade unions, etc. Figure 1.1 synthesises the positive and negative impacts from university settlements borne by the main urban stakeholders. Each of them has well-defined aspirations regarding the university settlement and development policies. In some cases, they could be conflicting: for instance, residents could be interested in a 'containment' of university development or in 'regeneration' issues, whereas the private sector could be interested in an expansion of university output. Students want to spend their education years in a lively urban environment, but this means in some cases utilising 'local' resources to which they do not contribute, and possibly subtracting them from the use by residents – e.g. 'street safety', culture, public transport, etc. Both landowners and students would rather develop 'informal' student neighbourhoods downtown that promise a more lively and enticing experience (and obviously higher rents), but the generality of residents could be more supportive of a 'campus-type' settlement where the

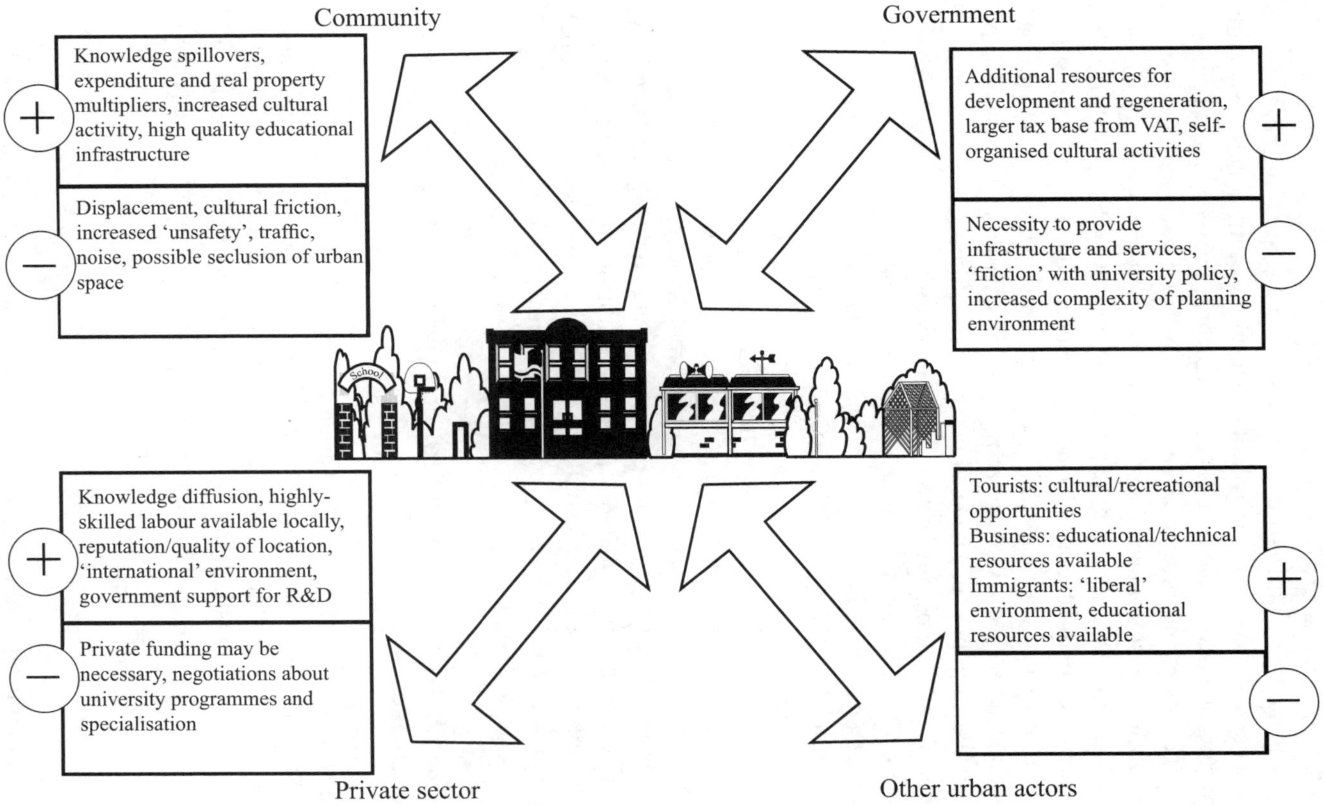

Figure 1.1 Urban stakeholders and impacts of university development

occasions for confrontation and resource-sharing with the guest community are limited; and so on.

These partially contrasting views should be accommodated in a comprehensive 'vision' of a university that enriches the local community without causing undesirable impacts like gentrification, seclusion, or cultural friction. This is by no means an easy task, because each stakeholder can influence the decision-making process to different degrees and through different means. For instance, voting residents, who have a direct and institutionalised control over the city policy through democratic representation, are the ones who might be more diffident towards university expansion in cases in which they do not clearly see the immediate benefits. Other actors who have less control over policy may have a more positive attitude towards university settlements – e.g. the private sector and some sectors of the population, such as young people and non-governmental organisations.

Local and regional governments should therefore integrate and recompose these ambitions in their development strategy, building a shared idea of university settlement and implementing it with the help and in the interest of all concerned actors. The involvement of the student and resident communities in the planning effort through a participatory approach is certainly useful in this respect. Universities could become real 'leaders' in this decision-making arena, either formally – for instance, through the constitution of city-university commissions in charge of planning for the linkages and the physical development of the research institutions in the city – or informally, through a direct engagement of university personnel in city policy.

1.4 The EURICUR Research Project. Framework of Analysis and Methodology

The objective of our study is the construction of a set of guidelines that should assist decision-makers and all the other stakeholders involved in community development when planning for the development, relocation and organisation of higher education institutions. These guidelines should be generally valid in all situations, but should indicate specific courses of action to be taken in different contexts.

1.4.1 Framework of Analysis

Careful planning may allow communities to undertake a *virtuous cycle of development*, in which an attractive city works like a 'magnet' for higher education and research, and this fosters an even more attractive urban environment for citizens, investors, tourists, etc. In the end HEIs become a real 'growth factor' for the regional economy and society. The state of formal and informal networks of decision-makers, the spatial-economic characteristics of places, the character of university programmes, and the local traditions and know-how are all elements that may influence the patterns of development of higher education and require to be managed with different tools.

The sustainable relationship between the university and its environment is assumed to depend on the balance in its process of exchange with the place. The interaction between the actors involved in the development of higher education in a city can be analysed systematically to highlight the role of the student community in urban development, the existing points of weakness in this interaction, and the actions that can redress this balance. This idea is synthesised in Figure 1.2.

Higher education has direct links with the private sector, including the business community, their workers and organisations. Universities transfer their knowledge to the private sector, in the guise of research contracts, consulting staff and training programmes. The resources generated in this exchange are utilised by universities to expand the education function, for instance financing new facilities and base research programmes. At the same time, the HEIs involve private sector practitioners in teaching activities. In this way education 'opens up' to real world, delivering a practical expertise that firms increasingly seek.

HEIs are also in a direct relationship with their host community, as they are physically located in a place. Therefore they generate jobs and revenues, like any other urban industry, and in exchange they demand services and infrastructure that the community should deliver or finance. The exchange relationship between HEIs and communities also takes place at a less visible level, as universities may enjoy 'city brands' which give them prestige above their actual merits, while on the other hand they offer more competitive and dynamic cities.

However, HEIs are not important only as industries in themselves. They are also levers for a stronger, more competitive and more 'embedded' local economy. The role of higher education is indeed that of upgrading the human capital available, influencing the location decisions and productivity of local companies, as well as the magnitude of the impacts that are trickled down in the community.

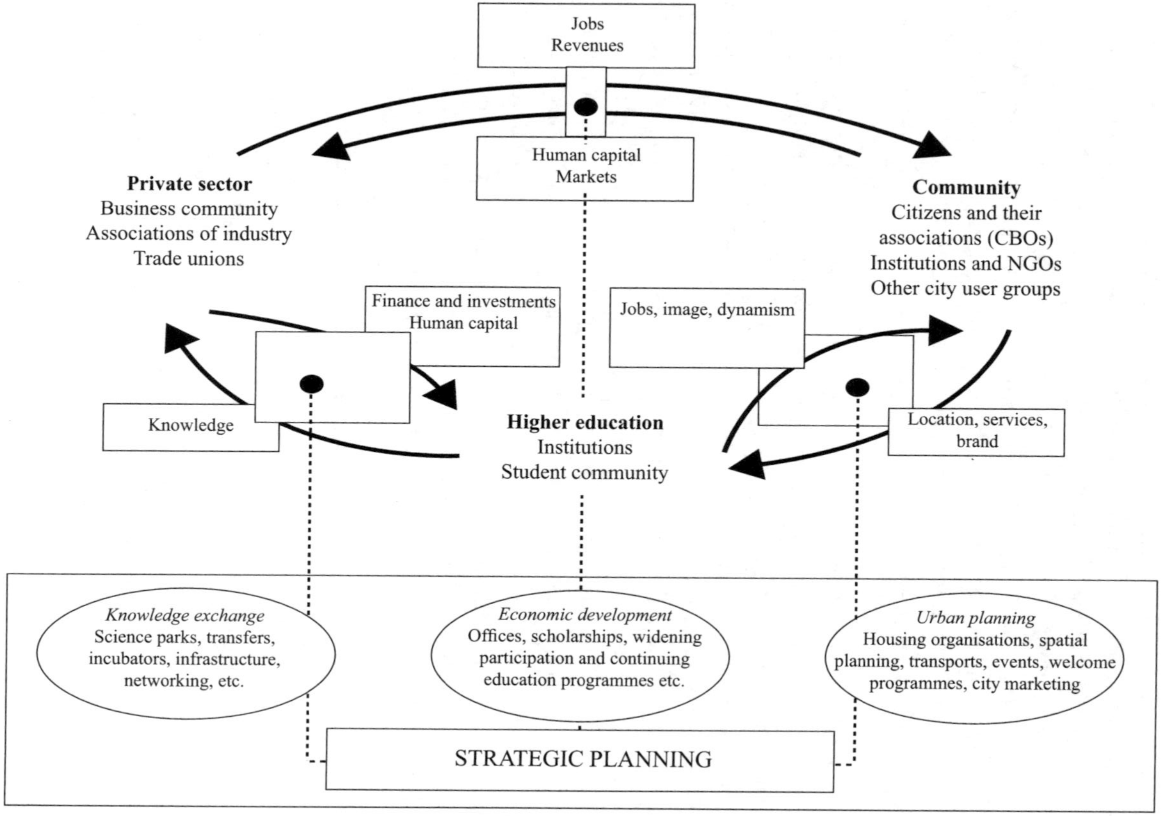

Figure 1.2 Higher education as an engine of sustainable urban development

These processes of exchange are not merely between institutions, but rather between 'groups' – students and the academic community at large, entrepreneurs, citizens – with distinct behavioural characteristics. Institutions do mediate these relations, for instance elected city councils would negotiate university development with the boards of the institutes, and chambers of commerce would stipulate partnerships with the same boards to activate stage programmes. However, the encounter between groups is more complex, as it relies on spontaneous, individualistic factors. Academic staff would comply with the partnership mentioned earlier, or students would switch residence location, only if they see a clear advantage in it, or in other words if an 'integration' is achieved based on mutual recognition, dialogue and exchange. On this level, the institutional arrangements that form the basis of higher education policies in cities are often partial and ineffective in bringing forward a real, sustainable development of the city.

Furthermore, some of these relationships are not supported by a market exchange, and therefore they are not self-regulated. They are also dynamic. If they are not satisfactorily developed, or they are not in balance, the consistency of the whole system decreases, and the city's competitive strength could be affected.

If, for instance, the university does form a valuable asset for the private sector, but the latter is not contributing to finance university expansion and to support quality of life for students, local universities may face decreasing public budgets for education and research. Conversely, if universities are irrelevant for the local business community, firms would migrate to areas where they find up-to-date knowledge and human capital. In both cases, higher education ends up losing its competitive edge and the 'knowledge input' that HEIs provide to the business community may be lost, leading to a decline in the economic performance of the region. Alternatively, if universities are not well served by the city in terms of services and 'sense of belonging', this will affect the integration of students into the community; conversely, if the university settlement model is one which minimises positive synergies with the community, the attitude of residents would be unsymphathetic to university development and social and political support to higher education would decrease. In these circumstances, the capacity of the city to tap into this capital and to grow would be affected.

To be sustained in the long term, the city-university relationship might therefore require some form of intervention from the public sector. An explicit, integral policy of cities addressing the student population can intervene at any of these levels. By actively promoting exchange initiatives between the

HEIs and the private sector, the potential for a development model based on 'knowledge' is improved. Students receive an education that is genuinely 'rooted' in the local environment, and the chances that knowledge will stay attached to the place are higher. By channelling resources into investments in quality of life for students, and into initiatives that increase their 'sense of belonging', cities make themselves more attractive and dynamic, open to the international environment, and visible. Finally, by supporting job market initiatives and continuing education, cities make sure that human capital is not lost, and that the firms stay attached to the place where they can find the best skilled people.

In this framework, it will be possible to evaluate systematically the state of the city-university relationship in all of the cities participating to this research, identifying critical points, priorities for policy and strategic developments.

In particular, the links identified in Figure 1.2 can be benchmarked through *qualitative indicators*. Each of the arrows in the diagram can be associated with a 'score' (++, +, 0, –, ––) ranging from very positive to very negative. Whenever there is imbalance at the level of each of the links, policy should turn 'minuses' into 'pluses', and/or progress towards an optimal situation in which all the links are 'double pluses'.

Finally, the overall score of a city can be derived, and the comparative analysis of points of strengths and weakness of different cities may yield a list of the most recurrent problems that any city is likely to meet.

Another objective of our empirical study is to identify a number of 'best practice' at any level:

- knowledge exchange projects;
- institutional arrangements and partnerships;
- investments in quality of life for students;
- events and campaigns;
- organisations models;

which contribute to the general aim of a sustainable city-university exchange, and could be imitated in other contexts. Possibly, new elements will be added to the picture, contributing to a deeper understanding of the relevance and issues concerning urban development in the presence of 'new' populations.

1.4.2 Research Methodology

The investigation was carried out in different stages. After a critical reading of the relevant literature and the preparation of the analytic framework exposed in the previous section, a preliminary desk research on students' communities and their impact on cities was conducted to identify the specific problems affecting each of the cities selected as case studies. Contacts with city officers and higher education managers were taken up for the collection of relevant documentation regarding student settlements (historical traits, settlement typologies and spatial patterns, magnitude and impacts, activities and organisations, behaviour after completion of studies), and other information on the structure of the local economy (e.g. job market, 'growth clusters', innovative attitude of entrepreneurs, etc.).

In the second stage, a team of EURICUR researchers travelled to each of the nine cities which participated in this research: Rotterdam, Utrecht and Eindhoven (NL), Munich (D), Lyon and Lille (F), Venice (I), Birmingham (GB) and Helsinki (SF). The nine case studies were selected in such a way as to represent a broad range of situations as far as city size, economic structure and geopolitical contexts are concerned, therefore allowing researchers to generalise the results from the study.

During the process they conducted an average number of between nine and 12 interviews with city officers, university managers, student associations, residents' and commerce organisations, and representatives of the business sector.

The information gathered was utilised as an input for the final comparative analysis presented in the last chapter of this book, which identifies typical profiles of student cities, as well as key elements determining the sustainability of the city-university relationship and cases of best practice in this relationship that can be transferred among the cities in this study.

Notes

1 For instance, Quadrado et al. (2001) find that regionalisation policies in Spain have favoured inequalities *within* regions with a strengthening of regional centres, instead of enabling a real decentralisation process.

2 In the UK, such institutional links between cities and polytechnic schools with a strong applied element was explicitly decided in the 1960s (Hall, 1997), but eventually this 'binary system' failed and polytechnic schools were converted into universities in 1991.

3 Felsenstein (1995) conceptualises the network of linkages activated by universities. Indovina (1995) has done the same, also exploring the 'intangible' nature of city-university ties.

4 For a formal example of such externalities in knowledge diffusion and the incidence of amenities, see Karlsson and Wei-Bin, 2001.

Chapter 2

Birmingham Case Study

2.1 General Introduction to the City of Birmingham

Birmingham, second city of the United Kingdom, is located in the West Midlands region, the oldest and most important industrial heart of Great Britain. It lies at the very centre of the country, at the hub of the country's transport network on road and rail. A leading manufacturing city during the industrial revolution and in the post-war period, Birmingham lost 10 per cent of its population from 1971 to 1988, maintaining the present million inhabitants to the present day, while the West Midlands region counts 5.2 million people. Fifty-one per cent of the population under 18 belong to ethnic minorities.

The urban region of Birmingham counts 2.6 million, and has maintained this population in recent years. It covers an area of 950 km^2 and is subdivided

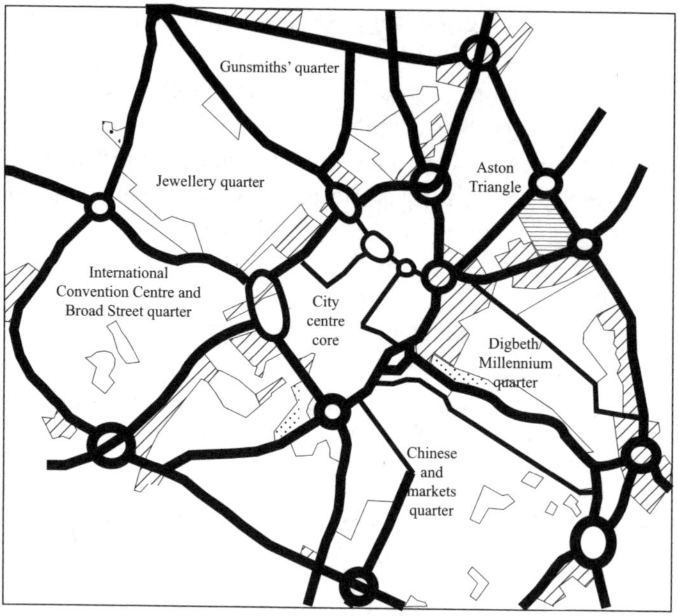

Figure 2.1 The inner city of Birmingham

Source: Birmingham Marketing Partnership.

into seven districts. Birmingham is the most populous, but among the others there are middle-size cities like Coventry and Wolverhampton. From 1974 to 1986 they were united under the West Midland Metropolitan Council. This level of government has since been dismantled (as everywhere else in the country), and competencies over infrastructure and public transport have been taken over by the individual city councils and independent organisations. Finer subdivisions of the local authority include constituencies and wards (see Figure 2.1). The city centre of Birmingham consists of nine wards.

The mid-1980s represented a turning point for the then declining and unattractive city of Birmingham. A strategy was set up to modernise the economic structure of the city. As the spearhead, convention and tourist functions seemed most promising, both because of the favourable location of the city and because of a certain know-how developed in the convention field. To this aim, a strategy to improve the city's attractiveness and infrastructure was badly needed. The Council also aimed at a more straightforward international orientation of the city.

A first giant step was realised with the renovation of the airport, through public and private investment. Its role was refocused as a national hub – a gateway to the whole region of West Midlands – and a basis for charter connections. In recent years, Birmingham has also invested in its internal accessibility. The extensive road network led Birmingham to be considered a real 'motown city'. However, this infrastructure has disadvantages as well, to the point where it becomes a main obstacle to regeneration. Efforts have been made to reduce the barrier effect of the flyover road links by building bridges across and sink roads below street level. In that way, pedestrian routes have been developed between areas that used to be physically separated. The spatial-physical structure of the city has been sensibly changing as a consequence. The inner city became ten times as large as before, yet retaining the mixed-functions flavour.

As a result of these changes, the present-day economic structure of the city represents a bridge between its historical profile as motown capital of Great Britain, and the recent diversification with the emergence of a strong service sector. The most important firms include MG Rover (car manufacturing, 6,000 employees), Cadbury (food processing, 4,000), Travel West Midlands (2,800) and others like Jaguar; HSBC banking services, and Cap Gemini Ernst & Young (IT). Though officially being part of the public sector, universities altogether account for more than 5,000 jobs.

Most of all, Birmingham put a great deal of emphasis on becoming a tourist destination. It now attracts 23 million visitors every year, who are for the majority day-visitors enjoying sightseeing, shopping, eating out and doing business. In 2000, Birmingham was the fourth most visited city in the UK by overseas visitors, after London, Edinburgh and Manchester. The flagship of the new strategy was the International Convention Centre (ICC), a congress and convention centre set up in Brindley Place, the top location of the inner city, adding to the existing event infrastructure.[1] The National Indoor Arena (NIA) seats about 12,000 spectators and can also be used for large plenary sessions of conventions staged in the ICC, which is within walking distance. A new hall with the capacity of 2,200 houses the prestigious Birmingham Symphony Orchestra. Theatres and clubs also sprang up in this area, contributing to the overall ambience of the place. The shopping and leisure quarter called 'the Water's Edge' developed into a remarkable tourist attraction in itself. Redevelopment efforts are today moving from the centre to the city edge, in the Eastside Regeneration programme which embraces a large, run-down section of the city formerly hosting manufacturing activities.

2.2 The Student Community and Knowledge Infrastructure in Birmingham

2.2.1 General Information about Higher Education in Birmingham

Birmingham is a relatively new city, hence its universities do not have the high-brow status of places like Oxford or London. Instead, the city has a tradition of large municipal universities. In 1890, Britain's manufacturing industry was falling behind Germany's, where excellent technical and professional schools had been developed. Therefore, the local government decided to invest in higher education. The new polytechnic schools were planned to realise a very close link between education and the job market requirements in the area. This link was revived during the 1970s, when universities had to play the role of a stimulus to the regeneration of the economy.

Birmingham's three universities are very diverse, appealing to different markets. Aston is the most specialised. The University of Central England, once a polytechnic, has strong local roots, and manages to attract minority groups, which places it at the centre of widening participation strategies. The University of Birmingham is the most 'generalistic' of the three but also the one with the highest reputation in fundamental research. However, in an age in which 'small is beautiful' and performance is measured by knowledge application, it is the one which faces the most serious challenges. The government predicts a 13 per cent increase in the 16–19-year old student population for the period 1999–2009. Competition in the market for full-time HE students and staff is likely to be fierce in the future, not only among local athenaeums, but also regionally and nationally. Birmingham also hosts eight vocational colleges.

2.2.2 *The City's Knowledge Sector: Actors and Spatial Patterns*

Universities and other higher education institutions The *University of Birmingham* (UB) is a well-established, world-class university, one of the leading research-based universities in the United Kingdom. An emphasis upon research in all schools is one of the distinctive characteristics of UB. The university was founded in 1900 by the citizens of Birmingham, who wanted their own university to train and educate the people who would create and manage the burgeoning businesses and industries of the midlands. It was natural, given the nature of Birmingham's industry, that the university should focus on the main scientific and engineering disciplines. It was also the first university in the UK to establish a faculty of commerce and incorporate a medical school. The contemporary UB is equally distinguished in the humanities, education, social sciences and law.[2]

UB hosts 15,214 undergraduate and more than 6,600 postgraduate students, giving a total of 21,896 students. Each year there are over 3,000 students from more than 100 countries in regular or exchange programmes. International students are in total 3,682 (17 per cent). Thirty per cent of these are citizens of the European Union, and 42 per cent come from Southeastern Asian countries. UB offers them more than 80 undergraduate programmes in 28 areas of study, and a large number of taught and research postgraduate programmes. The recent trends show that the number of students in the faculties related to the declined industries is declining (chemical, physical sciences), while students are increasing in social sciences, business and accountancy. This could be a problem for the future, as Birmingham still needs top human resources for the manufacturing industry. The UB is fully committed towards continuing and professional education, offering all sorts of courses and support. The main university campus at Edgbaston is spacious, attractive – one of the first 'redbrick' universities in UK – hosts a prestigious library and is well endowed with state-of-the art facilities for students, such as sport halls and IT facilities. The nearby Selly Oak campus is situated on a beautiful garden site just five miles from the city centre.

UB has relationships in research and in other ways with a host of private and public organisations. The university offers a wide range of services both on a generic basis and tailored to specific organisational needs. Examples of the wide range of activities include consultancy to business, industry and the public sector; business support services; technology/knowledge transfer and licensing; continuing professional development; student projects and work placements; and hospitality and conferences.

UB is attractive for students all over Britain because of its reputation. However, almost 50 per cent of its students leave Birmingham after the completion of their studies.

Today, UB is committed to maintaining an international reputation for the highest quality of scholarship and research, for academic excellence, and for the quality of its alumni network. UB also places a special focus on its mission statement regaring the relationship with the local area, which is seen as a laboratory for applied research and a community in which the university is a committed actor.

The *University of Central England* (UCE) is one of the largest universities in the United Kingdom, with over 25,000 students enrolled in full-time, 'sandwich' and part-time courses. This number has been steadily increasing in the last ten years. In total, UCE offers some 100 courses in its nine faculties. The staff of UCE in 2001 stood at 2,700. Approximately half the students are part-time, coming from the local area, working and looking for professional advancement; their motivation is mainly vocational. With these numbers, UCE is one of the most specifically employment-targeted HEIs in the Midlands. Among the undergraduate full-time students, 65 per cent come from the region (this share is rising, due to the changing university funding system, which has implications for parents with children at university).

UCE is committed to research, carried out by 250 PhD students, mainly in arts, industrial design, multimedia, video communication. Most research at UCE is market-oriented and applied, and thus in some disciplines has a particular emphasis on wealth creation and technology transfer. This is entirely consistent with the 'community role' of UCE. UCE's teaching provision is largely professional and applied in its orientation. The budget of UCE (expenditure) is around €115,000m, versus a budget of €328,000m for the University of Birmingham.

The size and diversity of UCE, combined with its excellent record in widening participation, ensures an interesting student mix reflecting a broad range of cultures, nationalities, interests and age ranges. Thirty-three per cent of the students are of ethnic origin. This mix creates a stimulating learning environment that provides an exciting student experience. A high proportion of students have a South Asian background, and the majority are women. The university's teaching supply is flexible enough to account for changes in society. New specialisations being developed include e-commerce and information systems; health services; music and arts; new media; and innovation management. The international activity of UCE is developing, with an increasing number of students involved in exchange programmes and distance learning programmes. UCE has bilateral agreements for exchanges with 94 partners in 15 EU countries.

UCE defines itself an 'urban' university, and quotes a commitment for education towards the local community in its mission statement, as well as its status as an engine of regional socioeconomic development. It has extensive links with the economic communities of the region through its Corporate Development Centre.

To achieve these goals, UCE seeks to develop partnerships with partners of all kinds, such as employers, community groups, and other HEIs. In particular, UCE looks with iinterest at Advantage West Midlands, a driver for regional change which has set up a network in which UCE can play a role, engaging with businesses, industry, the professions and the public sector in building an 'Agenda for Action'. In the medium term, UCE aims at strengthening this relationship for strategic objectives, such as the economic regeneration and improved competitiveness of business and industry in the West Midlands. To this aim, UCE has already started a number of projects, such as the establishment of the Jewellery Innovation Centre – a start-up and business support centre right at the heart of the jewellery district in Birmingham, whose aim is to revitalise this historical cluster.

However, the Perry Barr campus is working at full capacity, in spite of extensive investment in the quality and volume of the existing premises. As such, it is inadequate for postgraduate education, research and collaborative work with external organisations in research, consultancy, and economic development activity. UCE is working with the Birmingham City Council to find solutions to the space shortage in developments in the City Centre and in especially in the East Regeneration quarter, which could offer plenty of space for education and collaboration in the new cultural industries.

UCE guarantees accommodation to first year students, students with disabilities, international non-EU students and some EU students. The others go to the private sector. Some students indeed prefer to live in the Selly Oak area, where they find a nice 'student-city' atmosphere and good connections with the UCE. The area close to the UCE is not popular among students as it is a relatively poor district with little to offer. However, the strategic plan of UCE recognises the need to give priority to future investments in residential accommodation in the proximity of the campuses, as it would aid recruitment and retention.

Aston was founded in 1895 as Birmingham Municipal Technical School, and in 1966 it received its Royal Charter as *Aston University*. It hosts about 5,000 students in full-time and other kinds of undergraduate programmes and some 670 full-time postgraduate students, and boasts excellent teaching assessment levels. Aston is an applied, business-oriented university and its strong research and teaching base provides opportunities in a wide range of areas. The education at Aston is organised around four 'schools' or faculties, offering courses and postgraduate in around 20 subjects, including an MBA, and the possibility of 'sandwich courses' and combined honours in a flexible format.

Over 70 per cent of Aston's students indeed choose sandwich or language programmes, gaining valuable experience in industry, business and commerce, and integrating theory with practice. The result is an employment record within six months from the graduation that is the best in the UK, with a rate of 80 per cent against a national average of 67 per cent. Less than 4 per cent of graduates were unemployed in 2000, against a national average of 5.3 per cent. Astons' Business School ranks second in UK behind the London School of Economics, according to *The Times*.

There was no international office at Aston until 1994, but now its international relations policy is in full swing. To facilitate and 'standardise' the integration of the international students' community at Aston, the university has developed two one year 'International Foundation programmes', as an entry route to honours degree courses, in a partnership with Coventry Technical College. Admission to Aston is automatic after successful completion of such programmes. The main focus of Aston's international recruitment policy is on emerging markets in the east, such as Malaysia, China, and India. Only 20 per cent of Aston's students come from the region; this university is more oriented towards the national and the international level. This creates a balance between Aston and the other universities in Birmingham, that are more 'local' in reach. Nevertheless, Aston has important strategic relations with the local colleges, which share Aston's resources like the library.

Research is organised in 40 centres. More than 85 per cent of Aston's academic staff undertake research activity recognised as being of outstanding international quality. The location of the university, between the city centre and the Business Park, within the eastern regeneration scheme, is a potentially favourable one to facilitate links and networking with private partners.

Aston is appreciated by students for its friendly, close-knit 'campus atmosphere', contributing to a favourable environment for academic success and social life, and for its urban location. Students have all the benefits of living in a city, but live in a comfortable and secure environment. Aston guarantees accommodation to first-year students. Moreover, priority for non-first year students is given to those in their final year returning from a sandwich or language placement abroad/outside Birmingham. International fee-paying students are guaranteed accommodation for the duration of their programme, provided that they apply before a certain date. Aston offers to these students 2,117 places in university residences, all on the green campus five minutes walk from the centre of Birmingham. They consist of modern, secure self-catering flats with 24-hour security and controlled access, in flats of three to 12 students. On the campus, a student-run Residents' Association (ACRA) is active in organising social and cultural activities, and providing facilities for residents – launderettes, satellite TV rooms, etc. Non-eligible students have a wide choice of accommodation in the city. The vast majority of students live in favourite student areas like Erdington and Edgbaston. The Accommodation Office runs a register of local houses, flats, beds and lodgings for those students looking for accommodation off-campus, and advises on tenancy agreements, tenants' or landlords' rights, and safety. The Student Advice Centre in the Students' Guild also provides information and advice on accommodation and has a notice board advertising student housing.

The good relationship between the institution and its students is also reflected in an intense communication effort. Aston publishes a monthly newsletter (*Aston Aspects*) that represents a precious source of information for staff and students.

Aston also faces a number of challenges and harsh regional competition. Due to its relatively small size, it is an easy target for a merging process with UB, and indeed there is talk that this may happen in the not so distant future. Moreover, in spite of its strategic position within the Eastside regeneration areas, Aston has not as yet developed such extensive linkages with the economic community that is leading the business park development.

Research and technology transfer As in many other metropolitan areas in Europe, Birmingham is today the home of a number of knowledge-transfer initiatives. Each of the three universities of Birmingham has a leading role in a 'science park', where research establishes contact with the business world and collaborations at all levels are established.

The UB joined forces with the City Council to establish its own research park in 1986. Based in Edgbaston, directly adjacent to the university campus and the Queen Elizabeth Medical Centre, *Birmingham Research Park* offers tailor-made accommodation to companies seeking to collaborate with UB in research, development and training projects. The aim of the Park is to encourage the formation and growth of technology or knowledge-based companies. Past and present tenants have built upon the university's own reputation in biotechnology, computer systems, language research, medical diagnostics and scientific instruments. Healthcare developments, in particular, have been prominent. Many companies were founded by academic staff from the university to exploit new technology arising from their research. Others have external origins, but are developing their own working relationships with university departments. The companies in the Park employ some 270 people. The Park hosts a number of institutes and centres (ICT, innovation, business development) which support companies. These have direct access to the technical, sports and social facilities of the university campus.

The *Aston Science Park* is an integral element of the Eastside Regeneration Programme, occupying a 22-acre site adjacent to the inner city, Aston University and other flagship ventures like the Millennium Point. The Science Park is managed by Birmingham Technology Limited (BTL). BTL's aim is to enhance wealth and employment in Birmingham by providing facilities for the establishment and rapid growth of knowledge-based companies which can benefit from business support services, the management skills of BTL and interaction with other companies within the business community. Compared with the Research Park at Edgbaston, the Aston Science Park is more business-oriented, and its mission is more geared towards the incubation and start-up of new enterprise. Despite the name, the Aston University does not play a major role in the park, nor does it own property that it can use to start its own activity within it.

The Aston Park is still expanding. A pioneering incubation centre specially designed to help new and young businesses grow, is set to expand twofold thanks to the latest development, now under way at Aston Park, increasing floor space from 50,000 sq. ft to more than 100,000 sq. ft. More than £9m is being invested in additional incubation space to cope with the demands of new business start-ups and entrepreneurs. The brand new centre is a partnership between Advantage West Midlands (the development agency for the Midlands region), and other British agencies, and receives funding by the European Regional Development Fund. The Business and Innovation Centre has been the home to more than 100 companies since its inception 17 years ago. Over half of these companies began their life at Aston Science Park and have since steadily moved into larger premises within the Park as their businesses prospered. The success of Aston Park is being used as the blueprint for new hi-tech habitats in China and Tunisia.

Millennium Point is possibly the flagship project that is likely to represent the new face in Birmingham in the years to come. The enormous building (more than 37,000 m^2 of floor space), located at the heart of the regenerating eastern side of the city, is one of the UK's landmark millennium lottery-funded schemes. Its explicit aim is to foster a culture of science and technology, in a learning and entertaining environment.

Millennium Point offers a great potential to the city to enhance its 'spectacularity' and to bring forward regeneration in a marginal area of the city (900 jobs are generated directly and indirectly in this area). However, at the time of writing it is not yet used to the full potential, which poses a threat to the financial sustainability of this ventures; moreover, it is still perceived as disconnected from the city centre, as it lies on the outer side of the 'donut ring-road' and it is hardly accessible to pedestrians. The value of its location is likely to increase dramatically after the infrastructure impediments will come to a solution. The City Council in its economic development programme indeed aimed at realising a whole 'Learning Quarter', joining the Aston University campus with Millennium Point.

Spatial patterns of higher education and research The three university campuses are not entirely relevant to the inner-city economy and social revitalisation strategies, except maybe for the Aston campus which is a key element within the Eastside Regeneration scheme. On the other hand, they play a major role as development poles in suburban metropolitan areas.

The Custard Factory, a former derelict industrial building, is presently being developed as an incubator for creative industries, in an effort to replicate the successful model of Manchester. This impressive project presents itself as a hub to redevelop a whole neighbourhood for young people, artists and students. A possible adjacent 'Millennium Point Media Production Village' is being planned alongside the Learning Quarter/Eastside/Millennium Point, together with an expanded Digbeth Centre for Media Arts, putting together social and entrepreneurial objectives. It is also expected that the 600 and more workers indirectly 'activated' by Millennium Point will live primarily in this area. In the Jewellery quarter, in a student-friendly environment, extensive links are being formed between institutions, schools, community organisations and small private firms in the traditional jewellery sector.

In these projects, housing developments take into account the needs and the status of the resident population, in order to be socially sustainable. Not everything is left to the market, with the danger of an impoverishment, rather than an increase in complexity, of the social texture. The 'tourismification' of the city centre did produce some functional imbalance in the urban region, for instance traffic to the city has much increased, generating the need for an expensive upgrading for the city's obsolete road and rail infrastructure. The peripheral areas are now under pressure for development as high-quality residential neighbourhoods, and indeed a regeneration plan at the neighbourhood level, including an explicit catalyst role for higher education, is still missing.

The issue with the peripheral district of Selly Oak is an illustration of these spatial tensions. With 10,000 students, this district close to the UB hosts the major concentration of students. It is a working class area of ageing people, with old property and low-quality housing, badly in need of economic and social regeneration, and not exactly central in the reorganisation of the productive structure of the city. Indeed, in this neighbourhood, student housing is particularly convenient. Rents (2002/3) for a room in a shared house are around £42 per week. As a result, 60 per cent of the properties in the proximity of the UB are shared among students. Bus services in the city are frequent, extensive and cheap, with a per-term Birmingham student travel card costing £86.

The very economic structure of the place is changing as a consequence of this high concentration of students (Groves et al., 2002). While to some extent this change is beneficial – inexpensive shops, banks and all sorts of popular urban facilities have stayed here *because* of the presence of students – the traditional way of life of local residents are increasingly challenged by this dynamic, noisy group. Recent years have seen a proliferation of prostitution and connected activities, relying on students as a customer group. This development has resulted in chaotic traffic and use of parking places during the nights. This and similar episodes have led to an organised confrontation, with an association of residents lobbying the City Council and the University Boards for some mediation in the settlement development, which has been missing for so many years. Local residents are especially concerned with declines in property value, due to the lack of maintenance by students, short-term tenants with peculiar lifestyles. Students – who typically do not have a high spending profile – leave the area anyway when they finish their courses.

The City Council prepared a revitalisation plan for the area that includes the relocation of the hospital as a catalyst for social revitalisation, new supermarkets and other urban infrastructure. Yet housing and related policies in the area are likely to affect the university's performance to some extent. Alternative solutions therefore have to be competitive with respect to other factors, such as easy accessibility to cultural and social meeting-points, proximity to the university, provision of high quality services and residential facilities. But they also need to be consistent with the university's development policy.

Today, the internationalisation and development plans of the HEIs put additional pressure on these neighbourhoods, and even in Birmingham, where it has seldom been perceived as such, student housing has begun to be a problem. The number of students is growing constantly, and there is no institution that welcomes and accommodates students with an eye to the resources available in the community. Assistance is organised only for the category of indigent students, but is barely strategically connected with the development policy of the city. This shortage of high quality solutions does not affect the recruitment policy of universities, which relies on their reputation, but could definitely influence the students' after-study choices.

2.2.3 The Student Community: Behaviour and Careers

It should be considered that the character of student communities in Britain is quite specific. Higher education represents a highly relevant, culturally lively sector of the society, with its rituals, its traditions, and its reputation of excellence. This of course is much more visible in small historical student cities, like Oxford and Cambridge. In the big cities like London and Birmingham, despite the large numbers, students tend to get 'dispersed'. In a working-class, no-thrills city like Birmingham, it would seem appropriate that students become a 'wedge' of change, both socially and economically.

However, the student population of the city is to some extent 'invisible'. Only the recent 'opening' of the city centre to the Eastside projects gives an opportunity to a sector of the HE students, those of Aston, to be seen as a group of city users. The rest of the students are in 'retreat' between suburban campuses and student neighbourhoods, and a large number of the students would continue to commute to Birmingham from their family home during the course of their studies – 40 per cent of the students in Birmingham come from the West Midlands. The interesting thing about Birmingham's student settlements is that, rather than being the result of gentrification of past errors in planning, the 'peripheral' pattern of student communities seems to be a choice of students who want to be together in nice, cohesive communities where they can combine study with leisure at low living costs – like Selly Oak or

Edgbaston. In these districts private accommodation for rent is easy to find and relatively inexpensive. Their good accessibility allows students to commute to the city centre easily, which is quite rare. Associative and cultural student life thrives in the campuses, where hundreds or groups, guilds and unions are busy making life easier and more interesting for their acolytes, Yet, it is hardly 'metabolised' by the local communities, who still look at universities as 'towers of Babel' rather than resources for community development, in spite of the commitment of university's plans and mission statements.

If one strolls along the main malls and shopping streets of the small but charming Central Birmingham, one does not feel as if one is in a city which hosts more than 50,000, ethnically and socially mixed HE students. The student community is only visible in Central Birmingham during the weekends, and this is a relatively new occurrence, probably a positive spin-off of the city centre regeneration projects, which have created new meeting places and a whole 'entertainment district' between Brindley Place and the canals. However, these projects are more specifically targetted at the international visitor class and the appeal for young students could be limited. The city is now thinking how to develop other parts of town, having in mind the specific requirements of creative and culturally curious communities, providing 'laboratories' of social integration and development rather than consumption opportunities.

Though we do not possess complete data on this matter, the local private sector and the universities still perceive that too many of the Birmingham graduates leave the city after the completion of their studies and look for jobs and housing somewhere else. Most of them still go back to their original hometown, or try to find a job in London after their studies. Recruitment in Birmingham is problematic for firms, though on average it is cheaper live there than to live in London. Birmingham's economic structure, now a booming one after a positively accomplished transition to services and new economy functions, is probably less important than its cultural and social characteristics in determining the choices of most new graduates regarding the location to build life careers. It is true that today there are 'gentrified districts' where house prices peak to London levels, and nevertheless are attractive to young managers and professionals. At the same time though, these areas are limited in size and hardly appealing for the graduate with little money.

For these reasons, today, one feels that the transition of Birmingham to an international world-class city is only partial. Birmingham definitely lacks the social and cultural capital that is usually associated to 'young', 'hip' cities. Other big British cities like Manchester, Liverpool and Glasgow definitely have more appeal in that sense, though their socioeconomic problems might have been more serious in origin. Even on the regional scale, Birmingham is challenged by smaller but more student-friendly cities like Nottingham or Leicester for business in emerging sectors. Healthcare firms, financial groups and ICT prefer to recruit graduates in dynamic, thriving communities, offering ideal conditions to start a life career there. Birmingham, the old motown capital and today the capital of liberal professions, is not as good an option.

2.2.4 *Challenges for the City*

Like in many other European industrial towns, the economy of Birmingham came under strong pressure in the 1970s and early 1980s. In the 1980s, the loss of the competitiveness of traditional industries and the stagnation of the strong sectors of Birmingham (such as the automobile industry and related sectors) caused a slump in the economic fortunes of the urban region. Between 1981 and 1993 almost 300,000 jobs were lost, with the unemployment rate reaching the figure of 20 per cent in 1985. The recovery in the service sector could not match such dramatic drain.

Since the second half of the 1980s the situation has been improving. This is due in part to the general national economic trend; however, in this period the regional economy entered a process of restructuring from manufacturing towards the service sector (Figure 2.2). Despite the positive trends, social polarisation grew in the next two decades, and this went hand-in-hand with the physical decline of the inner city. All 'marginal' groups were affected: the immigrants, the long-term unemployed, the elderly and young people. There were 8,059 young people unemployed in March 2002, determining a juvenile unemployment rate of 10.0 per cent.

The polarisation also happened on spatial lines. In some wards, particularly in the inner city, youth unemployment rates are over 15 per cent (Figure 2.3). The unemployment rate at April 2002 is at 8.1 per cent, in recovery but still double than both the regional and the national rates. The number of people claiming Job Seeker's Allowance has decreased by 2,300 units from 32,558 in April 2001. A 'university' neighbourhood like Selly Oak has a low unemployment rate (5.3 per cent), as does Perry Barr (where the University of Central England is located), while the Ladywood constituency continues to have the highest unemployment in the city. Also in the centre (Aston, Soho) unemployment is very high and above 10 per cent. Unemployment problems are closely bound up with the average educational level of the Birmingham citizens, which is lower than the British average. As well as there being a considerable number of people with an academic or comparable education

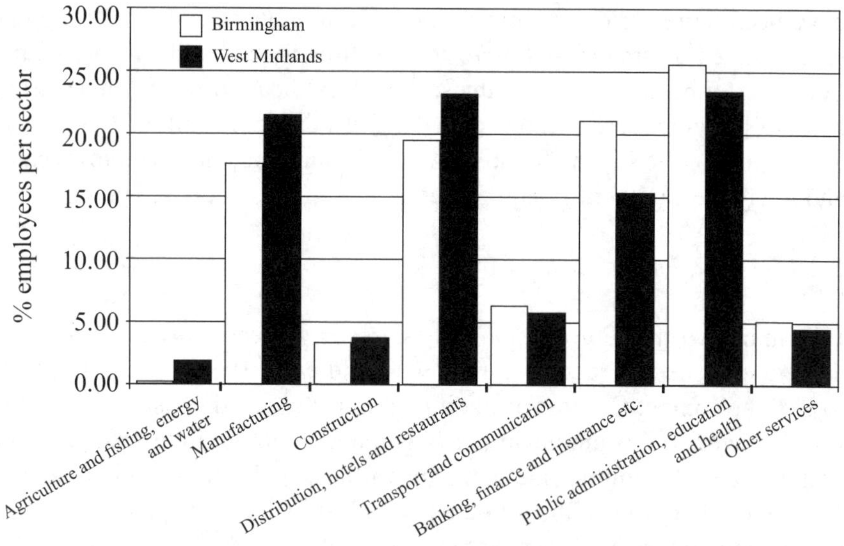

Figure 2.2 Employment per sector in Birmingham and West Midlands, 2000

Source: ONS/Annual Business Inquiry, 2000.

there is a large group of people without schooling or training (20 per cent of the active population). The number of long-term unemployed is especially high, at 35.8 per cent per cent of all unemployed significantly more than the British average (26.9 per cent).

To accomplish the transition to a post-industrial metropolis, Birmingham had to become more competitive and attractive. In particular, it was believed that the *image of the city* had to change, from an obsolete and rather dull industrial hub to a post-modern, visitor-friendly cultural centre. Other major issues concerned the city's economic competitiveness and the reduction of social polarisation, as well as the inadequate transport infrastructure and poor accessibility. The city had to reposition itself completely; regionally, nationally and internationally.

The city's development policy is focused around three key concepts: 1) supporting the growth of the information-intensive sectors; 2) enhancing knowledge creation and circulation in the urban region; and 3) creating an urban landscape that is attractive for the international visitor class.

These three ambitions have been translated into action programmes and, at the spatial level, have targeted specific areas of the city. Basically, the

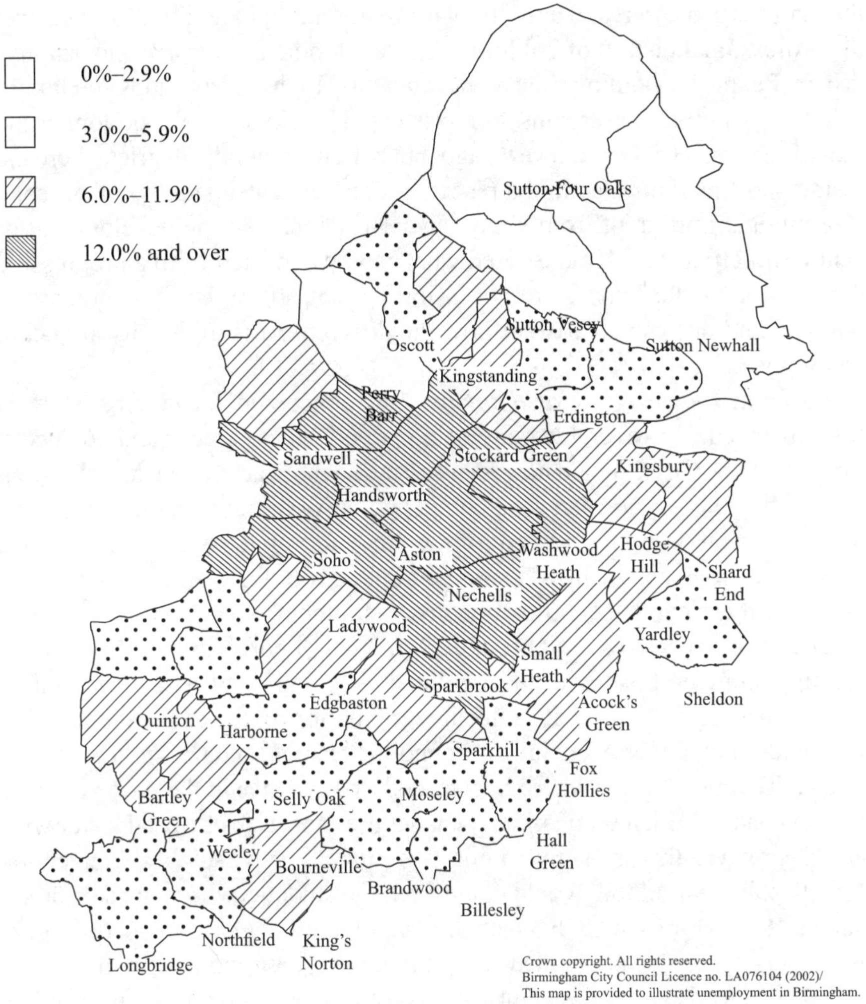

Figure 2.3 The municipality of Birmingham and subdivision into wards, with employment figures

Source: Birmingham Economic Information Centre.

visitor-friendly functions have been centred on the western-side redevelopment projects centred on Brindley Place and the ICC; these are seen as highly successful in changing the image and attractiveness of the city for a substantial share of potential visitors and for the local residents. The support for economic regeneration affects the city centre and the central business district, as well as the many urban centres in a multipolar urban region, including, for instance, the professional district of Solihull. This development is clearly opportunity-driven. People are coming back to the inner city; high quality shops and luxury apartments pull the re-urbanisation process. The city centre is no longer the monofunctional CBD of ten years ago, but a multifunctional district where the residential functions also find a place: an unprecedented situation that make sBirmingham different from many other British cities. The development of central functions has demonstrated what can be expected from a progressive diminution of the barriers to accessibility that still exist. The process is ongoing and one can clearly see how the different parts of the city are being recomposed.

As far as *knowledge-related functions* are concerned, the urban strategy is less advanced. However, ideas are being put on paper in plans and prospects. First of all, the City Council now wants to take a more active role in supporting and guiding the location and development policy of the universities. Until now, there has been little involvement of the university in the regeneration plans. The universities have enjoyed an autonomous role for their national status, but today it is clear that coordination in development, not only to contribute to regeneration objectives but also to exploit local opportunities – for instance, creating strategic links with local colleges – is a priority. The City Council sees the necessity to associate a 'local reorientation' of the education function with wider objectives of social and economic change, and tries to have 'more university' in the places where it is needed, for instance in the Aston Science Park and at the Millennium Point. The latter presents itself as a catalyst for wide areas. However, this and similar initiatives cannot survive solely as convention facilities. Its real 'blood' would come from the students who will take classes and use laboratories at UCE's technological faculty. Indeed, the City Library is moving there – it is planned to be the largest metropolitan library in the UK. The plan also provides plenty of green and urban amenities. The problem is that too much development has to be undertaken. It is a vast project that needs to overcome several existing barriers. Currently, the area has inadequate accessibility and infrastructure (there is an intention to build a metro).

Secondly, the City Council is now aware of the opportunities offered by the students' and in general young people's community for the regeneration of 'problem' neighbourhoods. Opportunities are being provided for a 'creative city' that may compete on equal grounds with cities such as Manchester or Sheffield in attracting the 'smart' functions of the knowledge economy. Research and education, albeit prestigious, are not sufficient to root the human capital in this area and to change the socioeconomic face of the city permanently; what is needed is a real integration and empowerment of the human resource base in the city's cultural texture. The 'neighbourhood level' of this process of integration is highly relevant; to some extent Birmingham has developed as a 'divided', secluded city, with wide areas of social marginality, mainly associated with low skills. These problems can be turned into opportunities. For instance, female and juvenile employment can be specifically targeted with HE programmes, for which local universities receive generous funds from the state. In this context, UCE has found a new niche market in Muslim women. Other projects, at the neighbourhood level see a direct engagement of universities and students.

2.3 Strategies and Integration. Birmingham as a 'Sustainable Student City'

2.3.1 Relations between the Students and the Community

The 'embeddedness' of the local universities into Birmingham society varies according to the profile of the institute and its history. Of the three, UCE has the most extensive social ties with the community. Such links are functional to the widening participation objectives in the HEI sector; access to higher education is facilitated through a 'planned transition' from vocational-oriented schools. The UCE also enjoys a long-term relationship with the West Midlands region through its Regional Development Agency. This relationship is strategic in character – the regional Education Department is looking to establish a medical school, and UCE, as the other universities, is intended to be the brain and training centre for this project. The region plans to develop a 'research corridor' on the axis between Oxford and Birmingham and the location of UCE in the southwestern part of the city will integrate itself fully in the development region. However, one can hardly say that the conditions exist for making Birmingham a 'valley' location, like Cambridge or Guildford, where students are welcomed in excellent structures, have all the tools to open their own firms and enjoy a high-class quality of life with their families. In big cities like Birmingham, the creation of a learning-city environment has to go through a thorough process of integration with the local community.

However, this does not happen easily, as the example of Selly Oak illustrates. Here, as well as in other student neighbourhoods, there is hardly any integration. The interaction between local people and students is limited to 'landlord-tenant' relations. Though the Guild of Students has been strongly involved in the negotiations regarding student settlements in this area, there is no sensible effort to join forces for a better future. The regeneration plans for this area are rather *top-down*. But what is even worse is that students are not users of the refurbished city centre, which remains empty at night and during the weekends. Aston is in the centre of the city, but it is small in size and it is a relatively inward-looking student community. This model of student settlement is less functional to an integration of students in the local community; in the cases of Liverpool and Manchester, student residences are right in the city centre and students are a social and cultural resource for the city.

The lack of cohesion between the relevant actors for a student-friendly Birmingham has historical origins. During the 1980s, the 'polytechnic schools', which in the UK were historically connected to local government, and evolved in corporate bodies independent from the educational department of the city. Eventually, they underwent a process of integration and became fully-fledged universities. Still, the city wanted to maintain its hegemony on these schools; this created a struggle for autonomy that resulted in very little collaboration on strategic issues. Through an arrangement by which the city appoints the chancellor of the university (UCE), this link has been partly restored, and today the state of relations between the universities and the City Council is improving. For instance, the city sees in the universities solid partners for the training of qualified personnel in social policy, and preferred consultants as far as planning is concerned. Aston, UB and UCE have been involved in studies and plans that brought to development of the science parks to fruition.

However, they had little chance of affecting the city plans so as to fulfil their own ambitions, for instance in relation to housing development policy. To meet the challenge of the Eastside development programme, for instance, more strategic relations are required in this respect, explicitly considering the 'soft' elements that would make the new facilities attractive, accessible, and functional to the development of the city as a world-class education pole. In general, these parties do not seem fully aware of the potential for synergies and of the possibility of joining forces to tackle the main problems. Secondly, the city and the universities do not strategically coordinate their respective international relations' policy. Apart from the initial cooperation within the Summer University programme, the international policy of the city is not concerned with higher education at a strategic level. The universities are not involved in the City Council's extensive city twinning and networking programme which aims at establishing cultural and economic relations as a precondition for economic partnerships; and a shared vision for the management of the foreign students' community is still lacking.

The City Council's Economic Development Department (EDD) is the main player in the city's economic regeneration and development strategies. It works in partnership with other key departments to meet the wide objectives laid down in the EDD programme, including transportation, planning and architecture, culture and leisure services and education. EDD is the main actor behind the development and regeneration actions that have almost completely changed the city in the last decade. The 'Economic Development Programme' document sets forward new challenges and objectives for the next years. The City Council has been especially active in forging a vision for the city, and translating it into an articulated action programme, involving partners and sponsors and effectively tapping funding at the national (e.g. lottery funds) and international levels, through EU regeneration and objective schemes.

The City Council, instead, has been timid in its attitude towards higher education and the three prestigious Birmingham universities in particular. The Council has to take care that the learning and widening participation objectives fixed by the Labour government are respected, a challenging task when the social and ethnic base of the city is considered. The context is not a favourable one either: the government has cut student grants, so the amount of funding per student is decreasing, while universities receive relatively more money.

However, it is commonly felt that the student community is not high on the Council's political agenda. Though the City Council hosts an evening reception for the almost 700 international students that come to the city every year, students are seldom involved in the local political and social life on any other occasions. Economically, students are not particularly relevant, being a relatively small-sized customer group for the city, with a markedly 'inward' consuming attitude and a generally lower-than-average budget. Instead, there is a growing recognition of the problems that this community raises. Universities are not considered by elected politicians as real stakeholders or strategic partners in urban policy. The relations are more advanced with the single departments of the city council and offices. Recently, the city has adopted a proactive attitude towards the issue of citizenship: students are explicitly invited to register as city residents so that they can vote, and the tax base of the city is thus enhanced. But most importantly, the strategic attitude of the city towards is student community is evolving in the new development documents and partnerships.

'Lifelong Learning' is a division of the City Council Education Department in charge of elaborating projects to tackle education issues at the local level, raising the number of students who access higher education. All future developments at university level will be guided by the principle of widening participation. An innovative project of the City Council is *Gallery 37*. This project targets people who left school early, who are particularly talented in the arts. Gallery 37 intervenes in the social processes that lead to low educational attainment, and at the same time, it is consistent with the city's ambition to market itself as a cultural hub.

A more noteworthy initiative has been taken by the international relations department, the main actor behind the creation of the *European Summer University*, which indeed has the potential to raise the profile of city-university relations in Birmingham. This school explicitly connects the city's ambitions to develop as a 'learning community' with more general marketing and regeneration objectives. It is a three-week pilot programme, built in partnership with all three of Birmingham's universities as a follow-up to a stage of common reflection on the possible contribution of the universities to the city strategies and the common benefits to be reaped. It targets postgraduate students and young professionals from all over the continent, but also from China and North Africa, offering special courses taught in technical English. These courses have a practical orientation and utilise Birmingham as a real-life 'laboratory' with redevelopment processes taking place. In this way Birmingham is making itself known to a wider arena and is trying to utilise this international outlook in an interactive way. All the participants are accommodated in the centre of the city, but their work is organised throughout the area, allowing them to be closer to the community which is the object of their learning.

The programme is promoted utilising the international network of both the participating universities and the city council. For the future, the ambition is also to involve the universities of the region, forming a consortium including Warwick, Aston, UCE and Birmingham's universities. Eventually this venture would become a private foundation. The secret of the success of this initiative was that – rather than being developed as a top-down initiative – it truly tried to look for cross-over in the agenda of the participating partners, who were encouraged to incorporate their specific interests.

Other projects started as ways to put together the interests of different players. *IDEA*'s ambition is to create a young citizens' college, bringing together people from all over the world in a course in citizenship that builds on public governance research capacity at the University of Birmingham. It will start in 2003, with 20 students coming to Birmingham for nine weeks. The city would also like to develop education and programmes in 'food tourism' as a support to one of the main hospitality products of the city, exploiting the global recruitment catchment area of the local tourism college.

The West Midlands region, rather than a fully-fledged regional council, hosts a decentralised governmental office in charge of developing at the local level the national programmes for education, housing, poverty reduction, business support, etc. The *Learning and Skills Council* is the national office responsible for funding and planning education and training for all the over-16s in England. It was instituted to articulate the national policy at the local level to respond to the local specificities and needs. The autonomous role of the West Midlands region in education policy is indeed limited. However, it has developed the *Connexions* programme, a new approach to advising, guiding and supporting young people aged 13 to 19. It is a service with the key aim of encouraging more young people to stay in education or training so that an increasing number have the qualifications they need for further education or work.

2.3.2 Relations between the Student Community and the Private Sector

The network of relations with the economic environment of the region is less problematic. Because of their historical origins, UCE and Aston have the best ties with the business and industry of the region. The 'sandwich programmes' developed by these universities are functional to a clear-cut market orientation of education. Pupils spend the central part of their study programmes in large national firms and learn to work in a business environment. In the last few years, students have been increasingly attracted by the number of small and medium-sized enterprises (SMEs) in the region – also because of the changing geographical composition – so it was in the interests of the universities to develop more solid links with these firms. In this context the universities have also started to provide services for firm creation and support.

The relations of UB with the business world are mediated by a number of affiliated organisations. *Research Support and Business Development* (RSBD) contributes to excellence in research and building firm partnerships with business and the community by providing assistance to all academic staff engaged in research and related activities, particularly with respect to the generation of research income and the development of partnerships and improved access for business, the public services and the community to the wide range of expertise and knowledge available in the university. *Birmingham Research and Development Ltd* (BRDL) is UB's technology exploitation company. Its main activities involve patenting and licensing new technology from the university, advising on the creation of spin-off companies and arranging consultancy and development contracts. The company is wholly owned by the university and its board includes senior Birmingham businessmen and representatives of the university and of Birmingham City Council.

The Corporate Development Centre of the UCE is an initial contact point for a potential partner organisation. In recent years it has made contact with organisations such as MG Rover, the City Council, Virgin trains, Cisco, Arthur Andersen. Through the Technology Innovation Centre, at the faculty of technology, a national centre of excellence for learning, innovation and technology transfer, it delivers education, support and research to local SMEs. UCE also has a Corporate Development Body that is concerned with contacts with the businesses, for internships and placement. The TCS (teacher company scheme) is a state-funded programme involving an arrangement between firms and universities to take undergraduate students to work on a specific project. Aston has used a substantial national grant to support the establishment of the Business Partnership Unit (BPU), a commercial organisation focusing on the development of links at all levels with the private sector in knowledge transfer activities, consultancy relations, and job market initiatives. The BPU offers easy access to academic expertise to the business partners.

This synergetic relation between local firms and HEIs is a strong point of Birmingham education function and a necessary precondition for the development of the 'learning city' concept. However, the Birmingham universities face a number of competitors, such as the University of Coventry, a resourceful research-oriented institution specialising in the information-intensive sectors of the new economy.

In the near future, there will be a large concentration of learning facilities in the eastern part of the city. Aston University, Millennium Point and the Science Park are already located there. However, the functional links between the 'new economy' firms that are expected to flock in this area and the education and research infrastructure have not been planned. There is a belief that a cluster environment will emerge, but the development and maintenance of these linkages necessitate a sound governance structure, with a clear leader, both of which are missing at the time. An example of the kind of initiatives that would be recommended is provided by the South Birmingham College, a pre-higher education institution that is planning to create a school for the creative industries in a partnership with some IT companies. At the higher education level this kind of initiative is more difficult to organise, therefore more strategic cooperation would be needed among all the relevant actors. The UCE is involved in the Custard Factory project, through the Arts, Architecture and Design Faculty, but this is not sufficient. The recent move of BBC and other TV stations out of Birmingham gives a negative sign to this respect.

Several organisations group the main private players of the region together, providing different platforms from which to coordinate strategies regarding human resource development and the interface with local institutions.

The *Birmingham and Solihull Chambers of Commerce* have merged with the region's *Business Links* to provide the most comprehensive range of business support services in the West Midlands. The new organisation aims at being 'the leading source of business advice, support and representation for all businesses' in this area, providing choice and quality of support services to business. One of the core strengths of the Chamber and Business Link is close strategic partnerships with key organisations in the territory.

In March 1998 the most comprehensive consultation in English regions began on the issue of shaping a *Regional Chamber for the West Midlands*. A consultative paper – *Making the West Midlands Work* – launched a debate amongst 400 regional and subregional organisations. Among the Chamber's main goals are widening participation and developing links between the business sector and education at all levels. The Chamber includes key members of the Birmingham business community such as Cadbury and Jaguar. *Cadbury's Chocolate* is an important presence because of its strong sense of responsibility towards the local community, which translates into philanthropic actions and support to social development. Continuing the tradition of involvement with education within its own communities, Cadbury has developed a School Links programme, working with designated schools and colleges around company sites, with many staff acting as co-opted school governors.

2.3.3 Influence of Higher Education in the Relations between the Private Sector and the Local Community

Contrary to many cities in the UK, Birmingham had a long history of good cooperation between local government and the private sector. An informal meeting of the main stakeholders in Birmingham worked to set up a strategic plan to lay the basis of the economic and social regeneration policy. A good deal of emphasis was given to investment in the quality of human resources and research ('matching people to jobs') and to the improvement and promotion of the city's image. The city's accessibility (internal and external) was also a critical point, as was the development of a modern infrastructure in the city to accommodate the potential cultural and business visitors.

The strategy note of 1985 and the development strategy of 1989 have served as blueprints for the policy of the 1990s. In those years, political support for the development strategy remained at the same high level; changes in the composition of the City Council at most somewhat shifted the policy accents. A series of aggressive pro-development strategic actions, such as the bidding for the Olympic Games of 1992 and the realisation of opera and ballet venues, marked a new stage of 'boosterist policy' to attract international investment and attention.

Since 1998, the Economic Development Partnership (EDP), the result of a joint venture between the city, the Training and Education Committee and the Chamber of Commerce, has been in charge of implementing the action plan. The action of the EDP was geared towards three main goals: the social and economic growth of people and the community; the development of a competitive and diverse economy; and the long-term investment in high-class infrastructure to stimulate the development of a lively economic environment. The new Economic Development Programme 2001 identified the short-term actions required to build a solid position for Birmingham in the network economy. It coupled actions for widening participation with the development of strategic sectors through channelled investments and partnerships, in an integral, decentralised framework. The Eastside Regeneration programme illustrates how, in this new approach, physical, economic and social objectives are pursued simultaneously, with a focus on the generation and multiplication of local resources and the establishment of synergies.

In spite of the huge improvements in the economic situation of the city, the 'trickle down' of the effects of revitalisation policies to the social realm is still lagging behind. This is acknowledged by policy-makers, who are now making plans for a new stage of development in which social development issues are given a new focus. The unemployment rate in Birmingham is still twice that of the rest of the country. As a matter of fact, this does not necessarily mean that the regeneration activity had no effect, if one considers what the trend could have been without it. There is, however, a certain difficulty in transforming economic growth into community advantages. The Brindley Place development provides for apartment buildings in an area formerly reserved for manufacturing industry, which implies a change in the social structure of the ICC zone. The same applies to the inner city, which is enjoying increasing attention as a residential area. However, the newly-created jobs are mostly in the lower scale of the job market. At the same time, social polarisation has increased in spatial terms. The spatial distribution of social problems has hardly changed in the past few years. Unemployment is still concentrated in the inner city. In the same period, the investment for housing and schools fell in comparison to other large cities.

So far, the attitude of the private sector in Birmingham with respect to human resource development is all-but cohesive. Large firms relying on a long community tradition are more proactive and forward-looking in their networking efforts with the education and research sector. Small and medium-sized firms in contrast are much less active in this area, and barely invest in the city's student-friendliness. This is a real problem for Birmingham, because the development perspectives are clearly there and not in the traditional industries. In this respect, it has to be taken into consideration that, for the skilled working class of specialised small firms, other cities offer a more attractive living environment. Thus the Birmingham governmental organisations have to take the lead in the process of bringing the new growth sector and the local universities closer. The construction of a 'learning city' environment necessitates more cohesion and a clearer vision in the private sector. West Midlands Advantage is a powerful platform for establishing such awareness, but universities are not represented in this development agency of the Regional Chamber.

2.3.4 Model Testing and Identification of Issues

The story of Birmingham is one of a city that has undergone a huge economic transformation in the last decade of the twentieth century, guided by a proactive City Council. It has also worked hard to change its physical face and to accelerate its transition towards a new organisation of the socioeconomic environment, a post-industrial one in which spectacle, *soft* functions, high-quality residence, culture and education are centre stage. At the time of writing, this process is incomplete. The economic and physical changes did not mobilise sufficient resources and interests for 'soft' functions to become central in the socioeconomic development model of the city. The old industrial system of values and relations is difficult to dismantle and permeates the development processes of the city and their governance.

In this context, the 'virtuous cycle' of development as a student-friendly city (or, to use the terminology endorsed by the Council, a 'learning city') does not take off. The satisfactory links between private sector and universities – which extend to the cultural level and are the legacy of the historical origins of the Birmingham universities – are the strong point on which the dynamics of the system should build. But the relationship is unbalanced. While universities provide the private sector with the knowledge and the facilities that they need to be innovative and competitive, they get little back – especially from the new small and medium-sized enterprises – in terms of financial support, expertise and facilities for students. Though there is room for improvement, utilising, for instance, the instrument of the 'sandwich courses' offered by the Birmingham universities to improve networking with the SMEs, the general impression is that the private sector is scarcely pushing for a reorganisation of their links with the HEI to meet the oncoming challenges from the 'glocalisation' of the economy.

This is reflected in a poor capacity of the private sector to bring forward a decisive step jump in the socioeconomic conditions of the city. Birmingham, despite being a centrally-located and diverse city, still has problems in catching up with other metropolitan areas in the UK in terms of human resource profile, social cohesion, and attraction capacity. In practice, it remains a divided city, spatially and socially, with high youth unemployment rates, large numbers of unskilled long-term unemployed, and little attractiveness for new citizens of the high-skilled working class. This in spite of the indubitable prestige and capacity of its universities and colleges.

The great risk is that all the new ambitious redevelopment projects will remain unused, because at present there simply is no customer group for the post-industrial, spectacular Birmingham that the recent and new plans are building. In the new economic environment, the universities of Birmingham, which were born as engines of local socioeconomic development, have problems in finding a new role. The elderly, the unemployed and the marginalised have little to gain from the universities, despite the widening participation programmes of the national government, because the cultural and functional links between the universities and their communities have faded with time and have been given little in the way of new stimulus. Instead, what makes the headlines in the local press is confrontations between students and residents in districts like Selly Oak. The lack of a proactive 'urban strategy' by the universities, including the community and the private sector as stakeholders in their development plans, is reflected in locations that are not completely functional to an harmonious relationship (apart from Aston), in problems with housing, and in an approach to widening participation that is to a large extent felt as a top-down policy taking away resources from the universities and their best students, rather than as a source of opportunities. The lack of communication and platforms between governments and universities, and within universities, is making things worse.

The links that would make Birmingham a student city are insufficiently developed (Figure 2.4), not only as regards the size of the resident student population, which could nevertheless grow if the city were more attractive and offered better possibilities to students. The lifestyle of students has not been put to good use and transformed into urban culture, which is still attached to

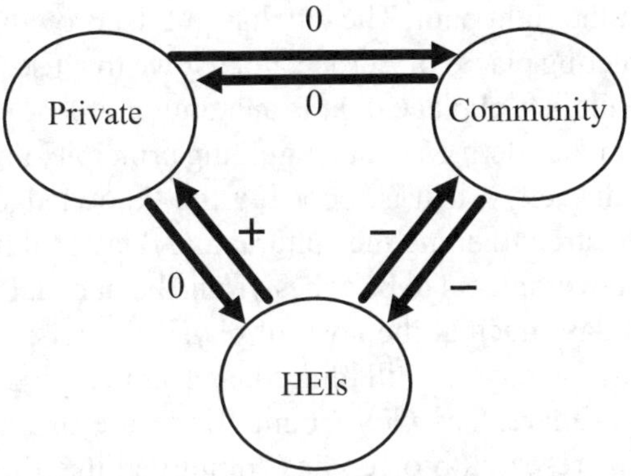

Figure 2.4 State-of-the-art: Birmingham as a student city

the legacy of motown city. The concept of a 'learning city' is no more than a guiding principle that has led to some outstanding physical planning but has not yet changed the face and soul of the community in the desired direction.

For the future, it will be necessary to plan this integration with a more cohesive attitude, striking the right balance between economic regeneration, sociocultural integration and physical development. Unless all the relevant players are actively involved in planning and implementation, in a genuine bottom-up approach involving the community, the development of the 'learning city' concept will inevitably lead to tensions, unwanted side-effects and more exclusion. In the context of Birmingham, with the general lack of cohesion between institutional actors, private partners and universities, this is precisely what is happening.

2.4 Summary and Conclusions

Birmingham's urban policy, with its accent on the functional mix of the inner city as a tool for developing a high-quality environment attractive for both residents and visitors, is not common among British cities. The formulation of a clear vision for the city and the capacity to transform it into policy action is a major achievement by the City Council, which could rely on a tradition of good cooperation with the private sector. Birmingham has been among the first European cities to succeed in anticipating a growth market: the market for conventions and fairs. The investment made to that effect is based on an integral strategy oriented to modernising the economic structure. Now, 15 or more years after the strategy was defined, the conclusion is warranted that

Birmingham is on the right path. The city has put itself on the map as one of the outstanding meeting places of Europe. For a town that used to be renowned as an unattractive industrial place, that is an admirable feature.

The City Council has defined another guiding principle of its urban policy: becoming a 'learning city', that is, boosting the 'knowledge' sectors in the local economy and strengthening the cultural and relational links in the local society that allow development to be cohesive and balanced. Progress towards this objective is today much at the level of physical plans; the *software* and *orgware* of the learning city are still blurred; a clear mapping of the available resources is not yet there. The City Council is aware of this and, with the participation of this research project, has committed itself to improving the strategic relations between the universities and the city, hoping to make a better use of the resource potential in the students' community to achieve its regeneration objectives.

There are at least two points where Birmingham can make substantial advances towards becoming a student-friendly city

Governance and Networking

For a city in transition like Birmingham, the accent in development policy on new functions and economic activities is crucial. Universities are already quite strong as individual institutions, but to become the engines of economic regeneration, they need to make better connections with the industries under focus, such as cultural and personal services, research and development, ICT, etc. Given the immaturity and fragmentation of SMEs in these sectors, building partnerships may be a necessary step towards involving firms clearly in the city's development programmes. In this respect, the approach of policy-makers has so far been limited. Instead, the role of the Council has steered development through partnership formation between private sector and research institutions is fundamental.

There is a lack of fora in the city where policy options can be brought into the open and discussed, guaranteeing the necessary social and political backup to key initiatives. This step could also mark a difference in the relations with the universities, which at the present stage is based on opportunities and not a synergetic, durable and bidirectional dialogue. The City and the universities are starting to look at each other in a more strategic way, but an institutionalised platform to discuss common problems and find solutions is still missing. This is also needed to accommodate the tensions between institutes. In its absence, destructive competition between organisations is likely to prevail instead of cooperation. Each university has its own charter and board, and there is no associative body of higher education that can speak with one voice to local government.

The lack of such bodies is clearly connected to the lack of a fully-fledged regional autonomy. However, Advantage West Midlands could be used as a platform involving the regional universities more closely and having them share the regional development strategy, with attention to the specific needs of the different areas included. The marketing of the region as a 'excellence pole' for education and research, as indeed it is, cannot happen without the direct involvement of the universities and in the absence of a strategic coordination of the single institutes' and the cities' international relation policies.

Image, Housing and Culture

Birmingham has to become more attractive for students, at least in three categories:

- beginning HE students who are in the process of choosing whether to become 'temporary residents';
- finishing HE students who have to make a career choice in or out of Birmingham;
- postgraduates and researchers looking for an attractive place to work and study.

In relation to the first category, Birmingham's universities have to extend their market catch. In this way they can become fully functional in the objective of the city to strengthen their national and international competitive position. The universities have no problems in attracting students or recruiting personnel, but their reach is local and hence they contribute little to the dynamism of the social base of the city. On the other hand, the city should start to market itself explicitly as an education centre. This objective has to be pursued in conjunction with the city, in a comprehensive marketing strategy to make the city more attractive and hospitable for young people. However, if the promises do not live up to the customers' expectations, the results are counterproductive. Therefore the city has to be aware that they have to invest in upgrading the city's cultural and recreational infrastructure. This is by no means the same as what will appeal to middle-aged conventioneers or tourists. Projects like the revitalisation of the jewellery quarter and the Custard Factory are first steps taken in the direction of establishing a 'hipper', 'sexier' brand for the city, which need to be articulated in a wide-ranging strategy for the identification and use of these assets.

In relation to the second and the third categories, Birmingham has to become an attractive place to live and build a life career. This is not the case at the moment; because of the proximity to London and other more attractive urban areas, new graduates leave and local firms have recruitment problems. This situation is linked to the image of the city, its social structure, its low-quality housing stock and the relative insecurity of its pedestrian passageways and malls. Birmingham should invest more in opportunities for 'life careers' in this area. To this aim it is required that the city takes a more active role in students' housing policy. By directly or indirectly managing part of the housing stock, providing good quality and dedicated facilities for selected users' markets, the City Council can influence location choices. Today, housing solutions are either completely centralised in-campus, and as such completely out of the reach of local policy, or left to the private market.

In this respect, we endorse the recommendations of Groves et al. (2002), who suggest that the UB and City Council join forces to find sustainable solutions to the emerging problems with the housing market at Selly Oak. They have to involve all the stakeholders in the area as well as the community and the students' organisations to manage the student settlements in the area better. Other actions could include a dedicated transport policy to reduce traffic in the area and, in the medium-term, the provision of alternative housing in university-owned properties in the proximity, or forms of student housing that reduce shared housing and hence do not affect the property value structure in the area and reduce the risks for tenants.

Notes

1 Other projects concern Broadway Plaza – a £70 million scheme to transform the old Birmingham's Children's Hospital into a leisure and entertainment complex; the Birmingham Alliance – a collaboration between three of Britain's top developers on Europe's largest city centre regeneration scheme; Five Ways Leisure Complex – the development of a complex of family-orientated leisure facilities including a multiscreen cinema, tenpin bowling and restaurants; Arena Central – the transformation of a six-acre site around the former Central Television Studios into a glass-covered leisure complex. Proposals also include the creation of 'City Tower', a multistorey skyscraper which would be Britain's tallest building. The most interesting project however is the Millennium Point, a key catalyst in the regeneration of the Digbeth area. This enormous building will be the base for the Discovery Centre, a Technology Innovation Centre, the University for Industry and the Hub, a cluster of places to eat, drink and shop.

2 In the latest (2001) UK Research Assessment Exercise, the University of Birmingham was the fifth most successful university in the UK for research excellence. Forty-five research areas were rated at the highest levels indicating research of national and international excellence, including a number of cross-disciplinary research centres. The figures also show Birmingham has an average of 78 per cent of its staff working at this level. The national average is 55 per cent.

Chapter 3

Eindhoven Case Study

3.1 General Introduction to the City of Eindhoven

The city of Eindhoven is situated at some 100 km east of the Randstad, the cultural and economic centre of gravity of The Netherlands. At the beginning of the twentieth century, the city of Eindhoven was no more than a small agricultural town, with some 5,000 inhabitants. The foundation of Philips in 1891 marked the beginning of the rapid development of the city (Adang and van Oorschot, 1996). The city has grown, incorporating the area of five smaller towns. Nowadays, Eindhoven is the fifth largest city of The Netherlands, with 203,000 inhabitants. It lies in the province of North-Brabant, and stands at the centre of a region, Southeast Brabant, which is often referred to as the Greater Eindhoven Area, or the 'Eindhoven region' (see Figure 3.1). This region comprises 22 municipalities. Like the city of Eindhoven, the population of this region grew quickly in the twentieth century. Today it counts some 709,000 inhabitants. Contributing 40 per cent of the industrial added value in The Netherlands and almost one-third of the industry jobs and half of the research and development expenditure by firms, the Eindhoven region can be considered effectively the true 'industrial heart' of The Netherlands.

The landscape of Eindhoven is strongly determined by twentieth-century urban design. In the city centre, buildings and housing estates offer a wide sample of modern residential architecture. The city's fast growth and its focus on technology have left a heavy mark on its appearance. However, apart from the impressive Philips complex of Witte Dame, the architecture is hardly memorable, unlike other modern Dutch cities such as Rotterdam and The Hague. Although Eindhoven is the fifth city of The Netherlands, up to the 1980s the urban environment did not live up to that status. The town centre looked chaotic, and there were no special attractions, apart from the futuristic *Evoluon*, which was closed down in 1984. Since the 1990s, Eindhoven has tried hard to bring its urban environment in line with its economic, social and cultural position. Today, the cultural climate in Eindhoven is enriched every year by some 200 exhibitions and thousands of performances in the mainstream and alternative music and theatre circuits. A high quality shopping centre (*Heuvelgalerie*) and a concert hall (Frits Philips) have been added to the city's infrastructure.

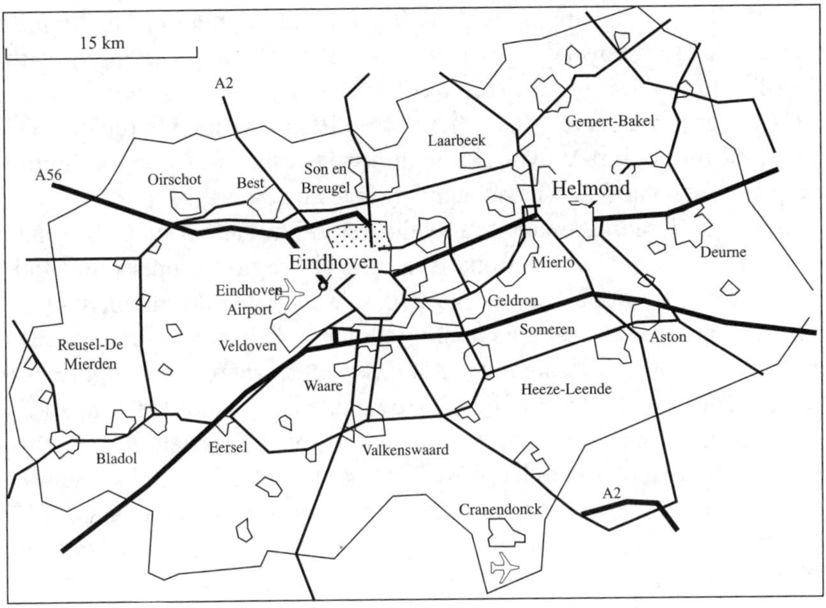

Figure 3.1 The greater Eindhoven region

Source: FPER, 1997.

The location climate of the region on the whole is very favourable for the firms in the area and their employees. By Dutch standards, the Eindhoven region offers plenty of housing space. Additionally, much natural beauty and open space surround the city of Eindhoven. Such environmental amenities score highly as attraction factors. Even foreign employees are generally pleased to live in the region. However, despite the afore-mentioned cultural improvements, a metropolitan atmosphere is definitely lacking. This means that certain types of hi-tech economic activities in the creative industries, such as multimedia firms and design workshops, are under-represented in the region.

During the last 15 years, the economic fortunes of the region have varied considerably. Up to the mid-1980s, the region did relatively well, with growth rates higher than the national average. From then on, however, regional employment growth was below national average. Between 1986 and 1991, the yearly employment growth rates dropped from 5.3 to 0.9 per cent. Things worsened in the following years, with the region starting to lose jobs for three years on a row. The main factors behind the economic problems of the region reflected a downturn in the fortunes of the two leading manufacturers. In the early 1980s, Philips employed 35,000 people in the region. This figure had dropped to 21,000 by 1993. In 1993, the DAF company collapsed, with the loss of a further 2,500 jobs. Additionally, the large regional supply network of DAF was also severely hit. After a few difficult years, the region strongly recovered in the mid-1990s. The region hosts other fast-growing high-tech companies such as ASM Lithogaphy, Simac and Neways.

Today the region is one of the wealthiest in The Netherlands. Incomes per capita are well above the national average. The region counts some 350,000 jobs (in 2000) in a population of 709,000, as well as 43,600 enterprises (mainly in the technology sector). The unemployment rate in the region in 2001 was lower than in other Dutch cities, at 4.6 per cent, whereas this figure rises to 6.0 per cent when the city of Eindhoven alone is considered (Etin, 2002a).

The technological focus of the local economy places a natural emphasis on the educational and research infrastructure of the city. Especially after Philips scaled down its operations in the city, one wonders what could be the role and the capacity of local institutions to support and renew the city's development efforts. One critical question arises: whether the city's lack of attractiveness is getting in the way of the city's strategy as a 'knowledge community'.

3.2 The Student Community and Knowledge Infrastructure in Eindhoven

3.2.1 General Information on Higher Education in Eindhoven

On the whole, about 18,000 higher education students are hosted in Greater Eindhoven. Coherently with the region's prominent position in the field of advanced technology, technical education is present at all levels, from secondary education to the university level. This makes Eindhoven a dynamic research centre, and an interesting place for technical students and scholars to develop first-class skills.

Arguably the most important player in the educational field is the Eindhoven University of Technology (TU/e), one of the three technical universities of The Netherlands. The city also hosts the Fontys Institute for Higher Education, one of the largest HE institutes in The Netherlands, and the Design Academy.

However, one should not forget the role of giant conglomerates like Philips and networks of small firms in creating knowledge, through dedicated research and exchange programmes and as a training infrastructure for higher education students.

3.2.2 *The Knowledge Sector of the City: Actors and Spatial Patterns*

Universities and other higher education Institutions The *Eindhoven University of Technology (TU/e)* was founded in 1956 as a polytechnic school. The presence of Philips in the region, in particular the NatLab, was an important stimulus to founding a new university in Eindhoven. Active lobbying and promotion of the local politicians in Eindhoven were decisive in attracting the university to Eindhoven.

Today, TU/e has grown into a fully-fledged institute offering a complete range of education in many technical disciplines, and carrying out a broad range of research, both fundamental and for the industry. TU/e is organised in nine departments, hosting 12 BSc and 19 MSc programmes and three academic teacher training programmes in mathematics, physics and chemistry. In addition, TU/e's Stan Ackermans Institute has ten postgraduate technological design programmes. A specific two-year postdoctoral course is offered in mechatronic design. In these courses, graduate students in mechanical engineering, electronics and physics are trained to develop skills to combine these disciplines and to think and work across disciplines. Education is based on the universities own research activities, with a focus on design. Lecturers and students make use of state-of-the-art information and communication facilities. In 2000 TU/e had approximately 3,000 employees, 300 professors, 6,000 students, 200 postgraduate students, 450 PhD students, 20,000 graduate engineers, 1,000 Masters of Technological Design, and has awarded 2,000 PhD degrees. Every year, more than 1,000 students graduate.

As with its research, TU/e education programmes have an international orientation. Students are encouraged to spend part of their study career abroad, or in collaboration with foreign partners. TU/e has close ties with a limited number of institutions, not more than four or five European technical universities (including Munich). Its Bureau for International Activities (BIA) is responsible for the domains of international scientific collaboration, development of cooperation and staff and student exchange. Practicalities such as accommodation, insurance, support and information on exchange programmes are taken care of by BIA. TU/e also leads prominent Dutch research schools and institutes, and holds a strong position in international networks. Nationally and internationally, TU/e holds a prominent position in the research fields of communication technology, materials technology and catalysis. Moreover, TU/e also aims at diversifying its activities. The cooperation set up with Maastricht University in the field of bioengineering is a clear example.

TU/e is an excellent partner for companies strongly oriented towards technology. The campus serves as a foundation for new enterprises. On the campus site, at the time of writing, there is lack of an incubator-type of facility. However, TU/e is very interested in developing this aspect. Currently, the start-ups which originate on campus (38 new firms were started in 2002 by TU/e graduates) can only stay there two to four years, after which they have to move to a business park. TU/e also seeks for optimal knowledge and technology transfer to industry and other organisations. Due to the increasing significance of contract activities for third parties, the Executive Board of TU/e decided to set up a holding structure, so that the exploitation of TU/e's scientific knowledge can take place in a businesslike and clear fashion by means of subsidiaries. TU/e *Holding BV* has some subsidiaries in its care, like the UTI and *CEBRA* (Centre for Electronic Business Research and Application). This institute applies and disseminates knowledge in the field of e-commerce, which is made available and is generated by TU/e, so that Dutch enterprises and government organisations can benefit. The centre carries out all sorts of applied research in collaboration with industry and other centres of knowledge.

The university campus of TU/e can be considered as the largest 'wired community' in Europe. Since 1996 every first year student at the university has been equipped with a notebook computer on favourable financial terms. Students can plug in anywhere on the campus, and have broadband access to the Internet, e-mail services, and all kinds of faculty facilities. A critical mass of users with broadband access is therefore in place. Recently, an e-commerce professor has taken the initiative a step further and is using the campus as test bed for e-commerce. A project named *VirTU/e*, realised by CEBRA in cooperation with 20 companies, completely funded by the private sector, will be finished soon. VirTU/e enables firms to offer state-of-the-art services to students, and to experiment with new technologies for very receptive customers. For instance, a supermarket seeks to test intelligent agent technology to improve its e-services. A virtual shopping plaza has opened with many on-line shopping possibilities. Particular attention is paid to the development of safe and secure identification and payment systems. The local government will also open a virtual desk on campus to experiment with the newest services.

TU/e is very active in the cultural sphere, with interesting repercussions for city life due to the dimension of the campus. The President of TU/e is also Chair of the association which brings together the city's cultural institutions. This association is making plans about how to improve the cultural life in Eindhoven. The aim is to boost cooperation between university and organisations with good ideas. Nowadays the university campus hosts three art ateliers. The initiatives are privately run; the support of TU/e is limited to concession in respect of land use. However, stimulating cultural activities can be a difficult achievement in a city that does not have the 'cultware' of other Dutch cities. Consequently, a more integral approach is needed. TU/e wants to upgrade and transform its status. The concept that is guiding development plans – 20 new multifunctional buildings are planned for the next decade, with a budget of €400m– is defined as 'Esklepion': the union of studying, living, evolving as individual as a holistic experience.

The *Fontys Institute for Higher Education* is one of the biggest HBOs (university of professional education) in The Netherlands. It is present in different other locations, such as in Amsterdam and Rotterdam, with 33,000 students, 3,000 staff academic, and 2,000 technical and clerical staff. Fontys was established in 1996, following a merger between a number of smaller institutes. It offers higher education to some 14,000 students. Some of its more than 100 professional education courses are also offered in English and German. These include courses in healthcare and social work, economics and communication, applied science and technology, visual and performing arts, logistics and transport, and education. In addition, Fontys is actively involved in business service provision, through the provision of Master's courses, post-higher professional education courses and training programmes, research, and consultancy. Between four and six meetings a year with the private sector are organised to discuss study programmes, and to make them suitable to the needs of the industry, according to the market evolution. The relations with the business of the region are therefore very good.

Fontys has established considerable networks with comparable schools, attracting students from Scandinavia, America, and Germany. The school carryies out an aggressive marketing campaign, distributing information packages in different languages to students and companies. Emerging markets like China and Eastern Europe are approached through the embassies and visiting groups are organised to show what the school can offer.

There are several other relevant elements in the knowledge infrastructure. The most important is the *Design Academy Eindhoven*, and attached to this, the *European Design Centre*, which educate students in the designing of industrial products. As the importance of industrial design in new products is increasing, this institute is an asset for the mechatronics cluster in the Eindhoven region. The Design Academy, established in 1947 as the Academy for Industrial Design Eindhoven, uses a conceptual and project-oriented approach to industrial design education. Its new name, coinciding with its move to the de Witte Dame, was adopted in 1997. De Witte Dame ('The White Lady') originally housed the Philips' incandescent lamp factory. For a long time a source of civic pride and symbol of technical progress, the building is now a symbol of human potential. As a centre for design, information and technology, de Witte Dame is home to the academy, the Eindhoven Public Library, the Mu-Art Foundation and Philips Design, thus representing an innovative cluster of education and production in the creative/cultural sector right in the centre of Eindhoven.

Research and technology transfer institutions The Eindhoven region is 'leading in technology' and in research and development activities. Public organisations and private firms contribute to establishing a top-level knowledge-intensive environment. Besides the research institutes and activities of the university, the campus also managed to attract top organisations such as TNO-Industry, the Eurandom Institute, and the Twinning Centre, developing into a focal point for technological research and business. The aim of the Twinning Centre is to offer space for young entrepreneurs in the field of information and communication technology (ICT), and 'twin' these firms with the expertise of senior businessmen to increase their chances of success. The initiative of this concept was taken by a senior manager at Philips and the Minister of Economic Affairs, with the aim of strengthening the position of The Netherlands in ICT. It is a public-private partnership between all the relevant actors in our study framework: local and regional governments, TU/e, and the private sector, aided with the European funds of the Stimulus Programme. There are only two other such centres in The Netherlands. The large firms in the region appreciate this centre, as they have an interest in maintaining strong ICT firms in their vicinity.

The research institute TNO-Industry, a branch of an important national research company, adds to the knowledge infrastructure in Eindhoven. This institute carries out contract research for firms, very often in cooperation with technical universities. TNO-Industry's focus is on mechatronics, combining existing technologies to solve production problems. It employs 400 people, most of them high-level scientists. In 1997, the institute decided to move to the TU/e campus. There were two reasons for this. Firstly, the majority of TNO's clients are situated in or very near to the Eindhoven region; secondly, it was felt that TNO could realise synergies from its proximity with the university labs. Cooperation would take the form of knowledge exchange, staff exchange, and joint projects in the institutes for metals and polymers. Furthermore, research cooperation is planned in the field of materials technology.

NatLab is the most important research laboratory of the Philips organisation, and the largest private laboratory in The Netherlands. With an impressive 1,700 staff, it provides for very high quality research. At the current site of the NatLab, Philips is building a large technology campus, where most Philips R&D activities that are scattered through the country will be concentrated. Other organisations of the region active in knowledge transfer are Interbaan, ECIS, PDNE, EESI, CCM, United Brains.

Spatial patterns of higher education and research TU/e is different from the other technical universities in The Netherlands. Twente, for instance, has its own campus, where many students are housed (20 per cent of the students, mainly first-year). This university has a real campus atmosphere, but is completely detached from the urban context. For its size, Delft is different; it is a real historical student city, and it is located in the Randstad. TU/e is practically situated in the centre of Eindhoven. The compact green campus does not include normal student residences, which are scattered all around the city, but an impressive number of university buildings, laboratories, and sports facilities, as well as a student centre, an auditorium and many restaurants. Most buildings are connected via walkways. The other higher education institutions are also located at small distances. In fact, Fontys has also expanded onto the TU/e campus.

In common with other cities in The Netherlands, Eindhoven has high commuting rates in the student population. The location of the higher education institutions and the transport links are good, thus commuting is not a problem. Only 40–50 per cent of students decide to live in Eindhoven. Fontys' students commute more frequently than do TU/e's. There are some differences between first year and older students, as the latter prefer to live in the city. Most first-year students stay with their parents, and possibly decide to find accommodation in Eindhoven later on.

The housing market in Eindhoven is changing fast. With its mere 40 years of history, it is still very dynamic if compared with cities such as Delft or Leiden. Still, Eindhoven ranks third in The Netherlands for lack of housing after Amsterdam and Utrecht. Apart from the quantitative scarcity, a planning regulation limits housing facilities per street – less that 10 per cent of houses per street are allocated to the use of students. Today this rule is applied loosely, and it can be circumvented easily as parents buy houses and make students live there. However, the number of students who seek accommodation is increasing very fast, and precise information on these numbers is missing, so that planning is made more problematic. With the introduction of the BA-MA structure, it is believed that the mobility of students will increase, and the pressure on the market with it.

The number of foreign students is also growing. They express a very different demand, as it is often difficult for them to arrange accommodation from their own country. Fontys, for instance, does not have its own accommodation facilities for foreign students.

As a consequence of the mismatch in the accommodation market, the scope for a fuller integration of students in the community is constrained. Moreover, the city's attraction capacity for skilled workers, also from abroad, remains limited. Opportunities for settling in good quality housing are missing, especially for those postgraduates and staff who are attracted to the prestigious hi-tech institutes of the region. This problem threatens to hamper Philips' plan to set up their own science campus; quality solutions will be needed for the thousands new staff expected to work there. Important stakeholders like the Mayor and the Chamber of Commerce are aware of this threat, and private developers have shown themselves to be willing to play a part in improving the situation, out of self-interest. Philips is moving autonomously, trying to buy or rent the houses all around the campus area, and in exchange they sell their own buildings in the centre.

A working group is elaborating solutions to the housing problem, not only for visiting academics, but also for private sector staff. An interesting TU/e project to expand the accommodation capacity is the *Fellow-tel*, a special structure that can host hundreds of graduate students next to the campus for short stays. This project has been sponsored by private parties, such as housing corporations and developers (with a contribution from local government), who expect to make revenue out of it. TU/e guarantees a certain occupancy rate. It was fully booked on an almost permanent basis in the blink of an eye. At the time of writing, another multifunctional project is on the agenda, combining in the same building housing, office, shopping and workplaces. Housing would be made available at three levels: undergraduate students and exchange short stayers; graduate students and young researchers; and visiting professors. In this way, it also becomes easier to find developers willing to invest, as market risks are reduced, and the profitability is enhanced. Another plan seeks to reuse the many existing first floors above shops in the city centre, to create accommodation units for short-stayers.

Vestide is the largest provider of accommodation for students and short-stay guests in Eindhoven.[1] It runs four complexes in the proximity of the campus, hosting almost 400 students of TU/e and Fontys. In total, Vestide owns 42 buildings within the inner circle of the city, and 1,700 residential units. Moreover, Vestide rents from other owners, housing corporations and private sector parties, catering for 18–20 per cent of the students who rent houses in the private market. Most foreigners are housed within agreements with TU/e. Two hundred flats are reserved by TU/e every year for exchange programmes; Fontys needed this service for almost 250 international students. In total 240 rooms are reserved for foreign guests, but the demand in 2001 was for 600; thus there is a shortage of about 300 units in this sector. The potential is four times larger than the number of rooms that Vestide has available. This is an important issue because the foreign guest market is leading this potential expansion. Vestide aims at increasing the number of rooms by 70 per cent, to 2,500, in the period 2000–2004.

Through a registration in a waiting list, students can apply for different categories of housing. Coherently with its mission statement – to make the living conditions for students as pleasurable as possible – Vestide guarantees a minimum quality of renting, for safety, space, use of certain common areas like kitchens, toilets, and garbage disposal, also offering to students a service for complaints and comments. Every flat has a cable connection. Vestide emphasises qualitative elements over quantitative aspects. In this way, they have an opportunity to convert the old-style accommodation into student housing. As regards the price policy, the basic rent is almost the 55–70 per cent of the rent calculated according to the government rules. Foreign short-stayers can get a furnished room for €200–300, postgraduates and staff of TU/e or Fontys for €350–900, offering also reasonable conditions for paying the rent. However, rents are on the rise in Eindhoven. The waiting time for a room for students is almost six months, which, despite being shorter than in other Dutch cities such as Rotterdam, is far too long for short-stay students.

In cooperation with another company, new housing developments have been planned within the inner ring or just outside it. Six hundred flats will be built in the immediate proximity of the campus and possibly in a few other places in the next 5–10 years. Moreover, Vestide seeks to expand the 'hotel' function to re-use the empty houses in summer. Vestide wants to involve the private sector and develop a semi-commercial segment in the accommodation market for visiting young people. As the biggest housing service, Vestide is the most affected by the 10 per cent rule. This rule seems to make little sense, considering that Vestide offers good level of quality, which is also beneficial to the community. The local residents are indeed willing to recognise a visible change in the tenancy attitude and mix, and they do not object to the presence of students. The 'costs' imposed on the community by this type of settlement are very limited. It is believed that the outcome of a regeneration process driven by student settlements is closely associated with a care for quality in student housing. The new city government has promised to relax the 10 per cent rule.

While Vestide caters specifically for the student community, the city is busy with more integral project to build a 'student-friendly', attractive Eindhoven. The restructuring of the old Philips facilities is a promising first step. The whole site of the Witte Dame complex has been purchased from Philips for €140,000. With a careful planning of its use, it can bring forward a real change to the image to the city. At the time of writing, the city is negotiating with private sponsors the redevelopment projects. 400,000m^2 of top-quality office and living space in the city centre will be recuperated with these operations. This high-density urban environment marks a neat difference from old-style city centre development in the city. It is definitely fit to attract the profile of new citizens that the city is looking for.

3.2.3 The Student Community: Behaviour and Careers

People appreciate the living conditions in the Eindhoven region when they are 35–40 and have a family, as the charming countryside is an ideal place for such households. Young people, however, tend to be very critical of the city. They choose Eindhoven because of the high reputation and quality of its higher education institutions, and they enjoy its student climate and atmosphere, not least because Eindhoven is also less expensive than Utrecht or Amsterdam. However, Eindhoven is not seen as an attractive city; indeed, it is considered one of the worst cities in The Netherlands. Still seen as the 'Philips city', Eindhoven lacks a historical heritage and atmosphere. Even if the region is rich in job opportunities and very wealthy, after their graduation students tend to go to work somewhere else. Both TU/e and Philips always have vacancies for PhDs and postgraduates, but these positions are not necessarily filled. It also depends on which kind of skills and which kind of jobs they get, but in general, the more mature students leave.

The student community is not very visible in Eindhoven; it is not a major group of city users. There are different kinds of people in the local higher education institutes. Many students come to Eindhoven from Brabant; this does not help to develop a 'student city' feel, because they go home at the weekends or even every day. Some would even go back home every night, just like high school students. This is not only due to the lack of housing, but also because it is widely believed that Eindhoven is not the place for having a good time. Others stay for longer periods, and they are active in other fields than their studies, such as organising activities, participating in sports events, and enjoying the university life from an active perspective. These sort of people are those who take full advantage of the student city. However, this is a relatively small number; many students preferring to spend their free time in laboratories and in front of PC screens. Moreover, the student community tends to be rather close-knit. Eindhoven's social life is peculiarly organised in hidden, close 'back gardens', rather than in the open, public places that – in the cold and businesslike city centre – lack appeal.

Student life in Eindhoven tends also to be highly 'organised' into student associations. A large number of students associate in clubs, debating societies, sporting clubs, etc.: there are almost 100 associations. Students are generally members of more than one group, and in this way the networks are strengthened. Being a member of an association may make finding a house easier, by exploiting informal contacts and sub-letting agreements. The Eindhoven Students' Union (ESC) is the oldest social club in Eindhoven. It is open to both TU/e and Fontys' students, therefore representing a mixed group of different backgrounds and interests. The ESC organises a large number of activities like sporting events – during which people from other cities come to Eindhoven (an important contribution to the city's marketing efforts) – forums, lectures, films and parties. The General Students' Union (SSRE) has about 300 members. Members can participate in activities like lectures, excursions and weekend trips. ESU Demos is the youngest of the three students' unions for TU/e and polytechnic students. Demos is proud of its democratic character, which frees it from restrictive procedures for admissions and hierarchies. Demos is actively engaged in projects for people in the community, like the disabled, the elderly, and charity activities, and also tries to involve international students, developing targeted activities and contacts.

All the student unions have their own equipped clubhouses with bars or TV rooms where members can meet every day. They are active in raising issues about the student population at the highest levels: the housing problem is the most immediate, but they are also concerned with the city's and institutes' communication strategies, and with the attractiveness of the city. An interesting initiative of the student union ESC has been the 'student party' (List 11) that ran in the last local elections. Though they did not manage to send an elected member to the City Council, the campaign was a good opportunity to start a communication channel regarding the problems of the student population in Eindhoven with an audience that was wider than just their peers.

These associations have a critical role in the university-city relationship. They are indeed busy trying to change the bad perception of the city. In the last few years, the community-orientation of the student population has become stronger. ESC has welcomed 50–60 people into their clubhouse to receive information on the union's activities. ESC also organises a large-scale festival annually on 10 August – citizen's day – organising exhibitions, computer games, and get-togethers of every kind.

These associations work like a platform, where active members can easily learn to be more entrepreneurial. At the same time, with their very role, they may be a limit to 'spontaneous' encounters between students and the community.

3.2.4 Challenges for the City

The partial demise of Philips, which moved its headquarter functions to Amsterdam some years ago, had profound repercussions for the city. For one thing, the social structure started to change, with fewer high-level managers living in the region. Moreover, it marked a cultural change. Philips, who so far had provided almost every facility in the city (culture, sport, infrastructure), could no longer be counted on. The city had to start looking for other partners and industry leaders. This marked a new phase of development for the city, one in which the city sought to create the conditions for a successful, sustainable development, acting proactively in anticipation of market developments, with an entrepreneurial attitude.

The city had to reinvent its economic texture, with an emphasis on small and medium-sized enterprises, and hi-tech functions. This required infrastructure to reinforce the network dynamics in the industry, and an improvement in the ICT alphabetisation of the local human capital. Along with the general recovery of the Dutch economy in the second half of the 1990, the thorough restructuring of parts of the operation chains of Philips and DAF generated positive spin-off in the form of newly-started businesses and the outsourcing of activities. Though the economic structure of the Eindhoven region is still dominated by the manufacturing industry, this has considerably modernised and evolved in its organisational structure, from a hierarchic, Fordist model towards a more articulated network structure.[2] The share of manufacturing in the regional economic structure (in terms of both employment and added value) is in this way one of the highest in The Netherlands, where the average employment in industry is 14.4 per cent. However, trade, producers and personal services are also large employment sectors.

Today, this process of change is moving fast and the industry is remarkably modern and knowledge-intensive. The Eindhoven region gave itself the title

Table 3.1 Employment structure, Eindhoven region and national average, 1999–2001 (%)

Sector	Eindhoven region		The Netherlands	
	1999	2001	1999	2001
Agriculture	4.3	4.2	5.3	4.7
Industry	21.4	21.8	15.2	14.4
Construction	7.0	7.1	6.6	6.7
Trade and repair	18.1	17.4	18.1	18.0
Hotel and catering industry	4.3	4.5	4.0	4.0
Transport, storage and communication	4.3	4.8	6.2	6.4
Financial services	3.3	3.4	3.7	3.9
Business services	15.5	15.3	12.4	13.7
Public administration and government	3.0	3.0	6.0	5.7
Education	5.3	5.4	6.0	5.9
Health and welfare services	9.9	10.2	12.5	12.6
Other services	3.0	2.8	4.0	4.1

Source: ETIN, 1999, 2001.

of 'leading in technology', and indeed it is one of the most innovative in terms of research and development: 50 per cent of total Dutch R&D expenditure occurs within the region (Kusters and Minne, 1992).

The NEI (1996) distinguishes a number of 'knowledge-clusters' in the region of Eindhoven/Venlo (see Table 3.2). The most important clusters in the region are electronics, automotive and office equipment. The dominant position of the knowledge-intensive industry in the region serves as an engine for other sectors, such as trade, business services and other services. In the high-tech sector, the number of start-ups in Eindhoven is quite high, and well above the Dutch average. Eindhoven has also proved capable of attracting firms: 200 of them originate from outside The Netherlands. With these figures, Eindhoven ranks among the top three of most knowledge-intensive regions in Europe, together with Paris and the Munich/Nürnberg region in Germany (NEI, 1996). The high knowledge-intensity of the region owes much to the large firms, such as Philips and Nedcar, which invest considerable budgets in R&D, as well as to the strong networks of their suppliers. However, the excellent higher education and research institutes of the city, such as the Eindhoven University of Technology, contribute crucially to this innovative environment.

Since 1994 the region has done quite well, showing the highest growth rate for industrial production in The Netherlands for five consecutive years. However, this impressive growth has stalled. One reason is that the human resource pool is insufficient to support this growth; either the rate of participation of the local residents in technical studies must be raised, or more skilled workers must be attracted to the region in order to keep growth to its potential. Presently the region of Rotterdam – which is far more attractive as an employment basin, but shows inferior education rates – is growing faster than Eindhoven. Science-trained human resources find this an attractive region for starting an economic activity, yet the labour force trained in other sectors, such as the creative industries and the financial services, decamp to other places like Amsterdam.

Table 3.2 Knowledge-intensive industrial clusters in Eindhoven/Venlo, employment in different categories, 1996

	Producers	Suppliers	Knowledge infrastructure	Total
Electronics	6,000	9,200	4,000	19,200
Automotive/transports	4,100	10,500	4,000	18,600
Office equipment	5,300	6,200	4,100	15,700
Transport/distribution	4,100	1,000	3,900	9,400
Environmental technology	–	370	3,973	4,343
Medical technology	5,100	2,000	3,600	10,700

Source: NEI, 1996.

Despite its peripheral position with respect to the Randstad, Eindhoven's location is by no means marginal, in particular in a European perspective. The region is situated within the rectangle formed by the Randstad, central Germany, the Ruhr Area and the Belgian cities of Brussels and Antwerp. External accessibility to the region is good. The region is well connected with highways and rail. Eindhoven's international accessibility is very important for the firms in the city, which often operate on European or even worldwide scale.

The region's internal accessibility falls short, however. The entrance roads into the city are not suited for the amount of traffic they have to carry. The road infrastructure of the city, a conglomerate of five former villages, is inadequate for current needs, though a project is under way to connect the region through a new system of fast and high-performing public transport. Eindhoven is a convenient location but not an attractive city. Having a lively car-free area with bars is already helping considerably, but the very ambience of the city has to change. An 'image restyling' effort targeting young city users is being undertaken, for instance, through projects like De Effenaar, a music club, which the city started and supports. Yet the attempt at achieving this upgrading through a richer cultural supply has been shown to be insufficient; more must be done to change the very atmosphere of the city and the image that it projects to the outside. Despite the bustling shopping district in the centre, with its cafés and restaurants, and the attractive typical Brabant landscape surrounding it, the city is convinced that it has to do more to bring itself to a higher status. For instance, its appeal as the of 'city of light' is hardly used to promote the city and create a mark of distinction.

To tackle these issues in a consistent way – the insufficient availability of human capital, its excessive specialisation, and the lack of attractiveness of the city for new talent – a clear strategy is needed, one that clearly identifies the interconnections between them and systematically tackles them. However, Eindhoven's experience is not of the best. In the early 1990s, the city set up a workgroup 'Eindhoven Studiestad' with the task of setting up a vision for an improved city for students. A paper with a number of goals was produced, but it had no real follow-up. The contribution from important players like TU/e decreased, as did that of private partners. Also little attention was given to the role of physical planning in this field. Clearly, a similar effort could be revived with new tools.

3.3 Strategies and Integration. Eindhoven as a 'Sustainable Student City'

The city of Eindhoven is an excellence centre in technical education and a major laboratory for knowledge transfer and business development. Naturally, it is confronted with the issue of how to sustain its status in the future. It is increasingly felt that the attraction factors that made the fortune of 'technical poles' in the past decades (research excellence, a large human resource pool, widespread scholarisation, a network culture) might not be sufficient in the future, as other factors related to quality of life, accessibility, diversity acquire greater importance. In this context, one expects public sector companies and programmes to play a decisive role as a propulsive factor for the economy; and indeed this is the case.

3.3.1 Relations between the Students and the Local Community

The city of Eindhoven has formulated a strategic document ('City Vision 2010'), in which four objectives are emphasised. Eindhoven wants to become an innovative knowledge centre focusing on technology; a city with an attractive centre; a city where people feel at home, as individuals and in their social environment; and an effective administrative centre. Local and regional governments are important in enabling and stimulating players in Eindhoven. A regional policy initiative stands out in this process.

Kenniswijk (E-City) is an innovative programme that aims at speeding up the development of the 'information society'. It started as an initiative by the Dutch government, which gave each city the opportunity to design a 'knowledge-neighbourhood' in areas characterised by mixed population groups. The project's goal is to improve the quality of life, work and education through the supply of innovative services to the local population and firms, providing adequate access and involving private as well as public partners. The Eindhoven project was rewarded for its soundness and articulation, and received €9m from the state, adding up to the €5m investment by Eindhoven and the city of Helmond, another centre in the Eindhoven region. The project also promises to generate substantial spin-offs in terms of accelerated time-to-market of new services.[3]

Kenniswijk has brought together the City of Eindhoven and the higher education institutions. In the past there were few regular contacts between the City and TU/e; these entities stood like two islands. Today the scope for cooperation is stronger and collaborations have extended at other levels on specific issues, under the pull of the City which organises regular meetings. However, until now the strategic attitude in designing a real student-city environment has been limited. For example, on the TU/e's website there is only a little information about the city. This is despite the fact that, in student decisions about where to go to study, the second most important factor – after the quality of higher education institutions – is the city where the higher education institutions are located. This sort of concern prompted all the relevant actors to get together in a city portal for education, covering all aspects of the local HE provision but also of the city life and of the facilities for students and young citizens in the region.

Another recent joint initiative in the framework of E-City comes from CEBRA, a TU/e research centre for e-commerce, which elaborated a project of a 'virtual debating room' for students. This project had the potential to bring different people together, but it also presented some threats and is not easily extended to the resident's community. It is instead believed that an evolution in the mix of students, with more international students, more women, and different disciplines, could achieve increased social dynamics in the student population, and, as a result, become more interesting to the locals.

TU/e is active in pursuing its own comprehensive vision of the role that the campus should have in the overall city development. For instance, the idea of a 24-hour campus life with facilities for staff and their families, which is also used as a pathway for citizens commuting to out of town, is consistent with the idea of bringing it closer to the community heart. It must be said, however, that such plans made little progress. It is possible that this outcome depended on the lack of a real strategic cooperation with the city. Had the planning department have been involved right from the start, the vision of TU/e might have been advanced further. However, this collaboration has been revived, though at a lower level of ambition. For instance, TU/e is an consistent partner of the city in cultural projects. The Opera organised at the university campus, as well as the Virus Festival, attracted many people who normally would not visit the campus. The spectacular event, with lighting provided by young architects, was an important occasion of encounter between the city and its university.

Strategic cooperation within the higher education sector and between it and the city is increasingly important for retaining the young talent and new graduates in the region. TU/e and Fontys cooperate in different fields. At the highest level, the two institutes send delegates to the respective management teams. Fontys runs a dedicated programme for students who were unsuccessful at the university. Furthermore, the institutes cooperate in specialised technological fields to increase the transfer of research results into educational products. One example is integrated circuits. The knowledge on this topic, developed by the university, is then transferred to Fontys; this enables Fontys to provide the most up-to-date education, and gives them a lead over other institutes.

A second initiative is called 'United Brains, the network for solutions'. It is a joint service for businesses and institutions offered by TU/e, Fontys and the Design Academy. United Brains is a joint portal to which an entrepreneur can turn for information. United Brains passes on questions to a specific source of information within one of the collaborating institutes, acting as a liaison. Moreover, it takes action to acquire, initiate and guide projects. Preference is given to projects in which more than one institute participates. The project relies on an 'information superhighway' among the three institutes, with linkages to firms. The aim is to develop and present integral solutions for firms related to design, a kind of on-line help desk. This project has tapped into the Stimulus funds, but it is planned to become self-supporting in a few years.

Working collaborations are also important. The city may finally start to be seen as a real-world laboratory by TU/e out of the mere technological domain. In the Urban Development Programme of the Ministry for the Great Cities, a cooperation has begun between the City and TU/e on two objectives. The first is the creation of a network of civic centres to be located in the centre of villages surrounding Eindhoven. TU/e is contributing to this project with planning design. Another project regards the reconstruction of the social networks in these places; research on behaviour is being carried out by Fontys, including the students' behaviour. Fontys, in the context of its educational programme, has educational programmes geared to problem areas: they offer to study these areas and find solutions, also contributing to strengthening their role in the community.

Despite the peculiar, 'closed' nature of student life, student associations have started to take a more active role in the community. List 11 was an unique chance to raise the issue of thes attractiveness and to present the disadvantages of the 10 per cent rule. After the elections, three parties out of five were in favour of a change in the regulation. Now the initiators of List 11 are watching progress closely, although they still consider change difficult to achieve.

However, spontaneous interaction of students and the community must improve in order to foster a higher quality social environment in the city. The question is how to encourage people who are used to spending their time in laboratories to be more communicative, to enter into relationships with different groups and eventually become citizens themselves. So far, these possibilities have been hampered by the peculiar nature of the local society. The peculiar genesis of Eindhoven through the 'conglomeration' of smaller rural communities might be an explanation for the small-town mentality that continutes to pervade community life. However, a typical characteristic of Eindhoven is the presence of social networks on a small scale, presumably the legacy of old districts or parishes (City of Eindhoven, 2000), which in a sense may be conducive to easier integration than in large cities, where close personal contacts with outsiders could be less frequent.

3.3.2 Relations between the Student Community and the Private Sector

Linkages between the business community and the knowledge infrastructure are relatively well developed, at many levels. In the first place, local firms take advantage of the presence in the region of technical higher education institutes providing a constant flow of ideas and intellectual capital. This has been the key to success in the development of the mechatronic cluster; the same experience is being now exported to other fields.

The second type of linkage between firms and knowledge infrastructure consists of research carried out by the university and Fontys for firms in the region. Contract research for the industry is important, but for the institutes it is increasingly a vital source of revenue. TU/e is active at both levels. Apart from the obvious transfer of human capital, the university is at the centre of network of laboratories and small companies carrying out contract research for the industry. The knowledge transfer agency of the university, where all contract research is channelled, has a turnover of some €14m. Fontys also carries out contract research. For example, in 1997 Fontys developed a new process enabling the creation of aluminium moulds. This project was carried out by teachers at Fontys and a firm, and tapped into funds from the Stimulus Programme. Fontys is also particularly active in the support of small and medium-sized enterprises. For this purpose, an innovation workshop was established, in which students and teachers offer a broad range of services and solutions for SMEs. The project also generates useful and interesting thesis projects for students. The turnover of the innovation workshop amounts to c.€0.19m. In the last few years, this figure has been growing by some 6 per cent yearly.

A third type of strategic interaction is the joint participation of education institutes and firms in strategic platforms. A recent example is the foundation by TU/e of the *Embedded Systems Institute*, which received strong support from the business community. This initiative strengthens the mechatronics cluster in particular, as this cluster seems to lack sufficient knowledge and capacity in this field. The business community is closely involved in setting up education and research programmes in this field, and plans to contribute financially in due course. The fact that the initiator of this institute is a senior researcher at the Philips NatLab and professor at TU/e is an illustration of the intimate links between the university and the business community. Schools and private partners also collaborate in the objective to increase the international profile of the city, extending the scope of international exchange programmes. All the actors are indeed convinced that, to survive, Eindhoven has to become an international city, and the student population is a good group to address first.

In the fourth place, higher education institutions generate dynamics by providing a breeding ground for new firms. For instance, the Eindhoven University of Technology offers different facilities for starting firms. Hi-tech starters can settle on the university campus at subsidised rents. The university policy targets two kinds of firm in particular: those that can benefit most from the proximity of the university because they use the same facilities, and those with which the university closely cooperates or with which it sees opportunities for cooperation. The importance of physical location to facilitate synergies and cooperation is confirmed by the fact that the Chamber of Commerce and the Regional Economic Development Association will soon be hosted on the university campus.

The *Chamber of Commerce* is indeed a very active player in Eindhoven. The cooperation set up with local government ten years ago to realise the Stimulus Programme brought plenty of new opportunities to the region. Thus the Chamber of Commerce was established as a leading actor in development policy. The Chamber supports the concept of Eindhoven as a place capable of retaining students, to promote cultural life and a high quality learning environment. The causes behind the demise of Philips lay in the quality of the local job market and in the attractiveness of the city as a place to host representation functions. This challenge is not over. A huge number of qualified staff will be needed in Philips' new science park who are just not available at the time of writing. Moreover, industry employers now demand a more integral education, for instance curricula that cover culture and human sciences, and they demand high-class cultural and recreational amenities for their staff. The quality of housing and of the environment then become more than ever crucial to local competitiveness.

The Chamber of Commerce started offering students scholarships and grants to do in-company training periods. It is now setting up a new strategic plan to tackle problems such as scarcity of housing and bad image, both in a international and a local context. It is lobbying with the national government to change the regulations of the job market to make it easier for foreigners to come to Eindhoven. For many private companies it is not easy to invest in 'no profit' projects, especially in periods of economic recession. The Chamber is therefore active, trying to facilitate cooperation among the relevant actors, and participating in strategic platforms. It is also working to improve the international orientation of the schools. After an exploratory study, an international school for children and boys up to 15 has been started.

3.3.3 Influence of Higher Education in the Relations between the Private Sector and the Local Community

Among the firms located in Eindhoven, Philips is the largest technology player in the region. Many other firms, however, can be considered engines of the local economy and in particular of its spearhead, the mechatronic cluster:

large, internationally-operating firms which act as a catalyst for other firms, not only by outsourcing production but also by transferring state-of-the-art knowledge and technology to SMEs in the region. This critical mass opens up space for specialised services such as engineering bureaux, technical design agencies, and consultancy firms. The cluster is therefore attractive to other firms and organisations who are considering relocation. This mass and diversity results in an environment in which the newest technical knowledge in different fields is rapidly dispersed and adopted, thereby contributing further to the cluster's dynamics. The 'network culture' of the regional firms is today well developed. They are eager to develop flexible networks in the supply chain, and to establish strategic partnerships with competitors in the region. There is now an explicit and broadly shared vision about how business should develop in the region, which emphasises network creation and the integration of business and research infrastructure. However, some important elements are still lacking to make a diverse and strong local economy. Furthermore, the region seems to lack main suppliers – firms with sufficient skill and capacity to codesign and produce complex modules for the large original equipment manufacturers.[4]

The process of rebuilding and strengthening the local economic cluster was accompanied by a host of initiatives where public and private have joined forces and where higher education and research play a prominent role. *Stimulus* is an economic regeneration scheme for Southeast Brabant, set up in the early 1990s by the local government and the Chamber of Commerce to fight the economic crisis in the region, followed in 1998 by a three-year new stage, 'Stimulus 2'. The European Commission and the Dutch government have largely contributed to the programme. One important role of Stimulus is to stimulate cooperation between different actors in the region (firms, organisations, educational institutes), by financing cooperative projects. One of the spearheads of Stimulus 2 is the improvement of cooperation between developers, suppliers and outsourcers in the region.[5] In this framework, *Syntens* – an organisation which emerged from a merger of the Innovation Centre and the Institute for Small and Medium-sized Enterprises – set up as its task to bring forward the conditions for economic development and innovation, and to build bridges between firms. Small and medium-sized enterprises are still the main focus. SMEs in Eindhoven have a natural problem, as they are not typical of this region and thus their networks are less extensive. These networks are not visible or formalised. There are quite a number of meeting places, but they are not open or accessible. Firms looking for a partner to develop a new product or process may thus seek the services of Syntens, which maintains relations with over 800 firms in the region, and has detailed databases about firms which can provide the required knowledge. In some cases, Syntens is the initiator of cooperation between firms. Syntens is clearly interested in integrating the human capital in the region.

Despite the success of this and similar initiatives, the relationship between the local economic environment and the education and research sector hides some 'black spots', and could be greatly improved. The mismatch in terms of human capital produced in Eindhoven is driven by the excessive specialisation in technology and the lack of jobs in other 'soft' sectors which are instead more easily found in the Randstad. But there are also constraints on the demand side, as technical higher education graduates are insufficient. In the last few years, this problem has intensified. Nowadays, more than 50 per cent of the new scientific staff come from foreign countries.

To attack these shortcomings consistently, the activities carried out by Syntens are extended in a whole new programme, *Horizon*, whose aim is to bring together, systematically, 'technology and knowledge, entrepreneurship and marketing' at a regional scale, extending over regional and even national borders (Commissie Regional Kansen, 2002). This ambitious programme, organised on thematic lines and projects by the 'Committee for Regional Opportunities' – a platform putting together local government, private sector and higher education – is geared to the mid- and long-term to bring forward the conditions for a sustained success.

The programme suggests a shift to the 'soft' sectors of design and culture; a despecialisation of the economy is encouraged, which implies enhancing the entrepreneurial spirit in 'new sectors' and offer competitive conditions for incoming firms. As the capacity to manage the knowledge created at TU/e labs and R&D departments is also low, since local higher education institutions hardly provide the entrepreneurial skills that are necessary to fully disseminate the knowledge and hook it to the region, Horizon sets out to redress this issue. TU/e is planning to organise a Master's course oriented towards forming such business skills. Links are already being established with Tilburg University. Despite the accent on internationalisation, one should note, as an aside, that the cultural regeneration of the city as well as the improvement of its image are not explicitly formulated as objectives of the programme, even if it is foreseen that Horizon will develop in full consistency with other initiatives. It is also not completely clear whether this programme will manage to get the funding that is needed, estimated at €10m a year for 3–5 years.

3.3.4 Model Testing and Identification of Issues

Eindhoven is not a particularly attractive city, despite the ongoing regeneration efforts, and this means that more and more of its precious human capital may not be willing to stay and contribute to the local community development, putting at peril the ambitious and wide-ranging programmes in which local stakeholders are investing to boost Eindhoven's status as a 'knowledge community'. In this light, the relations between university and community as described in Figure 3.2 are good but not optimal.

One crucial aspect for Eindhoven's future is how HE can contribute to the diversification of the local economy, on one side, and to social and cultural dynamism, on the other. The local economy is not yet at the stage of becoming more diverse, but the conditions for this change are definitely building up: good accessibility, more international exposure, and a local society that is today more attentive to quality of life factors.

A point of strength of the Eindhoven region to this respect is the good state of cooperation between government, business and knowledge and education institutes. The Eindhoven region has been shown on previous occasions to have a high level of organising capacity. For example, the E-City project is an effective demonstration of how public and private leaders in the region share the ambition to improve Eindhoven's position, and are willing to invest jointly in innovative projects. Key people in the public and private sector know each other well, are willing to cooperate and invest in innovative projects, and trust each other. The density of clubs and associations in the region is one of the highest in The Netherlands. As a result, when a firm seeks cooperation, it is

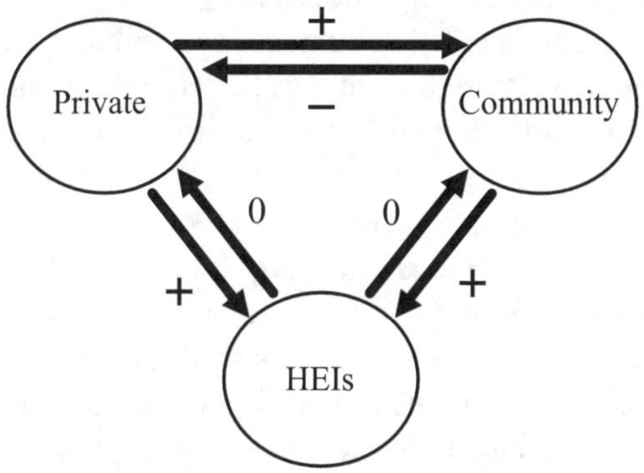

Figure 3.2 State-of-the-art: Eindhoven as a student city

not difficult to find a partner in the region. There is always a suitable partner, and the intermediary activities of Syntens may help to find one if necessary.

However, in more than one respect this situation could be improved. First of all, the city is not attractive enough to enlarge the pool of local work skills necessary to match the demands of the job markets which accommodate the growth potential of existing clusters and bring forward a more diverse economy. As a result, the private sector faces a shortage of opportunities in the local context. The lack of attractiveness is not only determined by the external characteristics of the city, which are hard to change, but also by the lack of social integration between students and the community, which greatly hampers the social and cultural dynamism of the city. Furthermore, the collaborative links between higher education institutions and the private sector should be more focused on strategic sectors in which the region is not strong, such as the creative industries and business management, and should encompass concerns about quality of life for graduates and students; the private sector must be led to invest in the attractiveness and hospitality of the city.

3.4 Summary and Conclusions

The capacity of the local higher education institutions to produce a wide range of skills necessary to the performance of the local economy has to be upgraded. This does not necessarily mean that higher education institutions have to expand, as they did with the new courses in industrial design, but they can pursue strategic collaboration with other institutes in the region and outside, as TU/e is doing to some extent with the neighbouring universities of Maastricht and Tilburg. Fontys can exploit its network throughout the country in this respect. These efforts have to be extended, taking into account that the range of topics taught at TU/e is quite narrow, and that a really 'student-friendly' city requires more diversity. This sort of cooperation could be expanded, dramatically increasing TU/e's utility for firms in the region. The education reform curricula increase the possibilities of enhancing this aspect. It is important that at this stage the streams of cooperation are investigated and selected with a strategic eye and in full agreement with the city's economic department.

The idea of teaming up two top local hospitals to start a medical school has succeeded in bringing together a certain number of partners, including, among others, TU/e, Tilburg University and Fontys, but is still waiting for government approval. It is important in this respect that the city fully endorses this initiative and guarantees it the necessary support.

The points of complementarity between university and 'vocational' education will have to be redesigned in the future, in order to achieve more integral, flexible study curricula in the region. In this respect, it must be noted that as a result of the technically-dominated orientation of TU/e, other studies that are nevertheless provided at Fontys have come to receive less attention.

Regarding the image of the city, and its communication strategy, Eindhoven describes itself as 'leading in technology'. This concept might have been apt when Philips was the main actor, but today, things have changed, as has the attitude of the society towards economic development and quality of life considerations. In this light, technology should no longer be the main focus. Technology is associated with cold landscapes, polluting productions, and lack of social contact and human touch. It has been suggested that 'leading *through* technology' captures better what Eindhoven would like to be known for: a human-scale environment where, thanks to technical capacity, everything is easier for firms and citizens alike. At the same time, there is no tangible expression of Eindhoven's technical excellence that the city can boost as a point of attraction. Technology should be given a face, and possibly it should be a human, familiar face. The E-City project is consistent with this idea, and it offers unique services that enhance the capacity of local government to respond to the needs of its citizens. If it progresses to more advanced stages, it may definitely build an advantage for Eindhoven over other 'technical regions' in The Netherlands. The E-City can gear the image of Eindhoven to the 'fun-technology' axis. However, the city should definitely highlight its environmental amenities more, as they are the only point of strength in the urban landscape, emphasising the social factors that make of Eindhoven a friendly, active city, for students too.

In this vein, the city has to overcome once and for all the ambiguity of its attitude to the process of building itself as a 'student city'. The 10 per cent rule on student housing is a reflection of this ambiguity. The city wants to protect itself from an excessive inflow of newcomers with distinctively different living habits but, at the same time, it would like to be more attractive and culturally stimulating. The challenge from the relaxation of this rule includes both aspects: give students an opportunity to settle in the city and become part of the community, and make integration bring forward an high-quality living environment for the whole community.

It is even more important to offer good conditions to postgraduates and young skilled workers, as a strategic policy to attract the right people. They should be offered 'Silicon Valley' conditions, but the urban quality is still insufficient. Despite the increased awareness of the problem, a strategic attitude in cooperating at the highest level to tackle this problem is not yet there. There was once a single agency (CEP) in charge of organising cooperation in city marketing projects, but it was not successful and could not convince relevant partners in the private sector to contribute. Today it is time to set up a new, empowered, participatory organisation to tackle consistently the issue of raising the status of the city. This organisation should implement a consistent communication strategy, based on the region's points of strength, but at the same time it should organise the conditions that can lead to a greater provision of new cultural and social amenities – ideas, sponsors, land, and jobs. The target groups would be young graduates and talent. They should be convinced to come to Eindhoven to study and to remain there. Then a virtuous cycle of development as an attractive city can be entered into.

The regional development agency NV Rede, which has proved to be highly effective, could be a trusted leader in the implementation of plans drawn up locally. It could be given more responsibilities, together with all other actors who have strategic importance.

In the operation which is leading to the restructuring of the Witte Dame complex, much attention will have to be given to the development of a high-class cultural environment. In this respect, 'intangible' factors such as urban design, street animation, and the regular staging of small but trendsetting events are more relevant than physical development. The location of the University of Arts and Design right in the middle of this complex provides optimal opportunities for synergies regarding this sphere.

A final issue concerns the innovation capacity of the region. As we have seen, this depends much on the network culture that brings different firms to have strategic collaboration on new projects. The research infrastructure exists and is of high quality. However, it has been pointed out that the role of Philips, and its willingness to become an integral part of this network again, is crucial in terms of 'critical mass'. The role of Philips would be different to the one that it took previously: it will be a supplier of knowledge and customer of products rather than a leading actor in a rigid hierarchy of relations. Formally, there is no technology and knowledge transfer from Philips to other firms in the region. In that respect, it is illustrative that Philips is not participating in the Stimulus Programme or in other regional initiatives.

However, the situation is likely to change; though it is not certain whether the networks activated through Stimulus will continue without these financial incentives, there are indications that many firms have discovered the advantages of networking and clustering. Philips' plans for a new science park in the southwestern part of the city provide the momentum for a new stage of interfirm cooperation. Most of Philips' R&D activities will be concentrated in the campus. According to the current plan, the site will only be for Philips divisions, not for other firms. In this respect, as a first step it would be beneficial if Philips would allow other firms on its technology campus to benefit from proximity. The Chamber of Commerce is also very much in favour of an 'open' campus. Public incentives such as the provision of additional infrastructure could be helpful in this respect. Next, a more formal structure of knowledge and technology exchange between Philips and other actors in the Eindhoven region could further enhance dynamics, both of Philips and its partners. The establishment of an independent intermediary structure, for instance led by TNO, could contribute to this, as well as a more outspoken role for those organisations active in knowledge transfer.

Notes

1 Vestide is a foundation (*stichting*), a public company like Stadswonen in Rotterdam, created by the university. It started as a small company, developed in a bureau within the university, then became an association of all student facilities; finally, eight years ago, it became an organisation with own staff. Since 2002, Vestide has been part of the Hertog Hendrik van Lotharingen (Hhvl) housing corporation.

2 An example of the change in the organisational structure of the industry is given by the success of the 'mechatronics' (mechanics + electronics) cluster. Large networked firms are engines behind the cluster's dynamics. State-of-the-art knowledge filters down from large firms to network partners in the Eindhoven region, resulting in a dynamic, learning environment, with open windows to the world. Suppliers' networks are located throughout the region, as firms generally prefer to outsource to companies in physical and cultural proximity. The knowledge-infrastructure is very much involved and part of the cluster, as it feeds the cluster with new talent, serves as breeding ground for new firms, and executes fundamental and contract research to the benefit of the firms in the region.

3 The E-City core area includes 38,000 households, the campus of TU/e and the polytechnic school. The main key concept of E-City is that every household and economic activity in the region should have broadband access. The city can count on a favourable position as it hosts 17 telecom providers. Main services that E-City delivers are: government services, community formation, consumer services like e-commerce, healthcare and education. Furthermore, 120 innovative projects will be launched exploiting the E-city infrastructure, regarding for instance safety for the elderly or disabled people, virtual reality, training, professional support. An interesting example which can be considered a 'pilot within a pilot', is the already mentioned 'campus notebook project' of TU/e.

4 Van den Berg et al. (2001) indicate that if the integration of the Philips organisation (in particular the large NatLab and CFT) with other firms in the region could be enhanced, the potential for further dynamic development would increase even further.

5 One well-known and innovative example of a Stimulus project is the KIC-project (Kennis Intensief Cluster; [knowledge-intensive cluster]) of Océ, the copy-machine producer. This

project's goal is to stimulate the joint development of new products and technologies at Océ and its more than 100 regional suppliers, for the purpose of upgrading the skills level and the quality of the suppliers' coproduction and codevelopment, and of keeping more work in the region. Very fundamentally, it has created an increased openness of firms towards other firms and knowledge institutes.

Chapter 4

Helsinki Case Study

4.1 General Introduction to the City of Helsinki

With a population of 560,000, the capital city of Helsinki is by far the largest in Finland. In 1998, the Via Finland belt – which refers to the urban corridor including Helsinki-Tampere-Seinäjoki-Oulu-Rovaniemi – had a population of 2.2 million, or 43 per cent of total population of Finland. Many of the fastest growing centres in the country are located in this zone: in recent years, its population has grown faster than the total Finnish population. According to Statistics Finland, this trend will continue. Table 4.1 gives an illustration of the social and economic dimensions of the Helsinki region at various levels.

Three geographical entities can be discerned in the Helsinki area: the City of Helsinki, the Metropolitan Area of Helsinki, and finally the region of Helsinki, including 12 municipalities.[1] The City of Helsinki, founded in the middle of the sixteenth century, remained small up to the nineteenth century, when it began to develop more rapidly. The metropolitan area of Helsinki, consisting of four municipalities – the capital and the surrounding cities of Espoo, Vantaa and Kauniainen – hosts now roughly one-sixth of the country's total population. The number of inhabitants in the Helsinki region is approaching 1.2 million. With an increase of almost 140,000 persons during this decade, Helsinki is in relative terms one of the EU's fastest growing cities. Although Helsinki has witnessed a noticeable influx of foreign residents during the 1990s, the city's growth is primarily based on its domestic attraction. The concentration of new residents in large university cities was a special characteristic of Finland's domestic migration during the 1990s.

The Helsinki metropolitan region is the main economic centre of Finland. The municipality of Helsinki alone counts more than 40,000 enterprises. Today, the regional economy is strongly dominated by services; some 80 per cent of jobs fall into that category. Manufacturing is much less important when compared to national average, and remains concentrated in the neighbouring cities, in particular Vantaa. Services and company headquarters concentrate in Helsinki and in Espoo. In Helsinki, there is also a concentration of talents working in creative industries, advancing the glorious tradition of Finnish design.

Before World War II, Finland was predominantly an agricultural country, with a rather narrow industrial sector. The subsequent development into a complex market economy took place very rapidly after the war. Alongside the wood industry, manufacturing areas like the metal industry, chemicals, textiles and construction gained weight. Today Finland is a very advanced country in the development of a number of specific post-industrial sectors, such as the production and application of state-of-the-art telecommunication services and new media services. Finland is a world leader in the development of mobile phones and their networks. It has one of the highest penetration rates of mobile telephones in the world, exceeding 50 per cent in 1998, and internet and new media services are also relatively widespread. A basic reason for the high penetration rates of new communication devices, the low prices and the success of Finnish telecom service and equipment producers can be traced back to the early liberalisation of the Finnish telecom market. At the time of writing, Helsinki is one of the most digitalised cities in the world. All community services have extensively endorsed IT systems. Science and technology have a strong role in the economy. The reason for this is also found in the social structure of the city: the population is young and well educated.

Table 4.1 Facts on Helsinki population (2000)

	Population	Employed labour force	% jobs in service sector	% students
Metropolitan area	945,725	489,861	83.4	6.9
City of Helsinki	551,123	280,773	85.2	6.9
Espoo	209,667	111,281	82.5	7.4
Vantaa	176,386	93,854	78.6	6.1
Kauniainen	8,549	3,953	86.6	8.2
Rest of Helsinki region	241,470	121,553	72.2	6.4
Finland	5,171,302	2,236,553	68.5	7.6

Source: City of Helsinki Urban Facts, 2002a.

By European standards, Helsinki is a peripheral city. However, this does not seem to hamper its dynamism. With its international airport and ferry links, the Helsinki metropolitan area is well connected to the rest of the world, though flights to and within Finland are on average rather expensive. Internationally operating firms can also count on first-class *e-accessibility* for operations and logistics. Within Finland, and within the Helsinki region, road, rail and air connections are good. Compared with Europe, congestion is very limited, because of the abundant space.

On an international comparison, Helsinki scores high as a city of culture. Together with a large student population, the highly-educated people of the region demand and appreciate top-class cultural and leisure opportunities. The local residents are indeed active consumers of cultural services. Helsinki offers an extraordinary wealth of classical music and it is the country's most important centre for education in the visual arts, a fact reflected in the numerous fine arts art museums and galleries that can be found in the city. The city centre of Helsinki, with its narrow streets, its busy waterfront, the large squares, the historical monuments, and churches and markets, is safe and small enough to walk around, with plenty of cafés, restaurants and shops. However, the urban environment is strongly characterised by the 'new': experimental art, first-class architecture and, of course, design. The media-city character of Helsinki is distinctively reflected in its everyday life and culture, with numerous shops and web-cafés as well as well as multimedia art galleries. A whole new vibrancy has been added to the city with new buildings and major renovation projects.

On account of this dynamism, Helsinki has hosted the European City of Culture event in 2000, together with eight other European cities, putting the city on the map of European cultural tourism once and for all. In this process, the city can count on special atmospheric elements, such as its diverse nightlife and its relaxed way of life, which to some extent provide a 'hidden face' of Helsinki in addition to the proud one reflected by the stately buildings in the centre. In fact, interest in the city is undoubtedly growing. In 2000 overnight stays in the Helsinki metropolitan area reached 3.3 million, half of which were by international visitors, marking a 6 per cent increase on the previous year. Finland is a popular country for international meetings and congresses, and its capital is the country's favourite host city for these. In 2000, 108 international congresses with 31,000 participants were held in different venues.

Finland ranks sixth in the latest United Nations' survey of quality of life. Indeed, the quality of life in its capital city can be judged very high. The region offers metropolitan amenities, and yet the countryside is never far away. These characteristics, among others, make Helsinki a pleasant place to live. On the other hand, the climate – cold and rainy middle seasons, long and dark winters – partly offsets these advantages. In general, the Finnish capital attracts many young people from other parts of the country, who prefer the urban, relaxed way of living in Helsinki. Four out of five Finns feel they are privileged to be Finnish, and have a very tolerant attitude towards foreigners and outsiders, which is why so many Finnish students – and an increasingly large number of foreigners – come here to start a life career.

4.2 The Student Community and Knowledge Infrastructure in Helsinki

4.2.1 General Information about Higher Education in Helsinki

The Helsinki economy is distinctively knowledge-intensive. Research intensity increased in Finland throughout the 1990s. The region of Helsinki is by far the most important locus for R&D activity in Finland: over 50 per cent of Finnish R&D expenditure is spent in the Helsinki region. Private sector involvement is high, accounting for some 62 per cent of R&D investments. The knowledge intensity of the Helsinki economy is also illustrated by the high educational level of its population: 40 per cent of the 25–64-year-old population in the Helsinki region has completed at least 13 years of education. Some 50 per cent of Finland's academics live in the Helsinki region, and the share of inhabitants with a higher education qualification reaches 40 per cent in Espoo and more than 50 per cent in the small city of Kaunianen (City of Helsinki Urban Facts, 2002a). In a comparative analysis of the 37 Finnish urban regions, the FUN project (Urban Facts, 2001) ranks Helsinki as the first by far both for diversity and for strategic specialisations; on this account it is one with the best preconditions for development, together with Tampere, Oulu and Turku.

Helsinki is indeed a city of students. In total, more than 60,000 students were registered at universities in the Helsinki region in the year 2000, and another 23,000 at polytechnic schools. The number of postgraduate students (third cycle) was 1,100. All these figures represent approximately one-third of the national total. Students come from all over Finland, and, thanks to EU exchange programmes and Finland's rising profile a place to study, also in growing numbers from abroad (1,865 in 1999). Table 4.2 provides a breakdown of registered students, graduates and foreign students in the main institutes. In the last ten years, the number of university students has increased by about 40 per cent. However, entry into universities is very competitive.

4.2.2 The Knowledge Sector of the City: Actors and Spatial Patterns

The supply of higher education facilities in the Helsinki metropolitan area is rich, diverse and qualitatively strong. The University of Helsinki is the

Table 4.2 Higher education students in the Helsinki metropolitan area (1999–2000)

	Attendees	Degrees taken	Foreign students
City of Helsinki			
University of Helsinki	37,244	4,034	1,190
Helsinki School of Economics and Business Administration	2,757	402	–
Swedish School of Economics and Business Administration	2,200	285	–
Sibelius Academy	1,700	147	–
University of Art and Design	1,600	244	240
Theatre Academy	–	56	–
Academy of Fine Arts	250	62	–
Espoo			
Helsinki University of Technology	14,264	1,053	790
Helsinki metropolitan total	**60,015**	**6,283**	**1,865**
Other HEI: Polytechnic Stadia	7,500	1,275	106
Other polytechnic schools	16,343	1,137	–

Source: various.

country's leading institute of higher education. Business studies may be pursued at the Helsinki School of Economics and Business Administration and the Swedish-language Swedish School of Economics and Business Administration. The Helsinki University of Technology is the most important institution for higher technical education and research. The Academy of Fine Arts, the Sibelius Academy, the University of Art and Design of Helsinki and the Theatre Academy of Finland provide higher education excellence in the arts. The City of Helsinki operates the Helsinki Polytechnic Stadia.

Most establishments are concentrated in the same area as top research institutes and knowledge transfer facilities (Figure 4.1). The area of Arabianranta is the location for advanced activities in the fields of new media, art and design. The large and already well-established Otaniemi Park in Espoo is a major site for innovative high-tech firms in the fields of information and communication technology. The Helsinki Science Park, for research and business purposes, is the centre of the new developments in the Viikki area, the east part of the city.

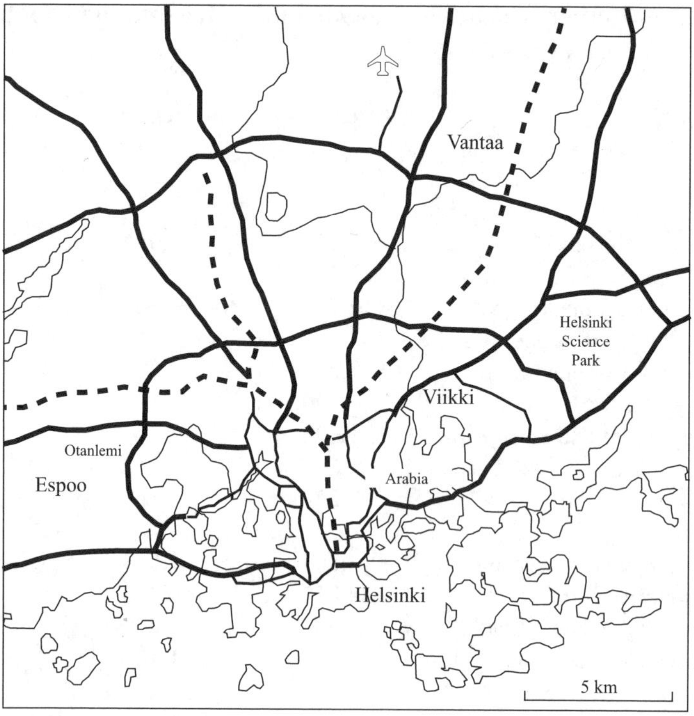

Figure 4.1 Main campus locations in the Helsinki Metropolitan Area

Universities The largest and most varied university is the *University of Helsinki* (UH). The UH is the oldest and largest university in Finland. It was founded in 1640 as the 'Academy of Turku' in the former capital of Finland, and moved to Helsinki in 1919. With nine faculties, the UH is Finland's most multidisciplinary higher education institution. These employ 3,339 teachers and researchers, 2,272 lecturers and another 3,607 employees: 36,822 students were pursuing a degree in 2001, of which 4,121 were obtained, and 390 doctoral degrees were completed in the same year. The UH is bilingual: 6.5 per cent of the students are Swedish speakers, and for 27 chairs Swedish is the statutory teaching language.

When the university moved to Helsinki, the UH premises were erected in the centre of the new capital. All higher education stayed entirely within the city boundaries until the 1950s, in the so-called 'City Centre Campus'. Today UH occupies four main campus areas, roughly divided by disciplinary field: the City Centre (theology, law, arts and social sciences), Meilahti (medicine), Kumpula (science), and Viikki (agriculture and forestry). A walk through the campus areas will show impressive architecture and affluent cultural history. As an old, traditional institution, admission to UH is difficult and the pace of advancement is usually up to the individual. Much of the instruction is in the lecture form, and relationship between students and teachers can generally remain quite official and hierarchical. However, some departments have now moved towards more student-to-student and student-to-teacher discussion-orientated lectures and tutoring, and most lecturers welcome active student participation.

High-level research is carried out in the departments of the faculties and department-affiliated research stations, as well as at independent research institutes in Helsinki or in other cities. Basic funding for research comes from the national budget; however, the volume of outside funding currently accounts for about 50 per cent of the total research budget. The most important sources of outside funding for research include the Academy of Finland, which has designated the 11 units of research of the UH as *Centres of Excellence* in 2000–05. The important position held by research at UH is also evidenced by the fact that, in 2000, the university allocated to it some €5.3m of its extra-budgetary funds. Moreover, each year, the *UH Research Foundation* awards about 30 scholarships to young scholars and scientists. A *Research Liaison Office* assists researchers in the acquisition of financing and in the management of the projects; it also assists new business creation and product development within the UH, and pursues the connections with the private sector and its network of affiliated organisations. Various departments and independent institutions are responsible for the *Centre for Continuing Education*, which runs a network of independent specialised institutes across the country. In 2000, the centre organised 1,381 courses for 30,487 students; 2,700 students took part in employment training.

The UH has cooperation agreements for scientific and student exchanges with almost 80 universities throughout the world and is active in European Community programmes in the fields of education, research and development. Furthermore, UH researchers have participated in special programmes promoting scientific cooperation with Eastern Europe and in projects funded by EU Structural Funds. *Student exchange* has gained in popularity in recent years, with the number of incoming exchange students, in particular, increasing considerably. The UH has proved a popular choice due, in part, to the rise in English-language education. In 2000, the UH welcomed 1,192 foreign degree students, 832 exchange students, and 324 foreign researchers and scholars.

The *University of Art and Design* (UIAH) is the largest art school in Scandinavia. Nearly 1,600 students were registered in 2001 (15 per cent of which are foreigners), and 400 teachers work there. Though it is already by far the most cosmopolitan university in Finland, its ambition is to become even more international. Nearly 3,000 persons apply to UIAH annually, and an average of 9 per cent are accepted as students. Since 1985 UIAH has been located in Arabianranta ('Arabia'). It provides education and research in the fields of design, audiovisual communication, fashion, art education, photography and arts. Eleven departments offer 11 undergraduate programmes and postgraduate education at Masters and doctorate level in 18 fields. The courses are held in English. The university is known to be business-oriented, and develops education and research programmes that comply with the needs of the business community. A *Research Institute* coordinates research and serves as a consulting body. It furthers cooperation with other interest groups and supports the university's departments in developing research and doctoral education. A business incubator is part of the university. It provides infrastructure, business, legal, and educational services. However, 100 start-ups created in the area have moved away for lack of premises.

Relationships with the private and public sector, other research institutes and universities can be considered very good and intense. The UIAH is also well connected to centres of art and design education throughout the world, offering scholarships to students for exchange or full enrolment in its programmes. Many activities and events are organised inside the university and across the city, like multimedia exhibitions of the students' works. Through the interesting magazine *ARTTU*, produced in-house, UIAH promotes and discusses the recent works of students and provides information to alumni and the general audience in the community. An important centre at UIAH is the *Media Lab*. Formed in 1993 as a separate department, it received full faculty status in 1998. The Media Lab provides education in new media design and production within UIAH and for its partners. It places strong emphasis on high-level project work, through collaborations with external partners; and it aims at preparing students for a leading role in the industry.

In technological and scientific research, *Helsinki University of Technology* (HUT) plays a significant national role; in many fields of technology HUT is the only place in Finland where university level teaching and research are carried out. It is the oldest and largest university of technology in Finland, dating back to the ninteenth century. There are about 100 laboratories in 12 departments at HUT and four separate research institutes. The HUT is located in the Otaniemi Park, in Espoo. The main language of instruction is Finnish but some basic courses can be taken in Swedish. HUT also offers a great number of courses in English. HUT is committed to preserving the balance between basic and applied research, and efforts have been made to ensure that research findings can benefit business and society at large. The Department of Architecture at HUT was nominated by the Ministry of Education as a national Centre of Excellence in Artistic Activities for 2001–03. The *Otaniemi International Innovation Centre* offers its services to HUT staff and students in tapping for research funding, commercialising the results, and managing property rights. Also HUT participates in extensive cooperation networks with Europe, the United States, and Japan.

Polytechnics *Helsinki Polytechnic Stadia* is a multidisciplinary institution of higher education in applied science, internationally acclaimed for the high standards of its educational units. It was established in 1996 by merging eight existing municipal HEIs. Stadia is a fully-fledged department of the Helsinki City Council. At Stadia, the open-learning environment produces professional experts, of whom 88 per cent become employed immediately after graduation. The unique needs of the capital region and the national challenges for economic and industrial life are explicitly taken into account as targets of the school's educational mission. The number of full-time students is 7,500, to which a number of students attending short and evening courses, open polytechnic teaching and customised teaching should be added. The number of full-time students is expected to increase to about 8,000 in the near future. Close to 30 different degrees are available in three faculties: Health Care and Social Services, Culture and Services, and Technology. Applications for the faculties of Culture and Health Care exceed the available places by far; it has been noted that technology studies attract a decreasing number of students. The activities of the Polytechnic take place on about 16 campuses across the city. A business centre of Stadia started operation on March 2000. The Faculty of Technology in particular has strong cooperation links with companies and different departments of the City of Helsinki, such as the Public Work Department and Helsinki Energy. Ninety-five per cent of the BA theses were undertaken for companies. Thesis preparation is a significant form of training for students, but at the same time it offers lecturers a channel for maintaining working links with the business sector. The Faculty of Technology also participated in an 'attraction project', through which the City of Helsinki promoted its industrial policy. This project aimed at inspiring young people to enter technology studies. Moreover, a two-year IT programme, in English, was started in the autumn 2000 as part of the internationalisation process of the Polytechnic. This programme aims at providing the third and fourth year students with an excellent technical language skill, helping them to get used to working in multinational work societies and at the same time orientating them towards today's leading technology.

The current international strategy aims at enhancing international activities, improving the level of professional expertise, increasing the cooperation both inside the organisation and with its partners, and developing the urban culture in Helsinki. Teacher and student exchanges and projects in several countries are run within the framework of Socrates/Erasmus, Leonardo and Nordplus programmes. In 2000, 40 teachers or experts from Helsinki Polytechnic Stadia taught abroad, and close to 50 of their international counterparts were welcomed to Helsinki. In 2000, more than 200 students arrived from various countries, including Europe, Africa and Asia. In the same year, the number of students undertaking different exchange and trainee programmes and bilateral agreements was 342.

Associations of HEIs The joint *Consultative Delegation of the Helsinki University and 12 polytechnics* represents the widest cooperation network in the Helsinki region. The main task of this delegation is to develop the regional development effect of universities in the capital region. In collaboration with the city, this body has launched a study programme for the secretaries working in the schools, geared towards vocational training and economics.

The Helsinki Polytechnic Stadia takes part in the *Consultative Delegation of the Uusimaa Polytechnics*, which, together with a cooperative group of the presidents of the region's polytechnics, develops cooperation between all polytechnics in the Uusimaa province. A new wave of discussion concerning the necessity of polytechnics and the level of degrees was started, and continued just as actively as that in the establishment phase. Universities in particular took a stand on how to develop the polytechnics. Stadia is now in close cooperation with the Helsinki University, the Art and Technical Universities and other polytechnics of the area: Arcada, EVTEK, Helia and Laurea.

Research and technology transfer The metropolitan region of Helsinki is characterised by a high intensity of research activities, with important implications for the urban economy. Examples of excellent combination of research, education and – notably – community life are promoted throughout the region.

The *Arabianranta* site in Helsinki[2] is possibly the most interesting because it offers insight on non-traditional ways to disseminate knowledge and use it to generate social capital. This district, located 6km north of the centre, has been one of the centres of Finnish design for over 100 years, but it declined, physically and socially, with the shift in importance from heavy industry to the service economy. A few years ago, the City of Helsinki and the other stakeholders of so-called 'Arabia' launched a large regeneration and development project, the *Art and Design City*, as a public-private partnership scheme. The main partners were the Ministry of Trade and Industry, the City of Helsinki, a large producer of consumer and design goods, the Pop and Jazz Conservatory, the University of Art and Design Helsinki and other educational institutions, as well as some other private firms. There was a clear effort to involve the area's residents in planning and as final users. The aim is to turn Arabia into one of the leading art and design areas in Europe, where the functions of living, working, studying and recreation are mixed.

An integral development strategy was adopted, providing space for such complementary activities as research, business, education, housing and shopping. This ensures that Arabia develops into a genuine, living community that does not turn into a barren wasteland at night. The project is due for completion by 2014. By the beginning of the next decade, Arabia will have grown into a community of 12,000 residents, 9,000 jobs and 6,000 students. The electronic infrastructure of the Art and Design city will be very sophisticated. Its bulk comprises an old pottery reconverted in a 'design factory' hosting many firms in the fields of new media and design are located. Already, 50 to 60 firms are present on the site, employing some 300 people. This number is on the rise as the rents are still low. Moreover, many of the small firms work for larger companies in the region. Nokia has begun to realise the importance of outsourcing selected operations to small hi-tech firms. One of the objectives of this policy is to let the SMEs benefit from the international market projection of the large firms, accelerating in this way their growth. IUAH is clearly one of the key players in the project, as it offers research facilities attracting many young people and it acts as a breeding ground for young talent and new activities. Arabia is also home to a real-world experiment in community networking. Sonera and its partners are creating the *Helsinki Virtual Village*, a wireless interactive community for the entire suburb of Arabia. The Virtual Village will be connected through a local area network on fibre-optics and wireless links, which makes available a wide range of services on a 24-hour basis. Large companies such as Sonera and Nokia could benefit from the superior quality of the electronic infrastructure by using the area as a test-site for new products, or as a showcase to display what is possible in the future.

The district of Otaniemi is the most remarkable concentration of research, education and innovation in Northern Europe. It is located in Espoo, in the

Helsinki Metropolitan Area, at some 20 minutes distance from the centre of Helsinki and the international airport. This area hosts the HUT, the Technical Research Centre of Finland, the Otaniemi Technology Park, and various other institutes of advanced and applied research, knowledge transfer organisations and incubators. Moreover, the headquarters of the international companies like Nokia are here. The *Otaniemi Technology Park*, opened in January 1987, is the largest and the most interconnected technology incubator of all the Nordic countries. About 5,000 researchers are employed in nearly 200 companies. Its main task is to function as an incubator and a business generator, offering business development and incubation services, financing services, and naturally premises and related services. *Innolinko* is the new early stage business incubator. It helps students and researchers to transform business ideas into new start-up companies, as a concrete starting point for ideas emerging from courses and research projects at the HUT as well as business plans and idea competitions. It offers furnished premises, mailboxes, Internet connections and hands-on help in developing business for nascent entrepreneurs. The strong cooperation with the universities enables the students to earn academic study credits while developing their own business.

Next to Otaniemi is *Innopoli*, one of Europe's larger commercial science parks. Its mission is to speed up the commercialisation and internationalisation of the results of investments in Finnish research and product development. Innopoli offers premises to more than 100 companies. Any kind of service can be found under one roof, including marketing, financial and legal services, technical assistance, and personal services for its workers. The *Spinno Business Development Centre* (Spinno BDC), established in 1991, is part of the Innopoli High Tech Centre's activities. Spinno BDC's mission is to promote the commercialisation of hi-tech and knowledge-intensive products in order to foster the creation, growth and internationalisation of new companies in the Helsinki Metropolitan Area and to generate new jobs in these fields. Its activities generate 20–30 firms every year, most of which continue their business activities within Innopoli or elsewhere in the Otaniemi Park.

The *Culminatum Ltd Oy* organisation is another important intermediary structure. It was established in 1995 to carry out the Centre of Expertise Programme in the Helsinki region, a national scheme aimed at stimulating knowledge and technology transfer between research institutes and firms. Shareholders in the company are the Uusimaa Regional Council, the cities of Helsinki, Espoo and Vantaa, and the universities, polytechnics, research institutes and business community of Helsinki region. This organisation aims at accelerating the technology transfer process from ideas through research to saleable products by coordinating the collaboration between companies and scientists, and by undertaking a part of the financing for the projects. The city councils of the Helsinki Metropolitan Area have assigned responsibility to Culminatum for preparing a science and technology strategy for the region, and a common development strategy for the universities and associated business operations. Among others, Culminatum runs the *Centre of Expertise on Digital Media, Content Production and Learning Service*, seeking to improve the profile of the cultural industry, the operating conditions in the cultural sector and to establish networks of experts and support services to improve the cultural sector business environment.

The *Helsinki Science Park* is located at the centre of the Viikki district in Northern Helsinki, a new development area in Helsinki. Viikki is called 'the green valley of biotechnology', because it is one of the areas which make up the Helsinki green belt. It provides a spacious living environment close to nature with uniquely beautiful landscapes as well as excellent recreational and outdoor facilities. The Helsinki Science Park provides new space for the fast-growing campus of UH – the UH Biocenter, the Faculty of Science and the Faculty of Agriculture and Forestry – but it also hosts offices and company incubators, a new local commercial centre, various types of residential units for common residents and the UH's staff and students. By the 2010s, Viikki is expected to have a total population of 13,000. More than 1,000 research scientists and technicians already work at Viikki. Helsinki Science Park Ltd[3] is responsible for coordinating this cooperation between the university and the companies operating in the science park. The excellent location of the Helsinki Science Park, 9km from downtown Helsinki and 20 minutes from Helsinki-Vantaa international airport, and the adjoining housing and recreational facilities offer the companies the opportunity to establish an advantageous new home base in Viikki. *Cultivator*, the first business incubator in the Science Park provides facilities and business consulting to over 40 companies. The new business park for companies interested in operating in the immediate vicinity of the Science Park and a new incubator will be started in 2003.

One should not forget the role played by individual firms and their laboratories to carry out their research activities, which come to be to a large extent 'filtered' in the rest of the business community. The most important firm in Finland is *Nokia*, world leader in the production of telecommunication equipment, which has headquarters in Espoo. It is estimated that Nokia makes up the half of the total electronics industry in Finland, with a strong growth during the last few years. Nokia is very keen on the development of new products, and it spends some 10 per cent of its net sales on R&D activities.

The bulk of Nokia's research is executed in the *Nokia Research Centre*, which is located in seven places: two in Finland – Helsinki and Tampere – and five abroad. Nokia cooperates closely with local technical universities and with other R&D centres in Finland and all over the world. In spite of its transnational status, 57 per cent of Nokia's R&D jobs are in Finland, with Helsinki as the main centre. Although Nokia is the dominant player in telecoms, the telecom company Sonera is also important as far as research is concerned (especially as regards community utilities), as well as many smaller firms in the Helsinki region.

Spatial patterns of higher education and research An overwhelming majority of Finns own their own residence. However, in the Helsinki region the situation differs, as 40 per cent of the housing is based on rented accommodation (City of Helsinki Urban Facts, 2002a). Consequently, the rental markets are sometimes relatively tight. Most students prefer to live in apartments shared with friends. Usually, renting a large apartment with friends is relatively cheaper than renting a small one. If their parents live in Helsinki, some students prefer to live with them, because of the high rents. The *Helsinki Region Student Housing Foundation* (HOAS) is a private foundation for student-housing service, including accommodation for foreign students. Student flats are located all around the metropolitan area, within 20 minutes of the city centre, but they can be quite expensive, at a price per square meter that reaches €2,180 in the centre (City of Helsinki Urban Facts, 2002a). Cheaper rooms are located further away from the city centre: prices are 25 per cent cheaper in Vantaa and 40 per cent in the rest of the region, but then one has to allow for travel costs. Some differences in the location strategy of different student groups can be noted. Students of technology prefer to live on the Otaniemi campus or in the Ruoholahti area between Helsinki and Espoo. Therefore, the pressure on real estate markets is 'transferred' to the municipalities of Espoo and Vantaa, where the living costs are lower and the availability higher. UH and UIAH students prefer to live in Helsinki. This is due also to the fact that UIAH does not offer its own residences.

The problem of lack of affordable housing for students in Helsinki has not been consistently tackled so far. However, it could worsen in the future because of the increasing pressure from the internationalisation strategies adopted by the universities and by the city, and because of the rising trend for single-person households. The fact that students stay in Helsinki after their studies and the fact that the city is very attractive for young professionals will result in a severe shortage of houses.

Because of the increasing number of students and the scarcity of space in the old buildings, the UH has been forced to pursue a decentralisation strategy, which is scattering the city centre campus to four suburban locations. UH acknowledges that this is a 'second-best' solution to the need for expansion and the requirements of high-quality education. The future strategy of UH foresees that the Faculty of Humanities and Social Science will remain in the city centre, while the other faculties will be spread around the city. The biotechnology pole is already located in the Viikki area. The other HEIs have been also doing the same in the recent years. Altogether, there are 60 different higher education locations throughout the metropolitan area.

The on-going suburbanisation of higher education poses additional demands on public services like transport. Though plans are under way for the realisation of a high-quality transport link between the main 'knowledge districts' of the Helsinki region (Figure 4.2), the budget for this work is far from readily available. But the real issue is the interdisciplinary cohesion. At the time of writing there are good relations among the different faculties of the university; with suburban campuses, some extent of separation may arise.

The impacts of this decentralisation process are indeed being discussed. There is a certain preoccupation with the segregation of the students. Viikki, for example, will host 7,000 students. The HOAS organisation is building student housing in the area, which could lead to a wave of concentration of students and social friction, something that until now – thanks to the equilibrium between host and guest communities in the city centre – has always been avoided. It is also possible that the decentralisation strategy will result in a loss of the vibrancy and the diversity generated by student communities in the city centre, though no fewer than 20,000 of the central campus residents will probably continue to live there. There are clearly stronger emotional links with the City of Helsinki rather than with Espoo or Viikki. In other cities in this study, such as Lille, decentralisation brought about considerable changes in the urban landscape, with an impact on the lifestyle in the city centres. Helsinki is still attractive and the public transport services in Helsinki are excellent. However, there are some problems with Viikki, because at the present time there is no fast connection and it may be necessary for students to find accommodation close enough to the campuses. Moreover, Helsinki is facing an €8m cut in the transport budget.

4.2.3 The Student Community: Behaviour and Careers

Helsinki is definitely a student city. Some 40,000 students are enrolled at polytechnics, universities and other HEIs in the Helsinki metropolitan area.

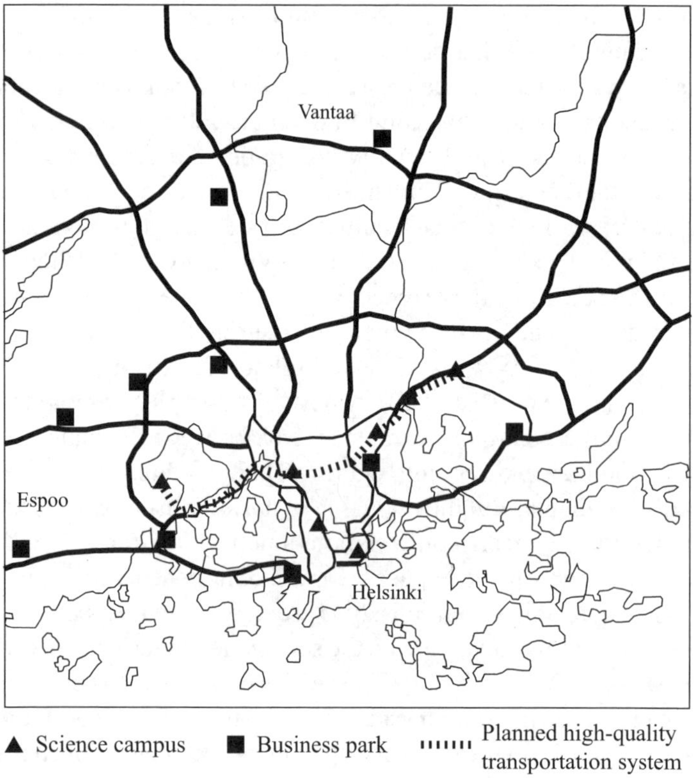

Figure 4.2 The new 'Science Corridor'

It is the biggest concentration of students in Finland and one of the biggest in the Scandinavian countries. In the last ten years, the number of university students has risen by about 40 per cent. Taking into account the age and the education of its population, Helsinki is also a young city. This generates increased attractiveness as a study location for foreign and exchange students. They choose Helsinki because of its image, the good reputation of its HEIs and the many possibilities for studying and working, not to mention the high level of quality of life. After their studies, new graduates look for jobs in the city, where the job market is by far the most developed in the country. The most attractive and well-paid jobs are concentrated in the area between Helsinki and Espoo. The main employers are the public sector and the telecom industry (Nokia and Sonera), the Municipality of Helsinki and many companies in the service sector. However, the chances of new start-ups in the SME sector are attracting more and more students.

Student life is very active. The peculiar, 'individualistic' structure of Finnish studies makes it important to establish strong social ties as a means of improving one's performance. There are many meeting places where social exchange takes place. In the numerous bars and cafés in the city or in the Kamppi – one of the neighbourhoods with the highest student concentration – students of different disciplines and different schools come into contact, and this diversity is the 'secret' of Helsinki's fascinating lifestyle.

There is a Student Union in each Finnish university, and membership is compulsory for all undergraduate students. The purpose of the Student Union is to represent its members and look after their interests in various academic and social issues. Founded in 1868, the *Student Union of Helsinki (HYY)* is a public corporation, with a long and glorious history that gained a high status in Helsinki's civic society. With 31,000 members, HYY is the biggest of all student unions in Finland, and gathers some 230 organisations under its wing. The membership fee (€62 in 2001–02) includes a healthcare contribution that entitles students to free or inexpensive healthcare services, as well as discounts at various stores and subsidised student meals. HYY is arguably the wealthiest student union in the world; until recently it ranked first among the real estate tax payers of the city. The Union's considerable patrimony is managed by the HYY Group, an independent company with four main lines of business: real estate, service, bookstore/publishing and the travel industry (Kilroy Travels). The turnover of the HYY Group is estimated at some €270m in 2000, and the profit account in 1999 was €6.2m. The Student Union owns buildings such as the Old Student House, the New Student House, and all the cafeterias in the UH. Hence, it is natural that the Student Union has come to be an important stakeholder in many urban issues, though the degree of involvement depends on the political colour of its board, which changes with regularity. One example of commitment that acquired general interest is the realisation of the present city-centre campus. This was strongly supported by HYY, which is still pushing for a rationalisation of student spaces inside the city, for instance through the pedestrianisation of a number of streets where student circulation is intensive. Their proposals for the use of public spaces are often taken into consideration by the city's decision-making bodies. HYY has recently started a programme of childcare for students with babies (three hours per day per student) with the financial collaboration of the university (1,000 members of the SU have children). The Student Union is also involved in more general cultural activities. Among more 'institutional' activities, HYY has published a 'Guide for Foreign Students at the UH', also available on-line.

Apart from the Student Union's activities, there are a number of active '*student nations*' or regional associations. One should remember that Finland is characterised by strong regional identities, and it is a bilingual country, both Finnish and Swedish being official languages. Student nations are among the oldest organisations in Finland, originally a fundamental support for incomers from different regions of Finland to whom they provided accommodation in their own private residences. Today, the 15 nations open their doors to every student; the participation in their parties and activity groups is indeed high. In addition to the nations there are over100 student clubs, such as department associations, interest groups and sport clubs.

This bustling activity has definitely upgraded Helsinki's profile as far as quality of life and dynamism are concerned. The benefits for the community are remarkable. For instance, creative new talents are highly appreciated by private employers, and not only in the dominating ICT sector, generating an almost perfect synergy between private and public interests. The city does not have any problems in attracting the best students, or retaining them after they have completed their studies. If possible, the opposite may be true: Helsinki could now face the problem of an 'excessive' attractiveness, which may produce imbalance both at the national level (a brain-drain problem for other urban and rural areas, though other education poles such as Oulu and Jyväskylä are also doing well), and at the local level, in terms of pressure on the local resources. In fact, the existing facilities and cultural services are sufficient to provide Helsinki students with the best possible conditions for a successful study period. However, in the future this situation may change due to the increasing living and accommodation costs, which oblige students to work during their period of study, inevitably lengthening their career.

4.2.4 Challenges for the City

In the last two decades of the twentieth century, the economic development of the Helsinki region was characterised by intense dynamism. Up to the late 1980s, the Helsinki region prospered. Income levels rose steadily, and the unemployment rate was very low, at only 3 per cent. In the early 1990s, the situation worsened dramatically, the main reason being the collapse of the Soviet Union, at that time a very important trading partner. In the period 1992–93, Finland's gross domestic product (GDP) dropped by 10 per cent and the unemployment rate rose to 18 per cent. Since 1994, the economic situation has improved. The service sector increased hugely under the influence of the telecom industry clustered in Helsinki. Recently, yearly GDP growth rates have been around 6 per cent, with regional growth in the Helsinki region somewhat higher, and the unemployment rate shows a clear downward trend. For Finland as a whole, the unemployment rate is 10.2 per cent, while the Helsinki region scores somewhat better, with 7.2 per cent. However, the fairly high number of long-term unemployed is a particular problem in Helsinki. It has proved difficult to bring these people back into active working life. However, there is a real chance to improve this situation on account of Helsinki's excellent and varied higher education: indeed, widening education opportunities geographically and among social groups, as well as providing continuing education to workers, are considered primary challenges.

The severe economic crisis of the early 1990s has proved an important stimulus to join forces and start to cooperate. As a result, in 1994, the key actors in the region – different levels of public administration and the business community – developed an integral and broadly shared vision, and drew up a regional strategy. Key elements of this strategy were to develop Helsinki as a Nordic region of interaction, with a focus on sustainable, knowledge-based development. Plans were developed to give substance to this strategy. Examples are the setting up of science parks, the Art and Design city, and the Innopoli cluster in the west of Helsinki, but also the important initiatives for cooperation between business and research. It should be mentioned that advisors' reports (van den Berg et al., 2001) in the past have underscored the difficult cooperation between actors within the metropolitan region as one of the factors hindering a development geared to the full potential of this region. Thus, a lot more is still to be done in terms of harmonising policies on a different spatial scale and governing Helsinki's sustainable development.

One necessary step in this respect seems to diversify the booming but volatile telecommunication industry, which now dominates, in order to make development more solid and balanced across sectors. In fact, the recent downturns in the ICT world have shown that putting excessive emphasis on only one sector exposes the whole country's economy to global crises. At the same time, Helsinki can use its information technology – largely available and familiar to all Finnish citizens – in creative ways as a support to the development of other sectors and to facilitate a dynamic community. ICT helps in handling an increasing number of daily chores, making everyone's life easier. However, it has been estimated that no fewer than 90,000 new jobs have to be covered in the Helsinki region up to 2015 (City of Helsinki Urban Facts, 2001: 14). Education excellence and top research facilities were key factors behind Finland's success with telecommunication. Today, this imperative is reproposed. The diversification of the economy and its strengthening depends much on Helsinki's capacity to develop a range of interdisciplinary studies, and to attract the best talents from all over the world.

Another challenge, which is linked to this, regards the system of social support. All Finnish students are entitled to financial support for their studies, as a means to guarantee equal opportunity in education and widen participation in what was originally a relatively backward country. On top of this, students can also get a bank loan, which is guaranteed by the government. The bases for the current aid scheme date from 1969. In the 1990s, the system was further extended: the amount of the study grant was raised in proportion to the loan, and the procedures involved simplified. The scheme is fairly comprehensive, covering some 75 per cent of all the students over 16 in secondary and higher education. The aid is granted for those who are admitted to an educational institution, study full time and are in need of financial support. Today, it is seen as a building block of Finland's excellence in education, and a tool for regional policy, as it may be used to give an incentive for one study location over than another. Yet the system hardly takes into account the rising living costs, and does not solve the problem of scarcity in the housing market. The system might also have to be reviewed in the light of the internationalisation strategies pursued by HEIs – the main goal of the government is to increase the number of international students from 2 per cent to 4 per cent – and firms alike.

Finally, a problem related to physical urban development adds to the previous concerns. The impending university reform (as everywhere else in Europe) will imply more mobility and flexibility in the study programmes, producing an increased need for effective physical connections throughout the metropolitan region the city. The present infrastructure may have to be upgraded to make the best out of the city's education supply.

4.3 Strategies and Integration. Helsinki as a 'Sustainable Student City'

Helsinki is a booming city, which serves an exemplary illustration of how far a 'virtuous cycle' of development involving excellence in education, urban attractiveness, and economic performance can go. However, we concluded the last section with a warning that some obstacles need to be removed to enable a new, more solid cycle of growth. In this section we analyse whether, and to what extent, the urban actors are ready to meet these challenges in a consistent way.

4.3.1 Relations between the Students and the Community

It was always considered the Ministry of Education's responsiblity to take care of the HEIs, and indeed the role of national government in boosting the knowledge economy is relevant, as Finland comes second in Europe after Sweden for the share of the GNP devoted to R&D. However, the City of Helsinki started to develop its own strategy. Until recently, the Municipality of Helsinki had very little formal contact with the universities in its territory. With recent initiatives, the climate has changed. Since 1994, the partnership between the University of Helsinki and the city has been a catalyst for shared efforts to build a community for 'Science, Technology and the Future', culminated in the 2000 European Cultural Capital Event; this partnership is expected to continue as advocated by the Research Programme 2001–03 of the city (City of Helsinki Urban Fact, 2002b).

Relations between the city strategy and the HEIs start to be relevant, as in the example of the key role of UIAH in the regeneration of the Arabia district. The municipality's investments in the 'Art and Design City' project sprang out of necessity and later proved a real point of strength in Helsinki's urban policy, following the development in Arabia and the recognition of the importance of the arts and design for urban sustainability. The UIAH has therefore proved a real engine in the regeneration process of Arabia and UIAH's students were involved in the European Cultural Capital programme, producing high-quality audiovisual promotional materials.

This new trend of cooperation between city and HEIs is also confirmed by the city's involvement in the development of the Helsinki Science Park in the Viikki area. Many other actors are now involved; Stadia is trying to transfer part of its school there. It could be more difficult to achieve a sustainable settlement in Viikki, as commercial developments do not always appreciate the market potential of student communities, who tend to be rather risky and low-profile. Also interesting is the relation between the Stadia polytechnic and the city. There is perfect dialogue with the city, which follows naturally from the institution's origins, but it significantly extends to decision-making. It is a special case that can be found only in Helsinki, Tampere and Turku. All the decisions about the school, like budgeting and location, are taken at the city council level. Moreover, in the near future the City of Helsinki will be one of the main employers of Stadia graduates, on account of the fact that a small but increasing number of students choose the Faculty of Culture and Services.

This kind of cooperation is consistent with the approach of the Ministry of Education on knowledge-induced regional development. At the time of writing, only UH has developed good networks with polytechnics schools (*Consultative Delegation of the Helsinki University and 12 polytechnics*) and other HEIs, like the University of Technology, the University of Economics and the Sibelius Academy.

A major problem in this respect is that the relationship between the universities and the separate municipalities at the metropolitan level is not optimum. For instance, the municipality of Espoo does not have as extensive relations with the knowledge network of Helsinki as it has with HUT. In other Finnish cites, notably Oulu and Tampere,[4] the situation is different: municipality and universities jointly build up regional economic and technical excellence. Culminatum's activities have brought about an improvement in this sense. However, there are again some problems with Espoo, because this administration opposes an extension of the metro connection with Helsinki, perhaps due to the fear of eroding its tax base as a result. Obviously this is an issue for more concerted planning at the metropolitan level.

Today, the City of Helsinki is promoting internationalisation as a strategy to compensate for its peripheral location. In the new geographic organisation of European trade flows, Finland could play a key role as a bridge to the east. The challenge is multifaceted; a key element is multiculturalism. In 1999, the City Council defined the policy priorities for the city's internationalisation, derived from the its strengths and opportunities, as well as setting the targets (Helsinki City Council, 1999). The first is to promote foreign investment in Helsinki, assisting the internationalisation of local business and developing Helsinki as an attractive city for international companies. The second is to strengthen Helsinki's position as a venue for urban tourism and a site for conferences and congresses. The necessity of marketing Helsinki as a centre of academic expertise and research is also highlighted. One initiative should be given particular emphasis: the City of Helsinki, UH and the Ministry of Education have jointly set up *six professorships in urban research* to strengthen the link between the city and its higher education in terms of problem-solving and capacity building. This five-year long project might be expanded to other universities in the Helsinki area.

However, not all the Helsinki HEIs fall under the city's authority, and this raises planning issues, for instance with regard to development of student facilities and transport. The Helsinki Metropolitan Area Council (YTV) is a joint organisation of the four municipalities included. Unfortunately, planning at Helsinki metropolitan level has not been very successful on account of the increasing competition between the four cities. Only two levels of decision-making exist: national and local. Hence, there is a certain difficulty for the HE sector in Helsinki in developing consistently with its aspirations.

As regards the students, relations with the community are good. The students can influence planning and other issues through their formal and informal channels, and though they are not directly represented in the city council, it is not hard for Student Union members to get in touch with the city's executive. The Student Union has a long tradition of taking part in the cultural and national politics. Students have the right to vote in Helsinki thanks to a change in the residential registration law.

Altogether, Helsinki holds its students and the academic community in high regard. Students are perceived to enhance the image, vibrancy, and atmosphere of the city. However, Helsinki has not been active in maintaining a balance or focus on students as citizens. The present favourable situation could be overwhelmed by problems of social exclusion and excessive spatial segregation; gentrification processes may happen *against* students' interests, chasing away one of the city's main important resources. With the campuses scattered throughout the Helsinki metropolitan region, students could be pushed out of the expensive Helsinki centre. Ring areas like Otaniemi are increasingly 'student ghettoes' that are largely disconnected from the local communities. In other popular districts, like Viikki, interaction may bring advantages, but the issue of a smooth encounter between residents and students remains.

4.3.2 Relations between the Student Community and the Private Sector

Universities and polytechnics have a strong network of relations with the private sector. However, the private sector's support to universities in sectors other than research is limited. Traditionally, they expect the state to take care of anything regarding education and infrastructure. Recently, there was an attempt to create a private university which not be exclusively for technology, with a greater involvement of the Finnish firms as regards education orientation and access policy. This ambitious project, which should become the spearhead of Finnish education, was strongly supported by the city administration. However, at the time of writing it had not obtained the agreement from the Minister of Education. It is remarkable that local and national governments have opposite views on this issue, despite their similar political colour.

The most developed networks of the HEIs with the private sector are those that are the principal players in the field of research and education regarding the new technology, the HUT and the UIAH. In general, linkages between business and education are numerous. In particular, large companies, like Nokia, are used to thinking of universities as research partners, but now SMEs are also exploiting the links with the universities. The private sector is today more willing to cofinance a major state programme aimed at increasing the number and quality of students of information technology and telecommunications, considering the fact that a labour shortage in this field is expected. Moreover, in Finland, science students are used to writing their theses on the basis of a practical research project in a firm. The advantage is that students are better prepared for their future jobs, and firms can benefit from their research and have some pre-selection of staff. In the non-science academic disciplines, thesis writing within a firm is much less common. Most students from the Art and Design Academy and UH do not write their theses on the basis of a business-related project. Still, these institutes do have relations with local industry. In some cases, university staff are hired for in-company training in large firms, and people from industry are frequently invited to lecture at the university, or are active in the composition of education programmes.

In the field of research, strategic links between universities and firms are more frequent. In particular the larger firms in the region closely cooperate with public research. For smaller firms, the situation is different. The intermediary organisation Culminatum Oy plays a major role in building bridges between public research and SMEs, facilitating knowledge exchange between universities and SMEs, and acting as a business incubator. As a result, the research and education agenda of these institutes has become more market-oriented, and the quality of education has benefited. The increasing level of contract research has strengthened the financial position of universities, whereas central government funds were shrinking. As universities and SMEs are culturally different, Culminatum Oy acts as a mediator and interpreter, raising the HEIs' awareness of their product. Despite this organisation's limited means, it has succeeded in initiating many projects, thanks to its fund-raising capacity. An element of weakness in Culminatum Oy is that large firms in the region are not directly involved in the technology transfer activities. In particular, the involvement of non-high-tech firms is very low. For the future, Culminatum's ambition is to increase further the number of SMEs involved in R&D and university collaborations.

The private sector, and in particular large firms such as Nokia and Sonera, take many important initiatives regarding the educational level of students and their quality of life, with some success. These concern in particular the effectiveness of job market and the support to students' career development. Though business support to universities and students is limited to the technical HEIs (and this holds true for Nokia in particular), ongoing restructuring and internationalisation in the ICT sector is pushing these firms to become more open. Increasingly, firms need people who combine several skills and are able to generate innovative integral solutions, not only in the technical sense. In this light, cooperation – both in education and research – between the several specialised universities in the region could benefit the local economy, resulting in broader, more integrated education programmes. For instance, the design of hi-tech products is a very important success factor in the market, and the UIAH's Media Lab offers a valuable asset to the business community in the field of integration of hi-tech and design.

Taking all this into account, stronger public leadership could achieve more. It is necessary that the government analyses the opportunities for a closer involvement of business partners in the development of a 'knowledge region'. So far, support from the private sector in student housing and infrastructure is still rather limited despite the emerging need to rationalise the presence of the students in the area. It would be up to the private sector to channel its efforts into long-term strategic actions. An important interlocutor in this process could be the *Chamber of Commerce of Helsinki*, which has over 4,000 member institutions. One of its key projects is aimed at helping members to make the shift to the information society, and this mission can be easily extended to the community as a whole.

4.3.3 Influence of Higher Education in the Relations between the Private Sector and the Local Community

Most firms in the new technology and new media fields are set up by young people, facilitated by incubators such as Spinno BDC, in which almost all the HEIs in Helsinki are involved. Finland as a whole expresses a large demand for this kind of services, as individual researchers (not universities) own intellectual property rights of their new inventions, which generates a great incentive to start up business. In the Arabianranta area, spin-off firms from the Art and Design academy abound. Their number is expected to increase with the completion of the Art and Design City project. Apart from formal arrangements, commentators would argue that the secret of Helsinki's boom is in the quality and intensity of its economic networks. The limited scale of the Finnish market and the non-hierarchic, informal way of working in Finland favours strategic networking. People know and trust each other, and common interests are easily recognised.

The aim of the recent development in Viikki is to create a knowledge hub in biotechnology, replicating in some way the success of the Otaniemi

Park, yet underlying a different 'urban theme' in which technology is no longer centre stage, and other issues – cultural and environmental – gain in importance. Through similar developments, the city is aiming at diversifying its economy and image. In this case, one could see a positive role for universities as 'agents of change' affecting the performance and development of the local economy, which comes to be more diversified in structure and more beneficial to larger areas of the territory. In the case of the recent development in Viikki, a stronger involvement of the private sector is required along the very lines that contributed to the success of other initiatives. At the time of writing Viikki is solely a residential location, scarcely attractive to private firms.

The public sector is paying special attention to small firms in this process of restructuring, through the *National Centre of Expertise Programme*. As a result of the programme, Finland now has a powerful and effective network of Centres of Expertise to meet the challenges of the twenty-first century information society. The aim of the *Helsinki Region Centre of Expertise Programme* is to ensure that the Helsinki Metropolitan Area retains its foothold amidst international competition and confirms its status as the country's principal engine of economic growth. The six Centres of Expertise in Helsinki, coordinated by Culminatum Oy, act as an intermediary body between the enterprises, business support services, universities and research institutions. The Centres reflect the needs and opportunities of local industries and innovation systems. Strengthening cooperation between enterprises and the research sector is a vital element of their work. They do this by bringing leading experts in research, education and private enterprises in the region or network into close interaction. Benefits from synergies in this knowledge-intensive sector substantially improve the environment for the emergence of new products, enterprises and jobs. The City is now trying to map the least attractive areas in Helsinki, where there is excess supply of jobs, and to promote them as job providers for new growth industries.

4.3.4 Model Testing and Identification of Issues

In the context of our research, Helsinki is certainly to be considered a success story. It is quite interesting then to see how strong points in one of the relationships sketched in the scheme of Figure 4.3 are reflected in the others. This case study also offers insight into how easily successes can turn to weaknesses, illustrating the importance of local government in anticipating changes and governing transition processes.

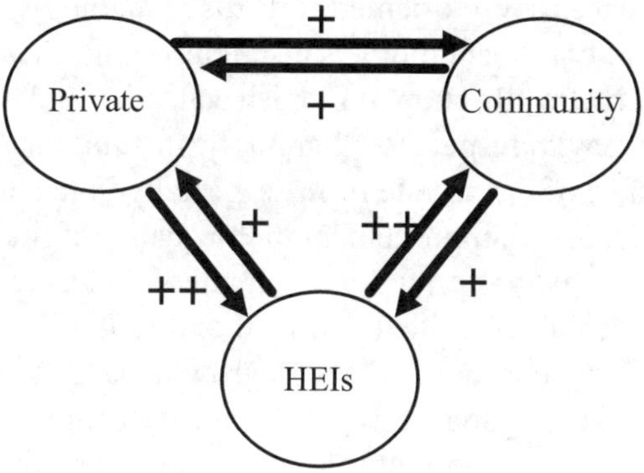

Figure 4.3 State-of-the-art: Helsinki as a student city

The private and public sectors and the universities work together relatively well. Examples of effective interaction are the Culminatum organisation, the Media Lab and more generally the Art and Design City, and finally the extensively-developed links between large firms and universities. These networks are to some extent dispersed, mainly because of the two municipalities that are involved. There are, however, some signs of a change for the better. One was the establishment in autumn 1997 of the *Helsinki Club*, whose members are influential representatives of the business community and public administration. The objective of the club was to seek opportunities which benefit the Helsinki region as a whole, among other things by instilling mutual trust among the actors, removing bottlenecks and coordinating the main actions in the area. One of these strategic areas is the preparation of a joint development strategy for scientific and art institutes, other university-level institutes and the associated entrepreneurial activity (Helsinki Club, 1997). This club could possibly be a driving force for new similar initiatives.

The link between city and community is strong enough, but it could face problems in the future to the extent that the City of Helsinki is not able to guarantee a high 'level of service' face to an ever-increasing student population and in the dispersion of higher education in the metropolitan region. This is also reflected in a possible loss of importance of students as agents of urban regeneration. The link between the private sector and universities is also strong, but it could improve by bringing together HEIs with different orientations for strategic coordination. In this way the restricted networks of each HEI could be shared in a wider metropolitan knowledge system. The private sector could benefit from the circulation of new talent in a local economy which has become excessively oriented towards technology. The project of a new private university of technology is also interesting, as it would greatly contribute to the internationalisation of Finnish education and research, acting as a wedge for change with respect to other public institutions. The enormous growth in performance of the private firms, especially in the hi-tech and ICT sectors, has caused a radical improvement in the quality of life for Helsinki's residents. For this success to continue, much attention has to be devoted to the research and education excellence. Social and spatial issues also affect this relationship. Balanced growth is is often growth that diffuses to all social groups and does not lead to spatial segregations and concentrations of costs and benefits. Universities have important roles in forging these spatial and social processes, and as such, should be at the centre of future planning initiatives, especially on the metropolitan scale.

However, as it has been pointed out on several occasions in this study, physical planning is not sufficient for a complete integration of the HEIs into the city nor for the development of a virtuous circle. Intangible, atmospheric elements related to the sense of integration of students in the community are also important and in this respect planning has to be more subtle and flexible. Also in this respect, though, Helsinki has proved to be at the forefront; but how sustainable is this model for the future? Helsinki enjoys the peripherality, the small numbers, and the familiarity in social relations which made this encounter easier; to this, an open attitude of the city administrations towards social and cultural innovation has to be added, to make Helsinki an ideal 'student city' environment.

4.4 Summary and Conclusions

Support for Start-ups

The Finnish private sector is dominated by a few IT giants, but it is apparent that this model has to evolve. The problem of Helsinki and Finland in general is to boost the growth of other sectors (even if Helsinki has the most diverse economy in Finland), diversifying the range of skills accessing the labour market.

One thing that should improve is support to the development of new companies, and this also implies a change in the traditional attitude of Finnish politics towards economic policy. A few leaders play an important role in the Helsinki economy, notably Nokia, Sonera and the Helsinki Telephone

Corporation. With their extensive strategic networks with smaller firms in the region as well as with other multinational firms over the world, they generate good dynamics and create spin-offs. Improved coordination of efforts could perhaps enhance the twinning of large firms with smaller counterparts and start-ups. In general, the entrepreneurial spirit has to be encouraged and supported. It is rather a question of encouraging creativity than lack of opportunities for start-ups. Higher education can become a powerful tool to instil this sense of entrepreneurship provided this issue spans all taught disciplines.

Strategic interaction of firms with the substantial public knowledge infrastructure is well developed. The large players in the region are very well aware of the assets that the R&D institutes constitute, and behave accordingly. Similarly, the universities see the firms in the region as partners for research and cooperate with them in education. A weak point is poor cooperation between universities. This is a serious drawback, as the firms urgently need people with a broad range of skills. The municipalities of Espoo and Helsinki could play a positive role in the stimulation of inter-university cooperation, but the large firms could also contribute.

Metropolitan Cooperation

It should be noted in this respect that in Finland the municipalities are the most important local governments in terms of financial and political power, as they have a high level of autonomy and generous financial means. Local income taxes amount to 17.5 per cent of the total tax payments, a very high figure in European perspective. Some 62 per cent of the budget of the City of Helsinki comes from local taxes (Holstila, 1998). This system of local taxation generates severe competition between municipalities about business companies and inhabitants with high income. On this account, the decentralisation philosophy of the 1960s has been followed by a stage of active policy for attracting new residents by the major cities. A large share of finance has been generated through the corporate tax system, but the state is now decreasing the share of this tax. This is deemed to hinder Helsinki's development (van den Berg et al., 1999).

Efforts to improve the *orgware* of the Helsinki metropolitan region in relation to economic development issues have not been overlooked. However, the situation with higher education needs more focus. Competition in the field of higher education, as in other fields, may prove destructive. It is estimated that to realise a metropolitan government, 10–20 years are needed. However, the pressure is increasing. On the metropolitan level there is no specific strategy aimed at the development of the telecommunications cluster. The municipalities of Helsinki and Espoo have their own objectives and goals. There is no vision as to how to develop the cluster in a metropolitan perspective, so possible synergies are about to be missed.

Due to the diversity of actors concerned and their partially opposing interests, this might need a new momentum of common dialogue and reflection. Culminatum Oy offers a good viewpoint from which to analyse the interests of the different stakeholders and to elaborate solutions. The municipalities of Helsinki and Espoo are partly aware of the possible benefits of better cooperation between universities, but up to now have considered the stimulation of university cooperation a task for national government. It might, however, be preferable to take a more strategic attitude and begin interaction in planning for higher education at metropolitan level. A first step could be to make a contribution to the total functional integration of the intranets of the Helsinki University of Technology on the one hand and the Art and Design city on the other, or at least construct a broadband link between the two nets. That would provide a platform from which to improve the integration of the Helsinki University of Technology and the Academy of Art and Design.

Housing and Segregation

The City of Helsinki is growing fast. According to Helsinki City Planning Department there is a constant need for 3,400 new housing a year for people moving to Helsinki, as well as for the more than 10,000 young adolescents still living with their parents. As a result, new residential areas are being built and planned. Also the nearby cities of Espoo and Vantaa are continuously growing. At the local level the big issues are housing and traffic. Helsinki is expensive and there is a shortage of land inside the city. The HEIs strategy is to decentralise facilities, but students do not want to live in the suburbs. As more students come Helsinki every year, the problem of housing will get even worst. Moreover, it will be necessary to improve the quality and a number of other facilities, in particular public transport.

At present, public transport functions well and dominates traffic in Helsinki and the region. Almost 70 per cent of daily trips in the city centre take place on public transport. The bus, tram, train and metro timetables are well-synchronised, the network is dense and the equipment is good. Traffic in the Helsinki region is smooth and traffic jams are rare. The trip from the fringes of Helsinki to the city centre takes less than half an hour; by public transport it can be even quicker. The presence of the decentralised campus of Arabia and Viikki, and the new housing developments in these areas, will require new investments to improve public transport connections. However, a potential problem may be decrease in the city's budget.

The number of exchange students is increasing on account of Finland's good reputation in the field of ICT. However, a weak point of this internationalisation strategy is that there is a decrease in the number of Finnish students who want to go abroad. Faced with high rents, students prefer not to leave their homes for fear of being unable to find another one when they return. The housing issue is also reflected in worsening quality of life conditions for Finnish students. Study grants are the same all over Finland, while it is evident that Helsinki students face higher prices, but the biggest problems regard the availability of cheap housing. There has been talk of reconverting a jailhouse into student apartments, but this project has proved too expensive. A solution may come from the ongoing relocation of the harbour to a new area, leaving free space for housing developments on the waterfront.

Notes

1 This should not be confused with the Greater Helsinki region, a new OECD-defined region called NUTS 3, encompassing over 30 municipalities, defining the functional urban region for Helsinki workers.

2 The name of the area, Arabianranta, means 'Shoreline of Arabia'. The historical reason for the name is the porcelain factory of Arabia still working in the area.

3 Helsinki Science Park Ltd is a joint venture of the Finnish government, the University of Helsinki, the City of Helsinki, SITRA and a number of industrial federations.

4 The City of Tampere can serve as an example: there, the municipality urged the two universities (one technical and one general, with 10km between them), to cooperate closely, and additionally invested in a science park in the proximity to the technical university to optimise research-business interaction. It is significant that in the last few years, research budgets (from public and private sources) of the Technical University of Tampere have increased faster than those of HUT.

Chapter 5

Lille Case Study

5.1 General Introduction to the City of Lille

The city of Lille is located in northern France, close to the border with Belgium, and lies at the centre of a densely populated, multinuclear cross-border urban region. The administrative structure of the area is rather complicated. The Municipality of Lille, with about 200,000 inhabitants, is at the core of a metropolitan area called *Lille Métropole Communauté Urbaine* (henceforth LMCU), the third largest in France after Paris and Lyon. Lille is the capital of the Region Nord-Pas-de-Calais, 4 million inhabitants, including two *départments* with a very high population density. LMCU is part of a larger *cross-border conurbation*, extending over the Wallonia and Flanders regions in Belgium (in some parts of the metropolitan area, Flemish is still spoken). This is the real 'functional urban region' of Lille, counting some 1.9 million inhabitants[1] (1.2 million in the French part and 0.7 in the Belgian part), 42 per cent of the population being younger than 25. As such, it is the most densely populated in France, and one of the youngest.

The urban region is economically unequal, with thriving areas especially on the Flemish side, and others that are lagging behind. The metropolitan area of Lille presents both faces in a large territory (600km^2). After decades of competition and rivalry, the cross-border character has recently been the object of transregional planning and extensive governmental cooperation. The region is characterised by a strong regional identity, tied to its industrial history, and at the same time, it is socially dynamic, with the highest birth rate in France. The emigration rates are high (5.4 per cent in the French part). The commonly quoted reasons are the lack of well-paid job opportunities (wages in the region are among the lowest in France), and the unattractive environment, in comparison to other French regions which did not go through industrial decline.

Unemployment in Lille was 12.7 per cent in 2000, compared with 9 per cent national average. The situation is slowly improving (cf. Figure 5.1), first at the metropolitan and then at the regional level, though the inner city of Lille is still losing jobs. The region is one of the most export-oriented regions in France, and trades a great deal with the United Kingdom, Belgium and Germany. Today, some of the oldest manufacturing industries are in poor

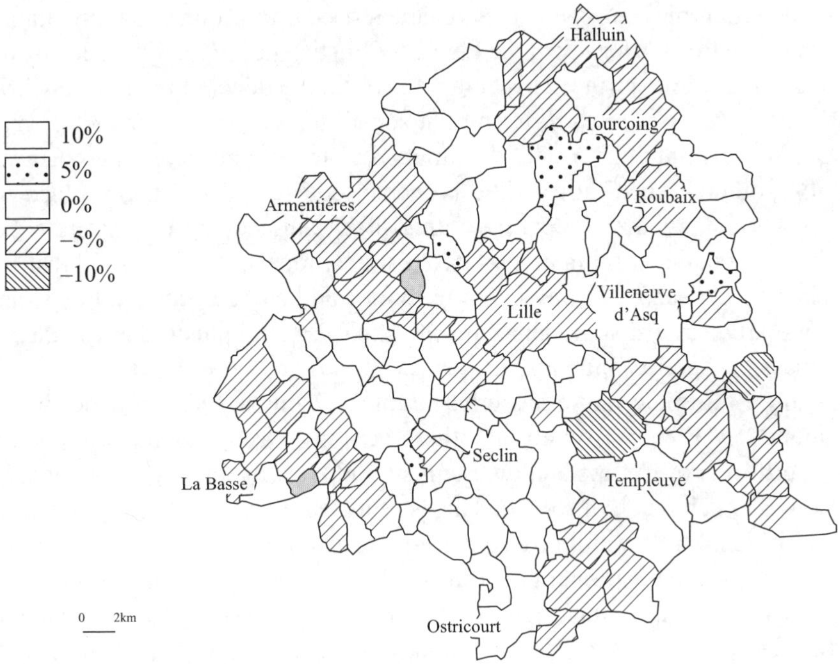

Figure 5.1 Employment change in various municipalities of LMCU, 1991–99 (% yearly change)

Source: INSEE, RGP 1990/99.

condition, after years of deep recession, and reconversion is progressing slowly. The industrial sector is still large, including construction, metal and textiles, but Lille has developed new economic specialisations, like research and education (about 6,500 jobs), and commercial services such as transport and distribution, banking, and communication. Other growth industries in the region are tourism, communication and international services. Lille can be considered a pioneer in commercial innovation in two fields: mass distribution and mail order sales, an area where the LMCU today ranks first in Europe. It has first class economic potential in the areas of fashion, electronic dataprocessing, printing and graphic industries. Ranking as France's third financial centre, and the ninth financial city in Europe, LMCU is attracting important foreign investment. Local businesses are strongly rooted in the community, but this is rapidly changing, adapting to Lille's new position. By choosing to settle in the region, Toyota has joined the ranks of numerous multinational firms that wish to be in the logistical centre of Europe.

The economic history of Lille is characterised by a strong social orientation by the local firms, a legacy of the Flemish Christian tradition, influenced by the social Christian doctrine. This context favours the development of networks based on *trust and mutuality*, and flat social hierarchies. These factors may explain the relatively peaceful climate and the recent economic success. Unlike other French cities, Lille is indeed considered a friendly place for public-private relations. Because of this, the social-economic environment of Lille is a peculiar combination of two distinct models, the 'industrial district' – characteristised by strong community ties and shared social values – and the hierarchic, Fordist model of the old Anglo-Saxon industrial cities, geared on mass production and large firm size.

Lille Métropole possesses other atypical features. On one hand, it is a distinctively French city; on the other, there is a strong orientation towards the northwest of Europe and the Flemish tradition (the city is also known by its old Flemish name of *Rijsel*). This is reflected in an increasingly 'outward look' in local policy initiatives, possibly a reaction to the lack of attention received from the national government. In fact, the decentralisation process that has taken place other big cities like Lyon or Marseille is in its infancy in Lille. The city thrives on other relations than the national – for instance, with Brussels, with which a certain 'complementarity' in terms of services is being considered as a planning principle.

The history of the old city of Lille is remarkable. Its legacy is in the Flemish, Spanish and Burgundy tradition, similar to that of Brussels and Dijon. Today, it has the status of a real regional capital of the north of France. It is a lively city offering numerous cultural opportunities: cinemas, theatres, an opera house and a philharmonic concert hall, the largest European bookstore, plus 33 classified museums, including the Fine Arts Museum of Lille - one of the most prestigious art museums outside Paris, a convention centre, and an international stadium. The city centre's exquisite grid of old decorated houses has been recently refurbished, and offers a first-class ambience for shopping and leisure activities. This enviable infrastructure is the result of a prolonged action in the field of culture, mainly led by the dynamic presidency of the LMCU. Under the leadership of a former mayor of Lille – Mr P. Mauroy, a charismatic political figure of national relevance – the LMCU managed to channel investments and resources into the development of a cultural capital. Thirty years ago the city was considered a cultural desert: its historical centre was badly in need of regeneration, and the urban atmosphere was considered dull. Today it has acquired a new vibrancy, to which students make a notable contribution.

Today Lille faces a considerable new challenge, the organisation of the 'European Capital of Culture' event in 2004. The programme will revolve around three major axes: the Lille's heritage, new cultural developments, and the work of associations. The 2004 event will be an opportunity to attract a larger than usual audience to the festivals and events held in Lille all year long. Though weather conditions tend to foster the 'grey' image of the region and the city, Lille also enjoys beautiful surroundings, such as tranquil rural areas and nearby sandy beaches at the Cote d'Opale.

Due to its strategic position at the crossroads of northern Europe, Lille is rapidly gaining status as a world trade and communications centre. The 'fixed link' between Great Britain and continental Europe with high-speed trains enhances Lille's location at the heart of a rapidly-expanding region. The flamboyant *Euralille* complex, however, is more than just a train station, but rather a large scale commercial development, built around and above the new TGV interchange, and an extension of the city centre, developed by world-famous and local architects. The ambition for Euralille is to become a catalyst to a different range of new urban functions, one of which is residence (Newman and Thornley, 1996: 189). The programme has led to the construction of 692 housing units, 204 of which are for rent, largely occupied by temporary residents, such as students and businessmen. Euralille is also an entertainment centre, both in public spaces, such as parks and streets, and in the numerous bars, restaurants and clubs of the area.

The high-speed train connection of Euralille puts the city at 30 minutes from Brussels, an hour from Paris and two hours from London, the real centre of a 'golden triangle' of European economic and political powers. Due both to ongoing European integration and to its optimal position in the HST network, Lille has changed from a city at the periphery of France to a city at the centre of Europe. Lille is estimated to become the most central city in Europe in 2020, from the eleventh place in 1990 (Face, 2001). Lille also expanded the capacity of its international airport in 1996. In the urban transport field, the Lille area was a forerunner when, in 1978, it launched the first automated subway in the world; since 2000, over 45 km of lines form the largest VAL (light automated vehicle) network in the world.

5.2 The Student Community and Knowledge Infrastructure in Lille

5.2.1 General Information on Higher Education in Lille

The higher education student population in the LMCU amounts to more than 100,000, to whom 6,500 researchers and 7,000 foreign students must be added, making it France's third largest higher education centre after Paris. With the exception of Toulouse, it is the largest student concentration in France in terms of students per resident. Most of these establishments are in suburban communities, such as Villeneuve d'Ascq. If the other establishments in the Academy of Lille are included, covering the whole region Nord-Pas-de-Calais, a further 50,000 students should be included. The metropolitan area, however, represents a 'functional region' for education on its own.

The French system of higher education is characterised by the coexistence of two systems, public universities and *grandes écoles*. Lille offers plenty of excellent opportunities in both spheres. Though the level of the *grandes écoles* of Lille cannot be compared with that of those in Paris or Lyon, they enjoy a strong reputation as engineering and business schools, attracting pupils from all over France. The varied and rich academic community of Lille is the legacy of its economic tradition, of a strong regional research potential and of a unique contiguity with the local authorities and the economic world. Seven universities are located in the region, four of which are in Lille, plus 14 engineering schools (the prestigious *écoles d'ingénieurs*, inherited from the industrial history of the city), five business schools, a journalism school and many other *grandes écoles* and higher education institutes.

Lille Métropole includes four universities and several schools (Figure 5.2). There are public universities, such as the the University of Science and Technology – Lille 1, the University of Law and Health Sciences – Lille 2, and the University of Human Sciences, Letters and Arts, which account for more than 80 per cent of the university students; and private universities, like La Catholic University.

5.2.2 The Knowledge Sector of the City: Actors and Spatial Patterns

Universities, grandes écoles and other higher education institutions In the fifteenth century, the university was in the old county capital of Douai, until the Flemish domination and then the French revolution, when it virtually disappeared, to be reopened in Lille. The university was restructured and named the University of Lille in 1896. At the apex of the industrial revolution,

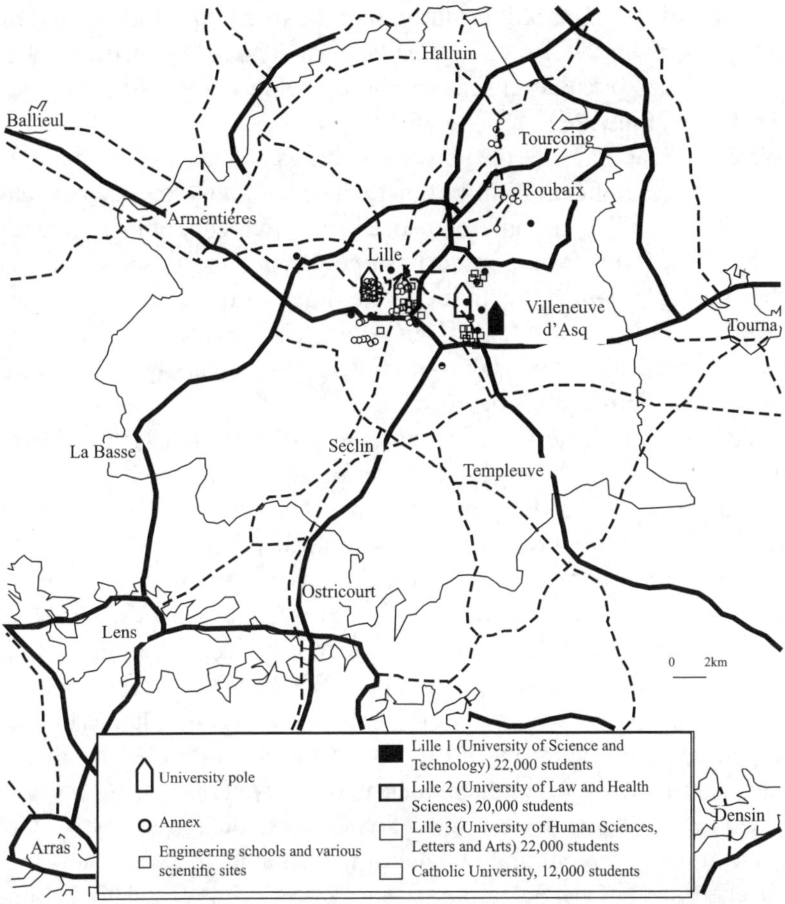

Figure 5.2 Higher education establishments in the Lille region

the big textile and metal industries saw in the local universities a potential engine in terms of knowledge and entrepreneurial capacity. La Catholic University in particular, founded in 1875, was at the forefront of economic and social innovation, due to the strong ties with industry leaders and the principles of social Catholicism. In those days the competition between public and catholic universities was strong, as they were both settled in the Lille city centre. In 1968, the separation with the transfer of the University of Lille to peripheral neighbourhoods also marked a stage of relative specialisation.

The *University of Science and Technologies – Lille 1*, located in the Scientific Campus of Villeneuve d'Ascq, is the main regional university as far as the range of education offered is concerned (120 national diplomas) and for opportunities in continuing education. It hosts 22,000 students and offers education in seven faculties and four 'special schools' or institutes. Lille 1 has an enormous prestige in France, and is regularly ranked among the top universities nationwide.

With its 48 laboratories the university carries out research in many fields. In particular, a significant contribution to the community has been developed through the research in the environmental protection. The campus includes the science park of Haute Borne, which hosts companies and all sorts of incubator and business development activities. The organisation *Cre'Innov* organises conferences and debates for students and teachers. Knowledge transfer activities are taken care of within the Enterprise-University Programme, an initiative of the university's board.

Since the 1980s, Lille 1 has developed an ambitious international strategy aimed at opening up the whole campus to the international environment; since then over 200 working partnerships have been sealed. There are 49 bilateral agreements worldwide, 220 European partner institutions within the SOCRATES programme, and over 850 exchange students and teachers each year. Despite this remarkable progress, the university's level of internationalisation is still low. This problem is being tackled in the present stage of reorganisation with education on a BA-MA structure, by organising a consistent supply of courses in English. In this way Lille 1 hopes to become more competitive, exploiting all the opportunities provided by increased student mobility and the complementarity of study curricula.

Life on campus may be boring for students, but the university is trying to cope. To improve the cultural climate of the campus, in 2000 a new facility was opened at the Cite Scientifique, the 'Espace Culture'. This includes a *café culturel* and an exhibition space for contemporary art covering 650m^2. At the end of 2003, a conference hall, spaces for workshops and art lessons, a photo laboratory, a CROUS office, a gallery and meeting rooms will be developed in this space.

The *University of Law and Health Sciences – Lille 2* is also characterised by a considerable international orientation, with 25 international cooperation agreements in European countries, Quebec and other Francophone countries. Lille 2 has 18,000 students. The university brings together two main faculties, Law Management and Health Sciences, which came together in the 1970s after a restructuring of disciplines between the universities of Lille. The two faculties, with 54 per cent and 46 per cent of the students respectively, are largely autonomous. Furthermore, there are two doctoral schools, one for each disciplinary group.

Lille 2 is located in the city of Lille, between the healthcare campus and the new compound in the Moulins quarter, an old working class neighbourhood at the edge of the city centre where the Faculty of Law moved in 1995. The move was seen as a positive development both for the university, which can count on a more attractive 'urban' ambiance, and for the city, which thus regained an important users' community. It was also part of a comprehensive regeneration project for the quarter. The hospital campus in the southwest of the city, Eurasanté, is the biggest in France, and includes a science park, a biotechnology centre, and many incubator facilities. This development, strongly supported by the LMCU presidency, is also considered a success in terms of physical development of a neglected part of the city. Moreover, the university maintains its headquarters and administrative offices in the old location in the city centre, where a centre for the promotion of research was recently established.

Research is indeed an essential function at Lille 2, and is associated to the national circuits like CNRS and INSERN. Lille 2 also recognises the importance of developing 'niche' specialities in order to be competitive. A new institute, the *Superior School for Business*, offers higher education in law and economics which became a source of specialised knowledge and human capital for local firms which have to plan their development in a dynamic environment. The university is also linked to different networks, national and international. One of those is IFRESI, which studies the development of societies and economies in the industrial age, with an important cross-disciplinary approach.

The *University Charles de Gaulle of Human Sciences, Letters and Arts – Lille 3* hosts almost 22,000 students. It is organised into 14 faculties and six special institutes and professional schools. Besides specialisation in ancient and modern languages, contemporary civilisation, music education, Lille 3 offers programmes in information and communication, tourism, business administration. It is also located in Villeneuve d'Ascq, in a campus halfway between the *Cité Scientifique* and Lille city centre. With its specialisation in humanities, Lille 3 has strong social and cultural links with the host community, which they explicitly try to nurture, being very active in the city life. They thus entertain a lively and healthy competition with Lille 1, but are specialised in 'new culture' that may be very attractive for students and may help them in the process of becoming new citizens, such as cinema, theatre and multimedia art. Lille 3 includes the most important centre of tele-education in France and a publishing house of national importance, *les Presses Universitaires du Septentrion*.

Today, Lille 3 suffers the problems of other schools of humanities and social sciences: how to maintain its mission between a 'market' for education that is not always willing to discover the potential of human sciences for their operation, and a state that, as always, privileges the 'central' universities of Paris. To face this challenge, Lille 3 is accelerating its process of internationalisation and networking within a northwestern European environment, which is believed to value the university's specialisations more. The region and other local authorities are seen as stakeholders and partners in this process, as it is believed that with their support, Lille 3 could become relevant and integral to the territorial development policies.

The biggest of Lille's private higher education institutes is *La Catholic University of Lille*, known as '*La Catho*'. It was founded in 1875 by the bourgeoisie of Lille, as a reaction against the danger of anticlericalism of the Third Republic, and that tie is still reflected in the strong liaisons with the local industry. Today *La Catho* is a leading centre for higher education in France, the largest private campus in the country with its 13,000 students, offering degree courses in almost every discipline, and 5,900 students following courses in vocational training. It consists of a network of polytechnic schools (six Engineering Schools, four Schools of Management), including six faculties and 30 GEs, all located in the northwest part of the city centre, at the edge of the historical city centre in one of the liveliest and wealthiest districts. *La Catho* offers, on one campus, over 50 degrees in five major sectors: humanities, law and economics, business and management, science and technology, medicine and theology. One of the essential characteristics of all teaching on campus is the importance attached to the humanities. All students, whatever their study programmes, follow courses in philosophy, religious culture or ethics. Studying conditions at *La Catho* are based on individual attention. Its originality lies in the fact that degrees are available both in the traditional style faculties and high-profile *Grandes Ecoles*. The university excels in teaching linked to vocational expertise, making students fully prepared for their future careers. Differently from the public universities, *La Catho* can count on a yearly research contract with the state.

La Catho has more than 250 international partners, and more than 1,000 students each year enjoy an international experience. Work placements and practical internships are integrated into the study programmes: seven out of the 13 students end up running firms in the area. Internships and practical placements in industry, commerce and healthcare services are often an integral part of the study courses. Though *La Catho* students are not exactly among the most 'creative' dwellers of the city, the inner city campus stands out for its social dynamism. Over 200 student associations operate in the area, mainly involved in sports, culture and charities. Highlights of student festivities include traditional student parties, various festivals run by part or all of the University, and important sports events like the famous EDHEC yacht race.

The *Polytechnic of Lille*, established in 1967, is part of the *Université Catholique de Lille*. The *School for Higher Industrial Studies* (HEI) is one of the seven schools that form the Polytechnic. It offers five degree programmes in engineering to 7,200 graduates, 1,200 engineering students and 240 graduates in continuing education. HEI also hosts 14 research laboratories, 160 full-time assistants/collaborators and 200 part-time teachers. HEI has also developed ties with companies and institutions, a legacy of its origin as a school functioning as a training centre for the network of local entrepreneurs. These links are an integral element of the education curricula offered by HEI, and provide HEI with teaching staff and knowledge. On the other hand, the research laboratories and staff often work for the companies which need this knowledge. The school is authorised to receive the Learning Tax, a tax levied on firms for professional training – its total is 0.5 per cent of the gross wage bill.

The *Lille Graduate School of Management (ESC Lille)* is one of the *grandes écoles* of management accredited as such, with 1,700 students at undergraduate, graduate, postgraduate and doctoral levels. The ESC was founded in 1892; its new location in Euralille dates from 1995. The decision to move was strategically driven by the attractiveness of the location, which provides plenty of networking opportunities. In fact, the school is still growing: a second campus is being developed for e-business, and another is located in Paris, at some 75 minutes from the school's headquarters. Through its close ties with the corporate world, ESC Lille offers three main paths to preparing for graduation: in-company placements, apprenticeships, paid day-release training. A special contract with the city allows students to be integrated into a company for two years, which exempts them from paying the school fees. Another noteworthy programme is *Alternance*: students can work part-time in the companies for one year. There is also the possibility to take a one year break in their studies for a prolonged apprenticeship experience.

ESC has developed a broad range of international activities, including student and faculty exchange, joint research initiatives, a full Masters in Management in English, and tailor-made intensive summer programmes ESC welcomes international students both through its more than 50 exchange programmes with its international university partners and through direct admission. Through the exchange agreements between ESC and its international partner schools students may study abroad in the second or third year of the ESC Lille programme for a period of one semester or one academic year. The strategy for the future is to develop programmes totally in English.

The *École Centrale Lille* (ECL) was created as an agricultural institute in the latter part of nineteenth century, as a partnership between the *Département du Nord* and the city of Lille. After long negotiations initiated in 1964, the Institute was transferred in Villeneuve d'Ascq (1968), entered the concourse that is common to all the *Écoles Centrales* in Paris and in Lyon, and was nationalised in 1975. In 1990, the *Écoles Centrales* in Paris, Lyon, Lille and Nantes signed a charter that defines the rules of cooperation between them. ECL hosts 1,000 teachers and some 20,000 students, plus nine research teams. The high-level scientific and industrial equipment enables these teams not only to develop research activities in France and with its international partners, but also to offer companies various services. What distinguishes the school most is the strong network with the private sector. ECL is open to companies who wish to profit from its rich and varied further education schemes through vocational training; some programmes, like 'Engineering and Management', clearly target engineers, executives, and corporate leaders, suiting the needs of the local firms. Moreover, ECL can provide technical assistance or valorisation projects. Each year, ECL approaches nearly 6,000 businesses which contribute to its funding through a *taxe d'apprentissage*. This financial support enables ECL to acquire the efficient equipment needed to give students state-of-the-art training opportunities.

Various types of practical training are offered in combination with the basic scientific courses, in order to prepare students for corporate life. To enable them to be in touch with companies, students have at their disposal the dedicated *Espace-Entreprises*[2] within the School. Finally, a newly-created Employment Observatory aims at assessing and analysing the professional insertion of young graduates from the ECL, and to facilitate their first 'job-hunting', as well as to contribute to the School's own planning. The international approach is also a priority for the school. Between 40–50 per cent of students go abroad on exchange programmes and internships for one or two years, also exploiting the possibility offered by the school to take a double diploma.

A school particularly devoted to the business world and with a strategy mainly focused on employment is the *Management and Commerce School Lille-Métropole* (*Ecole de Gestion et de Commerce* (EGC)), created in 1994 and located in Roubaix. This small school is part of the Management and Business Schools that make up the EGC Network, created by and for businesses, run by Chambers of Commerce and Industry and recognised by the French government. EGC offers a three-year degree course during which students are in constant contact with the business world, primarily via in-company training periods lasting at least 25 weeks. Studying at EGC school guarantees students rapid access to employment and successful integration into the business world (students spend an average of two months and ten days looking for a job after leaving the school). The school is close to the region's SME network and has an eye to the international markets. The teaching staff is partly made up of working directors and executives.

Other important HE schools in the region are the *École Supérieure de Journalisme de Lille*, one of the two most famous schools in France, the *Studio National des Arts Contemporains du Fresnoy*, located in Tourcoing, an ambitious educational project which tries to cross the visual arts and multimedia.

Associations of universities The Ministry for Education subdivides the French territory into 'Academies' as management units. The *Academy of Lille* includes all the higher education institutions in the Nord-Pas-de-Calais region, involving more than 150,000 students. The Academy works in close association with CROUS in offering services to the students, and with the *Pôle Universitaire Européen Lille – Nord-Pas-de-Calais* (PUEL). This association groups all the higher education institutes in the region, public and private, universities and *grandes écoles*, plus a number of research institutions, two Chambers of Commerce, and the main local government bodies. Created in 1991 and officially confirmed by the government in 1992, PUEL stands out in integrating the economic world inside its network of partnerships. It pays particular attention to the research sector through the association of the big national organisms. PUEL's main mission is to attract the resources for – and manage – large-scale projects of academic interest, like the strengthening of international cooperation programmes, support to emerging specialised research and education poles, the development of meeting places for foreign researchers and students, and the marketing of the local system of higher education. In this way, PUEL has become an important link between the educational and spatial-economic development of the region, and as such, it is an obvious partner for the LMCU in regional planning and marketing. For instance, PUEL is developing an ambitious international relations policy which is totally consistent with LMCU's orientation. This strategy started with a focus on The Netherlands, but is expanding to other regions of northwest Europe, in an effort to build a 'knowledge region' that stretches to northern Germany, Denmark and South of Sweden. For the realisation of this programme, which should take a decade, PUEL successfully tapped EU *INTERREG* funds.

At the local level, PUEL's ambition is a strategy to make up for the lack of communication between the regional universities. The seven presidents of the regional universities meet four times a year, and make plans for cooperation. At the time of writing, the trend within universities is no longer to be engaged with first cycles of education, but to focus on advanced studies, stressing the internationalisation of research. This, however, takes place in a strongly competitive regional context; PUEL tries to coordinate these efforts by developing consistent plans and strategies for the allocation of human and financial resources among the different players. In May 1992, PUEL created a statistical unit with the objective of improving the knowledge on the origin and development of the students, and in that way also the training requirements: the *Observatoire régional des études supérieures* (ORES). Thus, Nord-Pas-de-Calais is the only French region to set up annual files monitoring the life and study careers of almost all the higher education students in the region.

The *Reunion* (*Chapitre*) *of Management Schools* contributes to professional qualifications in management and determines the criteria for membership and the professional standards for accreditation. The Reunion comprises 28 accredited Business Schools in France.

Research and technology transfer Research as a sector of economic activity is making the Region Nord-Pas-de-Calais a model of scientific development. Research in Lille has a long tradition: in 1856, Louis Pasteur was the first dean of the Faculty of Sciences of Lille. Today this tradition is continuously renewed with investments and outstanding achievements in three major fields: biology and health, society and communication, and production technology. In the last decade, technology transfer became a focus of the regional development efforts, with the development of facilities like the *ateliers services* of the universities, the *pépinières d'entreprises* (incubators), and junior enterprises.

Research in Lille Métropole involves almost 2,000 researchers and 250 laboratories, 50 of which are associated to the CNRS. All around the traditionally strong research centres (engineering, life sciences, social sciences, and human sciences), there are other multidisciplinary poles which engage in national and international networks. Lille ranks second as a medical research centre in France after Lyon, with its *National Health and Medical Research Institute*, which focuses on genetic issues and parasitic diseases, and the world-famous *Pasteur Institute* The *Regional University Hospital* also runs a research unit and a technological innovation centre. The *Jeanne de Flandre Hospital* is the largest maternity hospital in France, with 4,500 births a year.

In total there are five business parks, 11 incubators and three centres of excellence. The main sites are:

- the *Cité Scientifique* campus in Villeneuve d'Ascq, which hosts more than 1,000 academic staff and 20,000 students. The campus is mainly devoted to education and pure research. The setting up of Lille 1 University on the campus in 1974 attracted other higher education institutes like the architecture school, plus various schools of engineering, like the *École Centrale*;
- the *Haute-Borne*, a Science Park contiguous to *Cité Scientifique*, whose aim is to favour the clustered localisation of different activities (research laboratories, education institutions and companies) with strong relations with the technologic and scientific vocation of the university. The park is located on a 130 hectare site beside the University of Lille 1;
- *Eurasanté*, which is a concentration of bio- and health-related firms as well as other companies wishing to diversify their activities. Eurasanté offers a wide range of services tailored to their needs, like consulting, assistance to start-ups, expertise for projects and technology transfer facilities.

These poles, especially Haute-Borne and Eurasanté, establish new and dynamic networks with the corporate research laboratoriess and with the regional small and medium-sized firms. Enterprise creation is given financial and logistic support by the Chamber of Commerce and the Region through specialised facilities like *North Entreprendre* and *Lille Métropole Initiative*. For example, the VAL, the first entirely automated metro in the world, shows the glowing success of a technology transfer initiative between a laboratory, and a private company, Matra. At Haute-Borne, the incubators support the creation of 350 start-ups per year, and the *Centre d'innovation et de transfer technologique du Nord* (CITTN) welcomes 20 advanced start-up yearly.

An interesting cluster of activities is organised around LMCU's 'technological arm', *Technolopole Lille Métropole* (TLM). *Digiport* was born from the merging and collaboration of different companies as a project in charge of developing the ICT strategy for the LMCU and a service centre for e-business, telecommunication, and digital reproduction. The initiative came from the big companies in the ICT sector, which were interested in developing an informal club to promote their strategies and create awareness among local customer firms through demonstrations and invited lectures. *Déclic* started half a decade ago as a spin-off of Digiport dedicated to e-business, mainly pushed by public initiative, performing different functions: 1) a monitor of the main world trends and link with local market strategies in the field of e-business; 2) an awareness centre targeting SMEs and institutions; 3) a 'nurse' for projects in ICT, matching demand and supply of specialised services, and assisting firms in developing their strategy at different stages. Déclic assists the LMCU in developing an 'e-vision' and attacking issues of bandwidth, infrastructure. Four or five times a year these companies open their spaces to students: although they are not a customer segment and the links with education are indirect and on a personal basis, these events are seen as good opportunities to familiarise students with the research environment, coherently with the development of a regional e-strategy.

Spatial patterns of higher education and research As in many other French cities, in the 1960s the public universities and private higher education institutes of Lille, like the School of Engineering, moved out of the city centre. In 1965 the university stated a decentralisation process towards peripheral campuses. The science campus hosting Lille 1 was opened, followed in 1974 by the human science campus of Lille 2. Partly this responded to education imperatives – the student population in the first and second cycles had multiplied by a factor of ten after the war – and partly was an outcome of the political turmoil of that time. Only *La Catho* stayed; its wealthy condition and smaller scale allowed this. In this way it maintained its image of a strong community orientation, a complete identification with Lille's image, despite the city's efforts to get rid, to some extent, of this very image.

The decentralisation brought about a big change in the urban landscape, which also implied the birth of a brand new municipality, Villeneuve d'Ascq, as a 'merger' of three smaller village towns. Since then, its population tripled, now counting 65,000 inhabitants, 35,000 of which are students, making it one of the most important *villes nouvelles* for education in France. Today, this town boasts a strong commercial infrastructure and a fully-fledged business park at Haute Borne. Before the decentralisation, students were accepted as an engine of socioeconomic development, and the 'university quarter' was a lively centre of integration. For many years afterwards, the *Cité Scientifique* has been isolated, with difficult transport links to the city centre; any form of interaction between students and residents was severed. Also, the university campus of Lille 3, at Villeneuve d'Ascq – which now hosts 22,000 students – started from scratch, without thorough planning considerations regarding its 'social' urban network. Urban managers have only recently started to be concerned with ways to bring this dynamism back into the city. An effort was made to make better connections with the city centre and the neighbouring regions, but the situation improved only in 1983 with the new metro system.

In the previous two decades, the location policy of higher education institutes entered a new stage of 'metropolitanisation', looking, for instance, at the other urban centres of the region, like Roubaix and Tourcoing, where plenty of space was available in old industrial and public buildings (VVAA 2001: 86). At the same time, technical research and knowledge transfer activities were increasingly 'suburbanised', creating specialised poles like Eurasanté. Even more recently, the city centre of Lille has become interesting again. New schools resettled there, such as ESC (at Euralille), CNAM, and Lille 2. In this regard, it should be considered whether the potential to revitalise marginal neighbourhoods through university settlements has been fully exploited.

An exemplary illustration of the difficulties that can be met is given by the move, in the 1990s, of Lille 2's Law Faculty to the *Quartier Moulins*, at the edge of the city centre, where a large factory complex had been elegantly refurbished for education. A historical middle class workers' area severely affected by deindustrialisation, Moulins had attracted marginal groups and petty criminality, and therefore expected the university development to bring a new source of prosperity. The plan foresaw that the factory would be moved to another suburban location at the university's expense, alleviating the social (and environmental) costs for the city. The students were supposed to find affordable accommodation and cheap living costs in this neighbourhood. However, things went differently. The local community did not integrate with 'upper class' pupils, and instead problems arose with the use of local services, for instance, parking places, leading to resentment and confrontation. Students were 'consuming' Moulins, commuting from other districts by car or through the nearby metro station, and hardly altering its social status. With time, this first negative impact was overcome, and the situation has improved, both in terms of liveleness and integration. The city has enhanced community services as a spin-off from the university redevelopment, with a new public library and sport facilities. Students attend 'neighbourhood fairs' and '*spectacles de rue*', sometimes they organise these events themselves. However, in conjunction, land values rose sharply, and many of the long-standing residents and foreigners were chased out. The same could happen in the future in neighbourhoods such as Fives or Hellennes, which host many students of the Villeneuve d'Ascq campuses because of their intermediate location between the campuses and the city centre.

The public residences found in the campus are mainly of the 'old style', small and uncomfortable, and students increasingly prefer to live in a lively city centre. In the past, this concept of student housing was the only response to the demand for accommodation, but since the private sector has started to invest (and speculate) in the student housing market, also the public sector has changed its orientation. This gave an opportunity to restore the building stock in the historical centre; some houses built in the early twentieth century were transformed into student residences. Though the extension of such housing development is still limited, it is not looked on favourably by local stakeholders, as the private developers tend to divide the original houses into flats, and in general to offer a lower housing quality, in this way influencing the market value of whole areas.

In general, the situation in Lille Métropole tends to be more favourable than in other French metropolitan areas. The 'mistakes' that populate the urban landscapes in cities like Lyon or Strasbourg have been avoided; social housing is more integrated into the inner city fabric, and the use of land is mixed, with no neat spatial specialisation. In this context, student housing does not generate excessive friction in less wealthy areas. An interesting project is the *Maison*, an international student house which arose from a cooperation agreement between Lille and Michigan State University, in a very central part of the city.

5.2.3 The Student Community: Behaviour and Careers

After high school, the most ambitious students apply for the best schools around France, and then choose among those universities which accept their application. Lille has good international accessibility, and the reputation of its higher education institutes is high: therefore it is attractive to students, despite the less impressive image of the city. All in all, with its rich cultural heritage and tradition of hospitality, Lille provides students with the best possible conditions for a successful study period. Moreover, living costs in Lille are highly competitive.[3] Job opportunities in the region, especially considering its centrality in the 'triangle' London-Paris-Brussels, are available for students wishing to combine university education with on-the-job training. This explains why students come to Lille from all over France. Engineering schools and business schools, as well as the prestigious scientific and medical faculty of the public universities, attract students from the old bourgeoisie of Alsace and Brittany. Up to the 1990s, the balance between Lille students leaving and students coming to Lille was negative for the city, but since then it has become positive. The fact that only 10 per cent of higher education institutes' students are from the region[4] indicates that the educational 'catch' of Lille is indeed very large. However, after the pleasant period spent in the city as students, it is likely that graduate students leave Lille for Paris or some sunny place in the south of France, or go back to their home towns. The city is apparently not attractive enough as a living and a working location to retain the enormous human capital that is formed in the local higher education institutes. This is also shown by the high share of commuters among the faculty staff, people who have chairs in Lille but reside elsewhere, like Paris, enjoying the fast train accessibility. This attitude is negative for the university, as these teachers obviously stay for shorter periods at work, affecting the quality of their academic offering; and for the city, which sees a high share of its working population 'leaking out' their income and tax base.

The highest share of students is accommodated in residences. Different private residences are available, otherwise students can refer to the private market or apply to subsidised housing if they are eligible. In a private residence, rents above €300 are charged. The rent can be reduced considerably by applying for a housing allowance, which are provided by the CAF (*Family Allowance Office*), and is calculated according to family income. International students can also benefit from one of these two allowances, provided the necessary procedure is completed upon arrival in France. This system has made private renting a profitable business, but the CAF is increasingly insufficient to cover the rising living costs of a big city like Lille, and more students apply for subsidised housing.

As in other French cities, *CROUS* (*Centre Régional des Oeuvres Universitaires et Scolaires*) provides accommodation and delivers scholarships, with competence over the territory of the Academy of Lille. It depends on the Ministry for Education, thus having the same status of universities but being an autonomous body. It caters for all students of public higher education (some 145,000) in the region of competence, plus a proportion of the students of *grandes écoles* and private schools. Its budget depends for a large part on national transfers; its own resources are diminishing as the number of students making use of CROUS facilities is falling off. For instance, some cafeterias and restaurants run by CROUS have been closed, as newer private residences offer own kitchen facilities. The university residences managed by the CROUS comprise traditional residences (with a common kitchen and WC) and '*conventionnées*' (with kitchen and en suite WC). Lille has seven such residences while another nine are in Villeneuve d'Ascq. CROUS hosts 9,000 students in its residences throughout the region, which means that 90 per cent of those entitled to be assisted by CROUS and applying for accommodation receive some support. However, their customer group is larger than the eligible students, as they run a kind of 'broker' service with the private market: through the service *logement en ville*, CROUS rents flats in the private sector and sublets them to students with particular needs, like the short-stayers. A *Service International* specifically assists foreign students in solving bureaucratic problems and facilitating their integration in the city.

CROUS is also concerned with other support to students, among which are grants like the '*Bourses au mérite*' and bank loans to low-income students, and it is responsible for distributing regional mobility grants to students who already have scholarships to go abroad. Significantly, Lille's CROUS ranks first in France for the number of scholarships delivered. Furthermore, CROUS provides an internship and placement service for students, with more than 5,000 offers per year, both part- and full-time, and all sorts of other services for students: placements in programmes of cooperation, social services, and on-the-job training. The action of CROUS in the cultural domain is becoming increasingly relevant. Established in 1989, the *Culture-Action* programme allows CROUS to subsidise and assist students in realising projects in all cultural fields. CROUS also promotes the enlivenment of both the campuses and the city by supporting the organisations of events, festivals, and contests, and financially backing the development of students' clubs in the campuses. Another actor with some relevance with respect to student support is the *Département du Nord*, which finances reductions for students in public transport, and is responsible for the concession to students of the '*Prêt d'honneur*', a complement to the Erasmus grants for students who go abroad, and in particular for students of art.

Students generally live for the first three years inside the campus, after which they move to more central locations. Living in the city or on a campus may have different advantages and disadvantages and generates different opportunities (stimulus to study, interaction with students of different faculties, reduced travelling times, etc.). The students of technical schools like the ECL have distinct preferences; they mainly live in the student house in the campus, which is relatively expensive but has better access to their schools' facilities and they are active in associations and clubs. These contribute to an 'atmosphere' based on small numbers, and the excellent facilities that the school provides to students. However, most of the 20,000 students who are educated on the campus of Villeneuve d'Ascq live in urban areas; thanks to the efficient metro system, daily commuting is not a real problem even at late hours. Therefore, unlike other cities in this study like Lyon or Eindhoven, even many technical students prefer to live in central areas. Today, students tend to give priority to quality of life considerations. As the metro system has facilitated student mobility, location is no longer a decisive element of choice, but quality is, and this contributed to make the existing public accommodation supply obsolete.

Foreigners are the most dynamic segment of Lille's student population and possibly the most problematic group to manage. A relevant share of them are short-stayers, for whom accommodation is not easily found in the private market; others come out of official exchange programmes, and clearly need additional assistance. Lille 1 maintains 150 flats in Villeneuve d'Ascq to accommodate foreign students, but this solution goes somewhat against the interest of the city's efforts to project a positive image and promote itself as a lively, comfortable historical centre. At the time of writing, other cities like Roubaix and Tourcoing are recognising this potential and trying to attract these students with dedicated facilities. Free-moving students include ethnic minorities, who generally cannot afford to stay in private housing. For them, old-style campus residence is the only alternative. However, the highest share of students comes to Lille within regular programmes; this group addressed 2,000 demands to CROUS in 2001, 1,500 of which were met (a 50 per cent increase over previous years).

In Lille, higher education institutes have agreements with the prefecture to distinguish the foreign students from 'normal immigrants'. They prepare the paperwork and assist the authorities in speeding up the process of registration of incoming students, which is an important element of the welcoming policy and a necessary condition to finding accommodation. Higher education institutes, have recognised the need to take an active role to this respect, although they are not formally responsible for this. Also, CROUS has a covenant with prefectures to facilitate the registration procedure for people applying for grants within cooperation agreements. They are also active in the effort to make this group feel more integrated during their stay in Lille, for instance organising trips and social activities.

There is a strong social network between students in Lille. In schools and universities, students have constituted a number of associations utilising national funds. Their aim is to welcome new students, provide information, discuss the study programmes, identify job opportunities, and organise conferences and social and cultural events as well as sports and humanitarian activities – in other words, linking studies with the outside world or the job market, often participating in the administration of their faculties and universities. An interesting association is *Once Upon A Time*, which is involved in several community projects of social inclusion and cross-cultural integration. Other organisations at the ECL are actively looking for opportunities for integration with the local community. Among these, the *Mongolfiades* organisation runs the 'Mongolfier' event in Lille-Champ de Mars, a popular event with the local community; *Lille Aux Echecs* organises a chess contest in the city. *Forum Rencontre* is a yearly meeting involving some 3,000 students and 100 firms from France and Europe, with the aim of matching the demand and supply of in-company education. The *Centrale Lille Project* is an association that develops privileged contacts among the businesspeople, local institutions and start-up services. A *Bureau des Sports* and a *Bureau des Arts* have been active for three years, to manage sporting and cultural life of Lille's students. The latter organises an art exhibition in Lille.

5.2.4 Challenges for the City

The regeneration of Lille has more than one dimension. First of all, the city is in the process of turning from a peripheral *French* city into a central *European* urban region. Furthermore, it is changing from a relatively backward and declining industrial centre into an international business metropolis. The first transformation is mainly due to European integration and the new infrastructure, such as the Channel Tunnel and the HST-network. The second is far from complete, but the direction of change is clear. Lille has to cope with its industrial past. The vision of a post-Fordist, knowledge-intensive region requires active urban management and knows ups and downs. With a new law ('Loi Chevénement'), the LMCU acquired competence for economic development. In collaboration with the Promotion Agency (APIM) and the Chamber of Commerce (CCI), priorities and actions are being defined, though the process is still at an early stage. A strategic plan for development until 2015 has been drawn up. Three ambitions are central to this plan. Firstly, Lille aims to become an international, world-class metropolis, developing a dynamic economy, targetting a few promising sectors in particular. Secondly, Lille seeks to develop the entire metropolitan area, with a complex multi-centred urban system in which the situation of each pole is different, with specific problems at the social, economic and physical levels. Thirdly, Lille aims to start a dialogue between the metropolitan area and its region. These ambitions are worked out in several projects. Clearly, further development of the already excellent higher education and research sector is functional to the achievement of these objectives.

Regarding *economic regeneration*, Lille is trying to expand its position as a logistics centre in terms of size and scope, building on its favourable central European position. At the Euralille complex, 20,000m^2 of office space per year will be made available in the next few years, and the quality of the labour is being enhanced with a focus on working in an international environment.[5] Regional headquarters of service companies such as banks and insurance companies are targetted, as well as innovative activities. A number of laboratories and research institutes are already located here, and dedicated technology parks are being developed for the start-up of new firms. Lille also targets the tourist market. Thanks to the high-speed train connection, visitors now come to Lille in unprecedented numbers, especially from the United Kingdom, making it a popular destination for weekend 'city breaks'. This made it necessary to improve its hospitality and cultural supply. The forthcoming event of Lille as European Cultural Capital in 2004 is a new stimulus to expanding and upgrading the existing facilities, at the same time increasing the city's vitality and the 'sense of belonging' of its users, such as students.

The master plan ('Schéma Directeur') has defined a major strategic and integrated objective for Lille Métropole, that is large-scale *physical and environmental regeneration*. Looking towards sustainable development, the aim is to 'rebuild the city on the city'. The *ville renouvelée* policy encompasses the scale of the agglomeration, and is articulated in a series of specific projects in the areas where decline struck hardest. Specific actions include the regeneration of Roubaix, a deprived area, abandoned by affluent households and left to decline, through an integral plan that associates retail development and cultural projects.

Despite these improvements, a number of issues that threaten to prevent Lille Métropole from realising its ambitions to the full are still open. For one thing, the bottlenecks to economic regeneration are not only on the demand side. The very factors that made up the social cohesion in Lille – the presence of large socially responsible firms tied to the solidarity tradition and committed to the provision of community services – have in the end discouraged young people in the region from engaging in business and taking the initiative with small, innovative activities. This lack of entrepreneurial spirit is symbolised by one of the lowest start-up rates of all French urban regions. The obsolete transport and industrial infrastructure scattered around the agglomeration creates discontinuities in the urban landscape and generates segregation. The citizens of Lille, including the low-budget segments of the society, prefer to live in suburban, green locations, which determines an unprofitable context for developers. On the other hand, the existing low-quality, working class housing is hardly attractive for the skilled workers of the 'knowledge economy'.

Indeed, the sustainable development of the agglomeration depends to a large extent on the capacity to refurbish public spaces, architecture and landscaping. The strategy should also lead to an upgrading in housing comfort, and to the construction of state-of-the-art educational, social, cultural and sporting facilities. In fact, the social situation in the region remains critical, with an unemployment rate well above the national average. Although the overall situation is improving, exclusion is emerging, as certain population groups and neighbourhoods are affected more than others.

The traditional image of Lille is not that of a vibrant international business city. Lille is a French city, but it is hardly seen as 'Latin', and this constitutes a huge image problem that drains resources and attention to the southern part of the countries. For this reason, city marketing is an important policy ingredient for a successful transformation. Lille has put considerable effort into city marketing, and the development of Euralille in the station area proved a good showcase for their city marketing efforts, as did its candidature to host the 2004 Olympic Games, and its appointment as the European Capital for Culture in 2004.

The administrative context to attack these problems is complex: two countries, three regions, four counties, six inter-municipal councils, 150 municipalities, and four chambers of commerce are involved. The urban fabric of some municipalities continues over both sides of the border. This demanded a coherent strategy of long-term cooperation, which is now present. Investors and territorial authorities from both sides of the border work together within the Permanent Cross-border Intercommunal Conference (CoPIT). A number of projects have been implemented and others are under way, thanks to the European Commission's funds to help bordering regions to get the best out of integration.

The Agency of Promotion (*APIM*) of Lille, affiliated to LMCU and Lille Chamber of Commerce, is directly interested in marketing the city as a knowledge-intensive, student-friendly community. Internationalisation strategies, such as stimulating the level of international services like language, open culture and foreigner friendliness, can support this development further. An influential player which could make a difference in the process of developing a vision for Lille as a 'student city' is the *Committee Grand Lille*. This informal club of 200 key decision-makers, acknowledged opinion leaders in both public and private sectors, debates and produces ideas for urban development, which are translated into concrete proposals through the personal and political ties of the committee members. For instance, the bid to host the European Capital of culture event was started in this way, as well as the unsuccessful bid for hosting the Olympics, which was an important occasion to draw people and ideas together.

5.3 Strategies and Integration. Lille as a 'Sustainable Student City'

5.3.1 Relations between the Students and the Community

In the case of Lille, the history of the relationship between the higher education sector and the local community is a long-lasting one. In the fifteenth century, the university was in the old northern capital of Douai, until Flemish domination and then the French revolution, when the university virtually disappeared, to be reopened after 1830 in Lille. In the era of the industrial revolution, the big textile and metal industries saw in the local university system an engine of development and entrepreneurial capacity. The Catholic University in particular was at the forefront of economic and social innovation. In those times, competition between public and catholic universities was strong, and they both were settled in the same 'university quarter'. The separation following the transfer of the University of Lille in peripheral neighbourhoods also marked a stage of relative specialisation.

However, the close tie between the university and its city was lost in this transfer. Paradoxically, whereas the national decentralisation policy brought 'university life' to cities such as Dunkerque, Valenciennes and Arras, the process of reintroducing education functions into the inner city of Lille has proved more problematic. The fact that the scientific campus is now well-connected with the city via the underground system takes away some urgency from this process, making it easier for students to commute to the campus from central settlements. In the era of the university quarter the university was an active stakeholder in the political life of the city. After the physical separation, the political class took a cultural distance from the public university, and did not allow universities to maintain a central position in the local political life. Even the efforts of the *Communauté Urbaine* could not reverse this trend. On this account, *La Catho* continues to be considered the only 'urban university' in the city, and its strong historical ties with the business community give it a favoured role. Public universities, instead, are proud of their independence, which is a typical feature of the French university system, and do not collaborate with outsiders or local institutions. Exchange with the local community takes place on the operational level but not at a strategic level, and is sometimes hampered by excessive bureaucracy.

On the other hand, the benefits that the universities provide to the local society are limited. In Lille, one cannot say that the housing shortage is a serious problem; opportunities are abundant and prices are low, therefore there is potential for a higher degree of student integration in the inner city economy and society. The real issues then are how to put this potential at the service of the city, how to use the student culture to improve the image of the city, and ultimately how to 'anchor' this human capital to the region, creating opportunities and reasons for students to become part of the community.

It is today accepted that students could play a pre-eminent role in regeneration strategies. Despite the stated wish to generate some social dynamism around the student population, so far the integration of students into the community has raised some issues, as the example of the Moulins quarter illustrates, or just indifference, as is the case among the residents in Villeneuve d'Ascq, where students are merely considered 'consumers'. In other popular neighbourhoods there is hardly any integration with the residents either. This is a demonstration that 'social engineering' is sometimes an illusion and it has to be actively supported with specific policies. To find some conviviality, Tourcoing and Roubaix offer better examples. In these cities students have repopulated run-down districts where cheap housing was available, contributing to the overall regeneration policy of the LMCU. In Lille, this happens to a lesser degree. Foreign students in particular – the ones who most explicitly could play the role of 'ambassadors' of the city's attractiveness – are often confined to campus residences for lack of information and alternatives.

More importantly, Lille is faced with the fact that the best students leave the region after the completion of their studies, despite the improving position of the city as far as the job market environment is concerned. LMCU is the leading institution in the process of bringing forward the conditions for a real 'student city' through its planning efforts. The driving idea is that, for Lille to become attractive as a living and working location and to retain its first-class human capital to its advantage, it must improve the quality of life conditions, offering new graduates a complete package of good housing, well-paid jobs and social and cultural amenities. In turn, this depends on consistency in the planning efforts of different players involved, universities included. The dialogue with all the partners in PUEL is so far good in this respect. It is recognised at various levels that a necessary precondition for a balanced relationship between the city and higher education is that the university system becomes more dynamic and less bureaucratic, and at the same time, that students are fully recognised as a target of the city's communication efforts. The output of the public university network should be given a more strategic focus regarding the 'local' level. Apart from healthcare and judiciary education, which are naturally 'embedded' in the community, other institutes of the three universities of Lille tend to have little local orientation, in contrast to the small but dynamic private schools.

5.3.2 Relations between the Student Community and the Private Sector

Private universities and schools in Lille share a well-developed network of relations with the private sector. This is illustrated by the high share of entrepreneurs among the teaching staff, reaching 40 per cent. The cohesion between enterprises and schools is sealed with the apprenticeship tax, a fee system inherited from the past industrial history of the region that firms pay to the university system, either directly or through the CCI. This system favours the private higher education institutes, as it is not likely that the CCI would distribute these funds back to the public sector. The tax income represents a negligible share of revenues for most higher education institutes, but is revelatory of the closeness between firms and schools in Lille. In practice, the higher fees charged by private institutions account for the closer links that they have with the private sector in terms of job opportunities – statistically, a graduate from a private school receives a 50 per cent higher salary in general.

It is not surprising that entrepreneurs prefer to maintain links and give their support to higher education institutes which focus more closely on applied research, like *La Catho*, *Arts et Métiers*, and less so on public universities, whose research programmes are not developed with a view to immediate utility. Firms require a critical return on investments in R&D, and therefore they are not used to thinking of universities as research partners. The system is in constant evolution, and today public universities are looking forward to a stronger engagement of firms as sponsors and customers; the increasing involvement of professional lecturers and consultants in higher education institutes is testimony of this process of convergence. In this context, many actors are now active in developing these synergies to the full. Today, universities are developing 'market-oriented' education programmes, especially in the new decentralised faculties, where teaching professionals do not threaten to over-run established hierarchies. This process should accelerate to enable Lille to catch up with more advantaged cities in Europe. A notable illustration is provided by Lille 2, which established a partnership with the private sector to grant its PhD students funding support while they write their doctoral thesis. This interesting initiative reveals that the private sector is willing to support public universities if there are good projects around which to organise these cooperative efforts.

The LMCU in particular is interested in developing focused links between higher education and the 'growth clusters' throughout the region. Small and medium-sized enterprises still need to be convinced of the potential that higher education institutes offer to innovation, and the Chamber of Commerce (CCI) has a key role in guiding these firms, for instance in terms of human resource development and training. To fulfil these ambitions the CCI set up a Development Charter, and has created an organisation, *Cepreco*, which focuses on higher and continuing education, and specialised training in the services sector and other fields such as administration, communication and business. Cepreco started a number of higher education institutes, for instance, a school for communication, graphics and multimedia created in 1999, and the *École de Gestion et de Commerce* (EGC) in 1994, in which strong ties with the corporate world are central.

PUEL is collaborating actively with the CCI in this field, for instance cooperating in the development of stage programmes. The next challenge, at a higher strategic level, is to define European priorities. The future of the project of a northern EU 'education region' depends on the quality of the network that will be built, which will serve to amplify the performance of the Lille region. On these grounds, relations with the private sector demands a more focused approach, with a clear research agenda agreed at the regional level. The regional government of *Nord-Pas-de-Calais* grants support for research activity, international mobility, and promotion, and organises some support to student life. Centres dedicated to research, innovation and technology transfer have been set up with Council grants. Through the *Inter-Regional Information Society Initiative* (IRISI), a cooperative effort with the French government which uses EU funds, the region supports plans and investments to strengthen the information society and its key industries, which today account for 23,000 jobs and 800 companies.

However, complete success in the effort of bringing firms and universities together is dependent on a real 'cultural revolution' in the higher education sector and to date this has not happened. Market awareness is introduced slowly, and most academics regard private sector lecturers with detachment, while schools like ESC and EC welcome this source of 'applied' knowledge. This cultural distance has so far downplayed the very importance of the scientific campus as a place for interaction between companies and knowledge producers. The few companies settled in the *Ville Scientifique* are not meant to incubate new activities.

Moreover, the campus is not attractive on account of its atmosphere, seemingly not an ideal one for informal meetings and knowledge sharing. The quality of life there may be good for a first-year technical student, but it is certainly not so for a young professional. Nor is the Haute-Borne site a 'Silicon Valley', where researchers freely develop their networks into business initiatives; rather, it is a successful example of a hi-tech business park, which does not really add to Lille's attractiveness as a working location.

5.3.3 Influence of Higher Education in the Relations between the Private Sector and the Local Community

The Lille region, like many other urban areas of northwestern Europe, has gone through a severe crisis with the progressive dismantling of traditional manufacturing industries. What allowed Lille to face this process without breaking down its socioeconomic texture was the strong tradition of social solidarity. This system of corporate social support has reabsorbed the worst effects of transition, and has avoided the creation of marginality and segregation. This explains why 'deprived areas' are less problematic here than in other similar cities, and the inflow of immigrants did not lead to severe problems as in other French metropolitan areas. However, this very model is at risk. The strong 'rootedness' of Lille's industry is increasingly challenged by the wind of globalisation, revealing a fragile model of organisation. Today, it is easier for outsiders to take over the local giants, the more so when the lack of entrepreneurial spirit among Lille's graduates is considered.

Therefore, politics is now looking at ways to strengthen the economic base, differentiating it and providing it with the necessary infrastructure and knowledge. It is also trying to link development processes with social care, once largely organised by the private sector through their affiliated institutions, and today at stake. In this light, education policies, as well as social policies aiming at the integration of students in the community, should be given extra relevance. For one thing, the private sector expects students to be well trained, and this is the case because the level of higher education institutes is high, but it is insufficiently 'practice-oriented'. Secondly, the best students tend to flee from this region. In this context, the danger for Lille is that new opportunities are not matched by an equal development of resources. An integral approach to local development is therefore on the agenda, trying to hook up equally with national and European government policy, and to link these initiatives to the private sector strategies. As a good example of this cooperation, *Creative Alleé* is an organisation set up in 2000 by some company leaders to stimulate the entrepreneurial spirit in the region through demonstration projects in schools, matching activities (for training and start-up programmes) and information management. The model is Quebec, where similar initiatives, taken decades ago, have stimulated a strong entrepreneurial fabric in the territory, helping its transition into a full-fledged 'information society'.

5.3.4 Model Testing and Identification of Issues

The general picture conveys the message that, despite the existing problems, higher education in Lille has potential both as a stimulus to regeneration processes (social and economic), and as an element of the city's attractiveness. This potential, however, is today not utilised to the full.

In Figure 5.3, we highlight the fact that several factors prevent the city-university interaction from developing better. The exchange between the private sector and the higher education institutes is not optimal. Prestigious but relatively small schools like *La Catho* and ECL have a history of strong strategic links with the private sector, but the far more relevant public institutes need to be oriented more than they are to the economic and social needs of the region. One clear sign of this lack of attention is the distance between firms and companies in the *Cité Scientifique*. The PUEL has been active in bringing together these realities and developing the local education and research system internationally, but the focus is today more on peripheral universities and less on the very local level, and it is more technical than cultural. Universities do not easily accept an integral involvement of practitioners in education, and this may make the process of orientation to business and development opportunities a slow one.

Even as far as the relationship between higher education and community is concerned, things could be better. After the 'seclusion' of the public universities from the city, their community focus has been downplayed as well. Universities are preferred research partners of the city, and fully involved in the planning process, though, arguably, with an insufficient strategic scope; the dialogue between higher education and policy-makers is confined to the stage of implementation. The initiatives where the two players combine their interests are fragmented, and in one of the major projects that led to the physical regeneration of the city, the Euralille pole, the higher education institutes played a limited role. Despite the share of educational activities

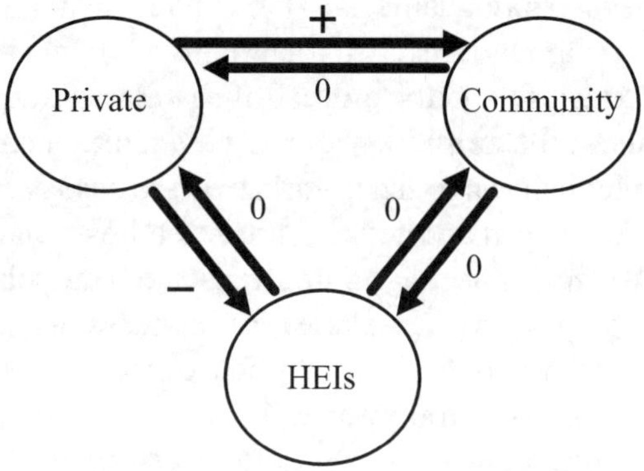

Figure 5.3 State-of-the-art: Lille as a student city

that occur in the city centre, higher education is organised in self-sufficient communities that do not use the city as much as they could; their staff prefer to live in suburban areas, and the students' interest in the city centre is limited to leisure activities, the more so since they can easily commute back to the campuses on the metro. In this respect, Lille is very similar to Eindhoven, another city in our research; but the potential for improvement in this case is much larger due to the historical atmosphere of the city which may facilitate the development of real 'student districts'.

These two aspects of insufficient development in the university-community interaction affect the very dynamism of the urban regeneration process. Firms boosted the importance of Lille after the city gained a new centrality in the economic order of Europe, and speeded up the transformation of Lille into a high-ranking European metropolis. But will the structural problems affecting Lille society ever be solved? The low salaries, high unemployment rates, lack of attractiveness for highly-skilled workers, and 'un-hip' image, seem local ills which higher education is currently unable to cure. The ambitious plans of the LMCU to regenerate the urban region, physically and economically, have to keep the contribution of higher education in a clearer focus. Today, the development strategy is opportunity-led, and hardly synergetic to the achievement of social and economic objectives, like the regeneration of marginal district, which has to be accurately planned, taking into account levels of integration that are not merely the economic regeneration effects on local shops and the real estate market.

5.4 Summary and Conclusions

Lille is a transforming city, and the process of change is full of problems. The enhanced centrality and accessibility of the city could become a drawback to the extent that the distance to more attractive regions is reduced, draining human capital away from it and reducing it into a low-ranked job location. Presently, students decamp to the South of France, Brussels or Paris, even though the balance between higher salaries and higher living costs there would probably be favourable to Lille. Lille Métropole cannot compete with cities of the rank of London or Paris, and anyway competition should not be the guiding principle of its policy in the emerging European network. Lille should instead become specialised in selected sectors and functions which are complementary to those found elsewhere, investing in its strong points – the excellence of its education, commerce and logistics, what is left of its manufacturing, and most of all residence, which is available in large quantities at low costs. But for this process to be effective, it needs to become more attractive and to provide better levels of quality of life to its potential new inhabitants.

Image and Attractiveness

Today, there are a number of factors which make Lille attractive to students. It has an excellent supply of educational facilities, and it is cheap compared to other educational poles like, for instance Paris, providing a good alternative to students who wish to concentrate on their studies and spend a pleasant period. However, the city's general attractiveness for those who finish their studies has to improve. Nowadays, good jobs are not hard to find in this region, but firms face the problem of attracting (or retaining) young managers with families.

In this respect, Lille does not have a problem with housing availability. The issue is rather whether the existing facilities are compatible with the development of a world-class metropolis. On one hand, Lille does not currently offer good opportunities for the best paid jobs; on the other, the firms which would offer these jobs do not settle here because they perceive that the location environment is not good enough. This problem could be solved by, for instance, offering sufficient good reasons to new graduates to stay after their graduation and generate an attractive enough job market for the best firms, upgrading the region's status.

In this light, the interests and ambitions of the society of Lille, its business community, and the city guests, could combine to a considerable extent. Universities and research centres, for instance, would be more competitive if there was a possibility of upgrading the accommodation stock, in particular for international students and visiting professors. If this upgrading went together with a proactive policy to favour 'residential' choices of new graduates who build life projects, a higher share of the human capital that it now only temporarily enjoys could be permanently attached to the area. It is also necessary to find ways to convince young graduates that they can profit from the city's attractive features and accessibility, for instance offering housing close enough to the main transport links so as to make it possible to maintain residence in Lille while commuting for work to Paris or Brussels.

There is, however, also a cultural problem to overcome: the perception that Lille is a declining city. The mechanism that could do the trick to change this perception is to make guests 'feel at home' and integrate them into the community from the earliest stages of their residence. The many initiatives coming from students' association and clubs need to be identified and linked in a network backed by active urban planning – in terms of location, financial support and marketing. Investing in *student-friendliness*, in an attempt to provide better opportunities to this substantial group for city users and making the city more visible to the international community of young entrepreneurs, visitors and firms, would be consistent with the development ambitions of the city.

In this context, fruitful connections between tourism development and the marketing of the city for students could be established. Lille has the advantage of being at a very early stage of its tourism development, one in which the value structure of the inner city has not been affected, and therefore it is still possible for students to play a central role in the transformation and animation of historical districts. The city's event strategy has to take this group of city users into consideration to a greater extent, as they are are at the same time important potential providers of cultural services.

Research and Education

The top-level universities of Lille and its *grandes écoles* may be real engines of development. However, the profile of the 'knowledge producers' is not sufficiently geared to the requirements for the region. For one thing, while universities provide sufficient education, research is under-represented. There are not so many researchers and research centres in the region as one might find, for instance, in Paris.[6] As a consequence, the innovative capacity of local firms is relatively low, and, in fact, Lille maintains an orientation towards traditional manufacturing and sectors with low capital intensity. On the other hand, its society is undeniably moving into the age of post-Fordism.

Economic diversification is thus a priority, and it needs to be backed up by more intense research activity in a selected number of new sectors, which in its turn requires a clearer, more focused involvement of local universities in regional planning efforts. The examples of smaller 'knowledge cities' in France, like Grenoble or Toulouse, indicates the way in which research and education can become spearheads of the local economic policy. Nowadays, the response of universities to the requirements of the local business community is to some extent inadequate. The existing research infrastructure and the many projects which have sprung from it, like CIEL at Haute Borne, Eurasanté and Euratechnologies, are not real 'communities' but rather science and business parks. To become real engines of urban attractiveness, they need to be expanded into their users' communities and linked strategically with education facilities, making it easier for these developments to generate a sustainable business model.

Today, occasions for strategic cooperation between universities and the city government have been revived within the ambitions of internationalisation on both sides. Yet, until now, the universities' international policy has been largely opportunity-based. The research agenda should instead be more explicitly set out according to local needs (community, city, private sector), and connected to what knowledge centres do. This could be done, for instance, within a simple project of a higher education portal, offering all kinds of links with universities and research centres as well as with the 'final users' of the university knowledge, such as firms and student and community organisations. Today, most university websites are not even available in English and comprehensive information educational opportunities in Lille is not easy to put together.

Integration with the Community

A more integral approach should be developed to build links between the host community and the student population, and in this way to bring again a 'student-city' feel to Lille. Presently, the government's education policy is dispersed over a host of regional decentralisation initiatives which are criticised for having transformed the higher education sector into a 'super high-school', with diminished potential for the sociocultural development both of the student population and of the host communities. The efforts of the PUEL should be more focused and more strategic. The large size of the student population, and the fact that the structure of the metropolitan region is varied and multipolar have to be taken into account. With one eye to the cohesion of the region and one to the strategic opportunities for revitalisation, efforts should be made to build 'cultural bridges' between the areas of the metropolitan region where the opportunities for student-led regeneration are stronger, such as Tourcoing, Roubaix, Villeneuve d'Ascq, the old historical districts of Lille, and the various campuses. A 'metropolitan network' of student settlements should be a key feature in the Lille of the future. The PUEL could even exploit the favourable climate with cross-border cooperation, strengthening the reach and articulation of higher education institutes towards Belgium.

This integration should happen with bottom-up rather than top-down initiatives. The unsatisfactory experience with the district of Moulins warns planners against the dangers of an 'imposed' regeneration of marginal areas through higher education institute settlements. For one thing, the main focus of these efforts should be on bringing together student residence and lecture facilities. If the function of a neighbourhood is only that of hosting the education function, the result may be a 'eat and run' way of utilising the local public spaces. Secondly, initiatives to bring student life into the local community should be actively supported with a more strategic perspective than the current one, which is rather limited to support for some old-fashioned events.

It is expected that a more balanced integration of the student community into the metropolitan region will bring forward better perspectives to contribute to the improvement of the quality of life. The city is now focusing on major events. This approach has the support of a population with a lively interest in the arts and an urge to get involved that gives cultural events a festive, participatory feel. The unsuccessful bid for the 2004 Olympics has certainly added momentum for a more cohesive approach to a metropolitan marketing policy. However, more focus should be given to low-scale, *soft* elements contributing to the city's attractiveness; the creation of points of aggregation for students, whose cultural and social profile in the city is negligible except from the proliferation of bars and student residences in the environs of Rue Nationale, could be a relevant step forward.

Governance

The actors involved in the French local development policy, and especially in higher education, are numerous. State, municipality, *Communauté Urbaine*, Region Nord-Pas-de-Calais and Department Nord, as well as the European Community are all involved in organising, regulating, developing, financing, accommodating and marketing the education function of Lille Métropole.

In spite of this complex environment, a real strategy to develop Lille as a 'knowledge region' does not exist, and the initiatives of these players take place independently. However, Lille's situation is clearly a more conducive environment for cooperation: in comparison with Lyon, for instance, there is an open dialogue between institutions at all levels. This makes it easier to progress in the right direction, but Lille has started this process rather late, when a vision was finally set and the opportunities from the improved international accessibility created the momentum for more integral action.

Therefore, if Lille wants to maintain this momentum, it is imperative to catch up with other cities that were more able to put together the interests of the city with those of its higher education institutes. Lille has a good and acknowledged leadership with a 360° vision of the opportunities. The formal (LMCU) and informal (Greater Lille Committee) circles of decision-makers act coherently to transform this vision into projects. However, the university world is not involved in these circles as a key player. Universities are jealous of their status and do not like to share development programmes with others, hence the isolation in which, so far, the PUEL has started to build up its vision for an 'international education region'.

In the future, LMCU, region and PUEL must work closely together. This might imply that the city is actively involved within the PUEL to influence its agenda in the desired direction, for instance making sure that high-level, ambitious plans for the internationalisation of education are matched by local planning initiatives to make Lille attractive and comfortable for foreigners. In this way, Lille can really make the best out of its enviable education and research potential. It is also critical that the general policy of PUEL is articulated in the activities of each institute; in fact, while the international exposure of the institutes proceeds quickly thanks to the support of PUEL in forging the foreign relation policy, the international orientation of university services (courses in English, assistance and facilities for foreigners, etc.) runs at a far slower pace.

Notes

1 CUDL /CCILM, 2002.
2 This is the main liaison organism between companies and the ECL, but also between firms and students. This centre manages internship offers, job offers, and is closely linked with the *Association des Ingénieurs*. Students are provided with free information on companies (brochures, videotapes, calendars, databank, etc.), as well as facilities for their own research (computer, video, telecopy, telephone, e-mail).

3 *La Catho* estimates a monthly expense from €547–712: a) accommodation: €210–375; b) meals: €153; c) books and equipment: €46; d) transport: €46; e) personal expenses: from €92.

4 25 per cent of HE students come from Paris, 20 per cent from the north and the rest from all over France.

5 'We choose Lille as the site for our head-office in France because of its excellent links with the United Kingdom and Paris and the quality of its human resources' (general manager, Abbey National).

6 Only 1.5 per cent of researchers in national institutes are in Lille, compared with 10 per cent in Lyon and 55 per cent in Paris.

Chapter 6

Lyon Case Study

6.1 General Introduction to the City of Lyon

Lyon is situated in the southeastern part of France, in the region of Rhône-Alpes. This region counts almost six million inhabitants, and covers an area roughly the size of the Netherlands. The Rhône-Alpes region includes eight *départments*, among which is the *Départment Rhône*, to which the city of Lyon belongs. This department counts some 1.5 million inhabitants. France is the first European nation for number of municipalities (36,000). In 1966, a law established the *Communautés Urbaines* (CUs) in order to remedy to the existing inadequacies in the administrative structures and the geographical reality of the agglomerations.

The 54 townships surrounding Lyon came together as Greater Lyon in1968, making it the second metropolitan area in France. CU Lyon is often referred to as 'Greater Lyon', an urban region extending across the jurisdiction of four departments. Table 6.1 illustrates the key population data at different administrative levels.

Table 6.1 Evolution of the population in the Lyon region

	1990	1999	% 1990–99
Rhône Department	1,963,941	2,087,351	+6.28%
of which Communauté Urbaine de Lyon (Greater Lyon)	1,134,687	1,167,532	+2.89%
Rhône-Alpes region	5,350,701	5,645,407	+5.51%

Source: Aderly, 2000/01.

Greater Lyon is one of the most important urban economies in France. The region ranks second in France in terms of regional added value and employment, and its economic growth can be compared to the fastest developing large European cities. The point of strength of Greater Lyon's economy is its diversity, both in terms of sectors and size of firms. These include a relatively high proportion of knowledge-intensive activities, with more than 3,200 hi-tech companies. Though civil and engineering industries maintain a leading position, among the advanced industrial sectors bio-engineering and life sciences stand out. These include many public and private research and development units, pharmaceutical productions, medical-technical companies, a major medical faculty, other educational institutions and an array of hospitals. Videogames and virtual reality technologies are specifically identified as a promising niche in the city's *Plan de Mandat* for the next few years as sectors to invest in, together with the more general 'knowledge industries'. The service sector is also well developed. The strength of the numerous corporate services in the urban area of Lyon lies in the city's strategic position at the centre of a powerful economic region. With some 60 banks located in the region, Lyon is also an important financial centre. The business sector is organised around major groups, like AVENTIS, and a pool of highly innovative small and medium-sized business in the areas of conventional manufacturing, advanced technology and services. Table 6.2 illustrates the economic structure of Lyon at the relevant regional scales.

Table 6.2 Economic structure in Greater Lyon and Rhône-Alpes region

	Greater Lyon*		Rhône-Alpes region**
	No. employees	Evolution 1996–2001	No. employees
Industry	135,484	–2%	491,028
Construction	39,225	–1%	128,457
Logistics	89,869	+15%	95,490
Retail commerce	41,751	+19%	290,356
Producer services	124,861	+45%	276,589
Other services	133,691	+10%	396,658
Total	564,881	+14%	1,678,578

Sources: * Opale 2002; ** CCI Rhone-Alpes, website, 2002 data.

Quality of life is a fundamental asset of the Lyon region. Firstly, the location of Lyon–near the Alps, not too far from the Mediterranean and only two hours from Paris – makes it an attractive place to live. Secondly, the city itself and its pleasant rural surroundings have much to offer. Compared to Paris, life in Lyon is not that expensive. This explains why most firms in the region find it easy to attract skilled workers. The Lyon's external accessibility is good, with frequent high-speed train connections to Paris. The Saint-Exupery airport, France's second airport, with a direct connection to the HST network, offers access to many European destinations.

Lyon markets itself as a cultural city, a meeting and congress destination, and a 'knowledge centre'. The city represents an interesting mixture of, on one hand, efficiency and modernity and, on the other, tradition and history. Now included on the UNESCO World Heritage List, Lyon has joined the club of cultural capitals such as Venice, Prague, or Saint Petersburg. In the city centre, the cultural heritage was accumulated over several centuries, and includes different elements: the Fourvière Hill, with its Gallo-Roman ruins; the district of 'Vieux Lyon', one of the largest Renaissance neighbourhoods of Europe; the central peninsula which was shaped by the great classical and imperial era, nowadays a bustling shopping and business district; the Croix-Rousse, a citadel within the city.

Huge investments have been made in the cultural sector, with renovated and brand new cultural facilities hosting prestigious events. The urban setting has also been greatly improved with the renovation and lighting of public spaces and promenades. Lyon has a point of strength in its 'cuisine', being the capital of French gastronomy. While the inclusion of Lyon into the UNESCO Heritage Site List in 1998 has accelerated the growth of leisure tourism, Greater Lyon has concentrated emphatically on business tourism with investments and marketing. In this tourist segment it registers strong and enduring growth rates, with 10 per cent annual increases since 1997.

Lyon is a relatively safe city, clean and tidy. The presence of night buses, very good illumination and a compact city centre reinforce this impression of safety, which is a typical of characteristic of the local culture. However, on the periphery of the city there are problem areas, and they are also close to the university campuses. Moreover, the city suffers to some extent from lack of vitality; the city centre does not live up to its status of second city in France as far as entertainment and liveliness are concerned, though Lyon's intimacy could be seen as an attractive point for a certain kind of visitor and most of all for its residents.

6.2 The Student Community and Knowledge Infrastructure in Lyon

6.2.1 *General Information on Higher Education in Lyon*

Research and education activities are a point of strength for Greater Lyon. The city hosts world-famous facilities and international organisations that make Lyon one of the leading cities in France for research, especially in the field of medical and life sciences. The average educational skills of the population are high. The knowledge base of the Lyon region is well developed: the city counts more than 100,000 students in higher education.[1] It is the second most important university concentration in France, after Paris. The establishment in 1992 of the ENS is seen as a particularly strong point for Lyon's knowledge infrastructure. The division of the total number of students in higher education is depicted in Table 6.3 and their disciplinary breakdown in Table 6.4.

Lyon's outreach as a student city appears more focused at the international level than that of Paris, which is very attractive for French students. Indeed the attraction capacity of Lyon for foreign students in higher education is rather strong, with more than 9,000 long-stay students plus about 2,000 short-stayers, some 10 per cent of the total student population from an impressive number of nationalities. This number is growing steadily (+6 per cent from 2001 to 2002). The most populous group is that of the African students, with more than 3,200 students, followed by Europeans (*c.*2,900) and Asians (*c.*1,100) (Michel, 2002). International programmes are an integral part of the institutions' strategies to promote themselves abroad. However, student exchanges are also important to promote the image of the city, and integrate its knowledge base. In this respect, a number of issues arise regarding the integration of the respective international strategies of city and higher education institutions.

However, the main strategic concern for Lyon continues to be the 'brain drain' from Paris. Paris offers a more attractive location to first-year students, though quality of life may be worse there due to high prices and long commuting distances. This is even more problematic for new graduates, who are attracted in the capital by the range of jobs available and the higher salaries paid there.

6.2.2 *The Knowledge Sector of the City: Actors and Spatial Patterns*

Universities and other higher education institutions The biggest and oldest university of Lyon is the *University C. Bernard* (*Lyon I*). The university covers almost every discipline, including three faculties: sciences and

Table 6.3 Students and staff in higher education in Greater Lyon (PUL)

Institution	Students	Staff	
		Teachers and researchers	Other employees
Lyon 1	27,072	2,430	1,200
Lyon 2	24,608	1,012	605
Lyon 3	19,674	590	420
ECL	1,214	220	200
ENS Lyon	919	211	107
ENSLSH	750	141	262
INSA	5,143	560	490
EM Lyon	1,675	83	190
ENSSIB	350	16	108
IUFM	4,273	245	146
Catholic University	9,284	226	105
ENV	483	93	260
Total	95,591	5,886	4,104

Sources: PUL, INSA, ENS LSH, Un. Catholique, EM Lyon, *Ecole Vétérinaire*, ENSATT, IUFM: October 2002 data. Lyon 1: December 2002 data. Other institutions: October 2001 data.

Table 6.4 Students in higher education in Lyon, various disciplines

Discipline	Absolute no.	%
Health and medicine	10,463	11.0%
Letters, social sciences/human sciences	29,026	30.5%
Law and political science	10,963	11.5%
Economics, commerce and management	12,998	13.7%
Sciences, technical disciplines and engineering	22,256	23.4%
Education	4,061	4.3%
Others	5,273	5.5%
Total	95,040	

Source: elaboration on Aderly data.

medicine (27,000 students, mostly in the Doua campus); social sciences (25,000 students); and literature (18,500 students). Research activities involve 2,500 scientists and engineers. The university laboratories – 95 research laboratories, 39 of which are associated with the CNRS (French National Scientific Research Centre), and 21 with INSERM (French Institute of Health and Medical Research) – are involved in close cooperation with high-tech industries. Faculties and the students' residences are spread all around Greater Lyon, while the labs tend to be concentrated at the Doua campus. Smaller campus sites are just outside the city or on the periphery, across more municipal jurisdictions like Villeurbanne, St Genis Laval and Oullins.

The research structure at Lyon I is to a great extent based on the quality of the partnerships with other educational institutions. Lyon I has some important tools to perform well in knowledge-transfer activities: a subsidiary company (EZUS Lyon I), a *Comité de Valorisation* and a Vice-presidency for Socioeconomic Relations. In particular, the EZUS company is an independent partnership between firms and universities, whose task is to execute financial and managerial handling of projects in which industry and university cooperate. The professional placement of the students is made easier through incubator activities and the creation of the knowledge-transfer centres.

Slightly smaller in size, *Université Lumiere – Lyon II* has an excellent reputation especially in the field of human sciences. This university has developed a strong international orientation, through a well-developed network of international exchange programmes with Central and Eastern Europe, Mediterranean countries, Eastern Asia and Latin America. However, Lyon II sends relatively few students abroad, a little more than 1 per cent of the total amount of students every year. Hence one of the priorities of this university is to increase the number of students who can benefit from internationalisation through the organisation of internships in non-EU countries. On the other hand, Lyon II receives 2,400 foreign students a year, 10 per cent of the total amount of students, among whom only 300 are official exchange students. Therefore, the need is felt to improve the quality of the information services, like the website. One of the most significant research institutes in Lyon II is the 'Institut des Sciences de l'Homme' in collaboration with the CNRS and Lyon III. It is part of the network of the *Maisons* of Human Sciences, and it is located in the old military school which the municipality of Lyon granted to the university. Other disciplines are located here (philosophy, arts, economics, etc.), thus favouring interdisciplinary and innovative research programmes.

Université J. Moulin – Lyon III is positioned in the city centre (Part-Dieu) and succeeded in remaining inside Lyon by moving the bulk of its structures to the recently renovated 'Manufacture du Tabac'. This university is well connected with its community and has a lead over the other public universities as far as economics and law are concerned. Its students like to live in Lyon and have transformed a section of the city centre in a real 'student area'. The campus inside the city is a stimulus and contribution to integration. Lyon III is developing special programmes in English, exchange programmes worldwide, and collaborations with the cities with which Greater Lyon has partnership links. For instance, it runs an MBA programme in Poland, in collaboration with the Rhône-Alpes region, attended by some 60 students each year.

The *Écoles Normales Supérieures* (*ENS*) are top-class *grandes écoles* of unique status. Apart from Paris, there are no other establishments of this school in France. ENS Lyon hosts 900 students, many of them specialising in life sciences. The school counts more than 300 staff, 150 PhD students, 62 post-doctoral researchers and some 500 technicians. ENS is located in the *Technopole of Gerland*, a prestigious area that hosts a mix of firms, and private and public research institutes. With its ten research laboratories of international fame, the school represents a scientific pole of international relevance. ENS is at the heart of a huge network of relations and partnerships with the business community. The acknowledgement of its sources led to the institution of affiliated ENSs. Recently, two other ENSs with human sciences curricula have settled in Lyon.

In order to become *Normaliens*, students have to take a highly competitive, multidisciplinary entrance examination. After obtaining their General Certificate of Education (the *Baccalaureat*) students must then spend two or three years in highly selective preparatory schools and then specialise in a field of their choice. Students are also required to spend one, and possibly an additional year abroad, although these placements cannot exceed two years; about a quarter of them spend a training period in a foreign country. In this way, the school has set up an impressive programme of exchanges with other centres of learning throughout the world. Finally, the ENS enrols PhD students from various disciplinary backgrounds under the supervision of the research laboratories of the School.

All ENS' laboratories are associated with one or more national research organisations, universities, or with big companies, SMEs, and public or private bodies, French or foreign. The number of research contracts is almost 100 per year, and a large share of these are in conjunction with the European Union. ENS Lyon is a member of *Crealys*, a network of incubators linking education and research institutions located in the north and west of the Rhône-Alpes region.

Another important *grande école* of Lyon is the *School of Engineers INSA* (Institut National des Sciences Appliquées), also located in the Campus La Doua. INSA is the biggest engineers' school in France, hosting a five-year course after the first level, with 4,200 students, 850 graduates per year in ten different specialised fields, 63 per cent of whom spend a significant length of time abroad each year. Particular attention is paid to the students' personal development: flourishing student associations, sporting and cultural activities, and courses in communication and humanities. It is an active centre for scientific research, with 690 postgraduate students, 31 research laboratories of which 13 are partners of the major research organisation. It employs nearly 1,000 researchers, and offers 24 nationally accredited PhD programmes. INSA-Lyon is also part of a *nationwide network of four GEs* (Lyon, Rennes, Rouen, Toulouse).

The teaching programmes are designed to facilitate the integration of foreign students. In order to get their diploma, the students of INSA must have a good working knowledge of English as well as a second foreign language. Each year, more than 535 students attend engineering courses abroad in one of the 180 partner universities spread across 40 different countries.[2] Each year, INSA welcomes more than 1,000 foreign graduate (first and second cycle) or postgraduate (third cycle) students. A peculiar characteristic of the INSA with respect to foreign students is that INSA selects its students abroad: the INSA, in collaboration with Rhône-Alpes region, set up an office in Shanghai to inform and promote the universities of Lyon.

The INSA model works very well and extends to private sector relations Apart from welcoming foreign researchers and trainees in its laboratories, INSA plays a role in supporting and spreading French expertise to its partner higher education technical schools through the TEMPUS programme of the EU. Furthermore, Eurinsa, Asinsa and Amerinsa are international first cycles that prepare engineering students to work abroad, stimulating their open-mindedness, and their autonomy as well as their ability to adapt and to innovate. These three organisations gather French students as well as students from other European countries, Asia or Latin America. The partnerships with the private sector are well articulated: the INSA is well-positioned in an extensive network of relations, both in the region and cross-border with neighbouring regions of Switzerland and northwest Italy.

Other important *grandes écoles* complete the enviable education infrastructure of Lyon, among which we can mention the *École Centrale* School of Engineering (ECL), the national veterinary school, and the National School of Public Works (ENTPE). The Lyon School of Management EM Lyon and CPE, the Lyon School of Chemistry, Physics and Electronics are supported and managed by the Chamber of Commerce. Finally, Lyon hosts a Catholic University and a number of business schools and technical higher education institutions.

PUL and other associations of universities The PUL (*Pôle Universitaire Lyonnais*) is a consortium of 13 public and private universities and higher education institutions in Greater Lyon, that offers services by utilising the resource base of its member institutions. In this way, the PUL involves 90,000 regular students, plus a population of 11,700 postgraduate students, 8,200 foreign students, 8,500 teaching staff, researchers administrative and technical personnel, 450 laboratories and research teams assessed by the National Ministry for Education as well as by major research organisations, and 815 PhD theses were defended in 1999. Within the PUL, each university is independent for what regards the internal organisation of research and education, but certain matters like international relations, cultural diffusion strategies, and the infrastructure, are catered for in a centralised way. It could be said that, after the establishment of PUL, the whole research and education sector can represent itself with one voice, despite the obvious differences for instance between public universities and *grandes écoles*.

AGERA is an association of universities and *grandes écoles* of the region, that works for the development of research and education in the region. The Conference acts in full agreement with the presidency of the regional Council on eight different topics of common interests.

Research and technology transfer Lyon enjoys a first-class research potential which encourages scientific and technological innovation. The players in local research networks participate in international centres, higher education, public national research organisations, hospitals, technical centres and private R&D centres. About 15 per cent of French patents are registered here. Research in Lyon covers a rich spectrum of fields: from health, life sciences and biosciences to physics (atomic and nuclear) and mathematics, from corporate sciences such as management, marketing, technology management to human and social sciences, from transport, energy production, infrastructures and resources to regional development, from the management of the environment to earth and space sciences. In the Lyon region there are 450 laboratories for public research, 155 located in Lyon (almost 4,000 researchers). The most important research centres are CNR (*Centre National de la Recherche Scientifique*) and INSERM (*Institut National de la Santé at de la Recherche Médicale*). Moreover, the private laboratories in the Lyon region employ 2,500 researchers.

Technopole Gerland is a relatively new example of a 'knowledge park', where firms and research institutes get together, and laboratories participate in projects together with the industry (mostly the giant pharmaceutical firm Mérieux). The close physical proximity plays a role in the interaction in Gerland; for instance, the fact that people from the laboratories, the firms and the ENS take lunch in the same restaurant is important, as it facilitates contacts and stimulates cooperation.

Spatial patterns of higher education and research With some 3,000 researchers and 20,000 students, *La Doua* – where Lyon I University and INSA are located – is the most important scientific campus in France. Apart from educational institutions, there are private firms established on the campus, some of them specialising in pharmacology or biotechnology. One of the most important science parks in Europe is also located in Lyon: the *Technopole of Gerland.* With 150 companies (8,000 employees, including 5,000 in the high-tech field), 26 public research laboratories, 1,000 researchers and professors – 70 per cent in life sciences – and 3,000 students), Lyon-Gerland could soon become the European pole for health and biotechnology and a real centre of innovation. Gerland hosts an impressive number of leading world organisations in biotechnology and health sciences. It also includes many locations for innovative start-ups or so called *incubators* of new entrepreneurs. In particular, *Novacité Oméga* is a welcoming centre for young entrepreneurs, with convenient conditions for office space, and consulting and engineering facilities.

The education and research sector has developed out of the historical centre of Lyon, in the Part-Dieu and in the suburban districts of Villeurbanne (a municipality in its own right), Bron, Gerland and other minor sites. The Doua pole is located at the boundary between the municipal administrations of Lyon and Villeurbanne. The fact that the public transport crosses the campus testifies to the fact that Doua is an integrated part of the city. However, this is not the case for the campuses of Gerland and, in particular, Bron. This pattern reflects the relatively recent institution of higher education in Lyon, but also a more general situation with French universities. Another 'eccentric' campus (but well inserted in an urban area) is the pole of Vaulx-en-Velin, further east from Doua, where around the school of architecture ENTPE prestigious institutions have opened their establishments.

The 'campus model' was explicitly chosen in France as a planning practice in the 1960s. It was believed that the concentration of studies in 'citadels of science' favoured knowledge transfer and innovation through the meeting of different talents, as in the model of the North American campus towns like Harvard or Stanford. In pursuing this model, the French government was also hoping to control the increasingly 'noisy' student population of 1968; there was concern about the role of students as agents of change and social unrest in cities. However, the results of this strategy have been arguable. In the majority of cases, campuses grew as lifeless and unattractive ghettoes from where students and staff would flee as soon as possible. After 25 years, awareness of this situation has led to a new wave of investments to 're-bring universities into the cities' (VVAA, 2001). Qualitative improvements in the study locations are also looked for.

This is also the story of Lyon, which, however, like many other cities in France, faces a shortage of space and resources in the city centre to (re)accommodate education functions. Thus, this 're-appropriation' happened only partially, through the re-use of former industrial buildings. The most significant of such interventions has been the refurbishment of an old tobacco factory, which now hosts locales for Lyon II and III, providing a nice atmosphere and state-of-the-art facilities in a lively and central district. This sort of intervention could continue. Other universities and *grandes écoles* are still confined in dull or inaccessible campuses, like Bron. The consequence of the disaffection of French society with the campus model is that a substantial part of the student population organises its life around the central, cheap districts that are 'buffer zones' in the development of business districts and tourist areas. An example is Croix Rousse, where housing is relatively cheap in spite of its centrality, and a lively student life may flourish. The neighbourhood has a mixed population of older citizens, young professional families, students and a large foreign population. Renovation work and social planning have given the quarter ambience and cultural vitality.

This, on the other hand, generates huge daily mobility flows in and out of the university areas, which the local governments cannot master effectively as budgets are shrinking. Hence it has been a long struggle for Lyon's higher education institutions to obtain tram stops within the campuses, and the available services are still judged insufficient. It is also important to note that students change their attitude dynamically, as price patterns and lifestyles are in continuous evolution. This determines frequent episodes of mismatch between supply and demand of services in given locations.

6.2.1 The Student Community: Behaviour and Careers

Lyon's catchment area for higher education students is mainly regional, though some faculties in the public universities have a national reach, and so have the prestigious *grandes écoles*. While in the brochures and promotional material Lyon is presented as an attractive and cultural city, well placed in international networks, in the end it is the quality of its education that matters the most, as it is revealed by an interesting research project ('The Conditions and Strategies of Youth Relocation towards Big Cities') conducted by a joint research team of Lyon II and INSEE, the National Statistical Institute.

This study reconstructed the student population's life and study patterns. Even if a large proportion of students choose to live near their university, this is also because they see their studies as a parenthesis in their social life. Very few of them, at least in the first and second year, take full advantage of the cultural and social opportunities of Lyon: their relationships and social life take place elsewhere. In the majority of cases, students return to their hometown as soon as their classes have finished for the week, and work at home during the holidays. For a smaller group, the social backgrounds vary, and the common denominator is parental mobility. As a result, students seek coherence, stability and independence from their parents. Relocation is much more difficult for them: they have no parental assistance in finding accommodation (temporary solutions with friends are frequent), and are often obliged to find employment throughout the academic year, regardless of the parental income. They are inclined to live in the city centre, close to the most lively districts, thus becoming the only student group in the most awkward rental market.

According to this study, arrival in Lyon is a particularly critical moment for foreign students. Lyon is a big city and the first 'face' that foreign students get to know of France – especially those who are not placed in regular exchange programmes and are not used to the European culture – is that of bureaucracy. They have to obtain a residence permit at the *Préfecture*, and this is by no means easy. Foreign students are not familiar with the organisation structure of universities in France, nor with the language in which lessons are taught, and they might feel completely alienated during their first months, which causes a regular delay of preparation. Apart from the welcoming activities organised by student associations and *grandes écoles*, and the psychological aid offered by CROUS, the support received is limited compared with other French cities, where higher education institutions themselves take care of the bureaucratic duties, and a more consistent familiarisation programme is organised.

The greatest proportion of students is accommodated in residences. Most of them are granted housing benefits paid out by the *Caisse d'Allocations Familiales* (CAF). The amount varies in relation to several criteria, such as price, income, family status and so on. The public university residences in Lyon as in any other French university city are managed by CROUS. Lyon has different kinds of accommodation possibilitiesranging from traditional 9m^2 rooms in residences to equipped studios, for a total of some 5,000 opportunities. Students in Lyon are not used to renting an entire apartment to share with other students. However, the majority of Lyon's 100,000 students lives in semi-private or private residences, some of which are located in the centre of the city. There, apartments can be rather expensive, but are a convenient solution for students who want to enjoy the life of the city. Quality is a prime requirement for most students. In this market, students face competition from young professionals, who are able to pay higher rents. This may become a problem in the future, because of a shortage of large apartments in the city centre which will push workers to look for houses in student areas where the supply is richer.

Other public residences are located within the Doua and Gerland campuses. As they appeal to different segments of the student population (the high- and low-budget), there is no real competition between public and private structures. However, given the new trend to develop university functions at the edge of urban areas, the public residence capacity should be expanded or at least reorganised spatially to guarantee good access to these facilities. Otherwise, there is a high risk of public residences becoming 'ghettoes' for low-income students, where all the benefits from the mix of vocations and origins among the student population is lost. In fact, though the university residences have improved in the years in comparison from the basic facilities that were offered some decades ago, CROUS did not build any new ones.

It should also be noted that in France, residences for students may be a very profitable business for real estate operators, who enjoy a substantial tax benefit. Private residences with services are also very successful as a form of investment because they can easily be reconverted to other uses off-season, for instance to host tourist groups or conferences.

Students generally live inside the campus for the first three years, then move to more central locations. Most *international students* are accommodated by CROUS within the agreements of exchange programmes. Finding a private apartment outside of regular exchange programmes is difficult for any student, and especially for foreigners, due to the intricate bureaucratic procedures, like having to find a 'warrant' for stipulating rental agreements. Therefore, on-campus housing is the rule for foreign students; in this way, as most of them remain only for short periods, they do not ever get to know the city during their stay. At the same time, there is an increasing demand for inner-city residences from French students.

Student life in Lyon is affected by the spatial and historical structure of the higher education institutions. As a result, it is rather fragmented or organised in poles that correspond to the different campuses and schools. Even within universities, it gathers members of the same faculty or association. In general, there are considerable differences between the students of the *grandes écoles* and the students of the universities, as well as across faculties, for instance between 'technical' and social science students regarding the attitudes towards the social and cultural infrastructure of the city. In the case of Lyon, as in others, student associations and groups are more inward-looking and self-sufficient in technical schools and *grandes écoles*. Other groups are more active in community life, for instance organising courses for children, and spreading academic culture at the neighbourhood level.

In schools and universities students have constituted a number of apolitical associations and fraternities known as *Corpos, Amicales* or *BDE*. Their aim is to welcome new students, provide information, discuss the study programmes, identify job opportunities, organise conferences and social and cultural events as well as sports and humanitarian activities – in other words, linking studies with the outside world or the job market, often participating in the administration of their faculties and universities.

The *Association Rencontre Française des Etudiants Internationaux* (RFEI) has developed a pilot project, 'Bilan Moral', concerned with welcoming students, both French and foreign, who are met directly on their arrival in the city at the railway station or at the airport. Private partners like France Telecom, the Airport Company and the Satobus carrier showed an interest in cooperating in this project. Up to the time of writing, almost 300 newcomers have been welcomed, but with more support of other actors the impact of this interesting initiative may extend.

6.2.2 Challenges for the City

The post-war years gave Lyon a European dimension with the development of the transport network, the municipal infrastructures (convention centre, the library, the Auditorium, etc.) and the creation a brand new central business district and administrative centre in the Part-Dieu, aided by the construction of a TGV station in 1978. The multimodal hub at Lyon Saint-Exupéry further reinforced Lyon's connection to the rest of world. Major development projects have been undertaken on strategic sites in the city (Cité Internationale, Champ du Pont, Gerland). Hence, it can be said that from the point of view of physical planning, Lyon is now a metropolis fully integrated into twenty-first century

Europe: it brings together a historical, medieval tourist area, a prestigious and elegant city centre offering all the cultural and personal services of a European capital, and a modern, well-equipped and well-connected business district, including the large education and research sector. It should also be highlighted that the city has been active in strengthening its higher education facilities. The idea is to bring forward the conditions for the development of an excellent education and research pole, creating a critical mass of top institutions and marketing the city as a knowledge centre with unique characteristics in France, as it puts together a human-scale city with the best that French education and research has to offer. The local government, through its intermediary Aderly, is the main actor behind the successful establishment of two *Ecoles Normales Superieures* in Lyon. The biggest challenge for the city is therefore how to *use* such enviable assets in order to stimulate further economic development and to produce diffused benefits for the community.

Today, Lyon is committed on many fronts as far as strategic planning is concerned. The first regards the further stimulation of a creative and innovative economic environment. The Plan for Economic Development of Greater Lyon or SDE (*Schéma de Développement Economique de Grand Lyon*) has fixed the main objectives of a global project for the agglomeration since 1997. Recently, the SDE entered its operational stage, based on 15 actions which are currently at varying stages of advancement. The main goals are the development of new enterprise, structural change in the economic structure with a boost to innovative SME, and finally the reinforcement of the most solid 'growth sectors', which the scheme identifies as biotechnology and health sciences, on one side, and videogames on the other; in both sectors, Lyon has the features of a world leading city.

In recent years the number of French firms engaged in R&D activities has increased dramatically, and the national context for improving knowledge transfer between science and industry also improved with the Law on Innovation and Research of 1999, which deregulated the opportunities for education institutes to develop entrepreneurial activities and granted academic staff the possibility of maintaining the intellectual property of their research, and also enhanced the mobility of personnel between the private sector and higher education. Lyon has a natural advantage in this sphere, given its top-class research and education infrastructure. In the framework of the 'Technopole Plan' (1998–2000) it has reinforced its innovation capacity with actions targeting research-industry links. This plan has been implemented with full cooperation between Greater Lyon and the partners in the Lyon University Pole (PUL), and is entering a new stage with the 'Lyon Innovative City'. This brand new policy document basically opens up the cooperation effort to a wider range of actors and disciplines than the pure sciences, and enlarges the scope of this cooperation to issues such as communication and internationalisation which had been left in the background in the previous stage.

One of the actions foreseen by the SDE targets higher education students and freshers in order to support and stimulate young entrepreneurs through an original approach that foresees a 'playful' path towards the discovery of the world of the enterprise, supporting fiscal incentives to facilitate starters. However, despite the merits of this initiative (which, by the way, is less advanced than others in the SDE), the 'quality of life' factors are not explicitly mentioned or cared for; and instead, these may still be pre-eminent in determining a lower-than-expected innovative profile in the region.

Lyon is attractive for students, both regionally and nationally (and to a large extent also internationally, in particular for medical disciplines) on account of its excellent infrastructure. However, the quality of life and services that it offers to its students is not high enough to convince them to stay for good after the completion of their studies, at least in numbers that match the requirements of the local economy and the development ambitions of the urban region. Instead, Paris remains an attractive location which drains many highly-skilled graduates away from the Lyon area. Other cities like Toulouse or Nice also offer a 'cosier' atmosphere, resulting in strong competition for Lyon. The issue for Lyon is then to improve the way in which it welcomes and manages its students, hoping in this way to retain the necessary human resources in an effort to make the local economy even more innovative and creative, keeping pace with the developments at the national and international level.

This issue is indirectly but substantially related to other challenges faced by the urban agglomeration. The first is the refurbishment of its image, which is still considered inferior to its potential. The perception of a 'working city', rather than a leisure centre, has proved difficult to change, though arguably visitors are willing to revise their view after an occasional visit (van der Borg, Rumi and Russo, 1999). The city lacks a real blockbuster attraction, a symbol by which the city can easily identified. Moreover, the city centre is not particularly lively, especially at night. The relative lack of animation is due – among other factors – to the specific character of visitors attracted to the city (business travellers, families) and possibly to the lack of a real 'student district'.

Finally, the city is looking forward to further strengthening its international network and overcoming in this way the 'second place' position that it suffers at a national level. Clearly, the advances that will be made on opening up the research and innovation strategy to the external environment are critical to a successful positioning of the city in the European context. They can add to the capacity of the city to attract and involve talents, investors, and a wide community in the development of its education and research environment. In other words, a research community completely opened to the 'outside', be it the global market forces or the international academic networks, could be the spearhead of a new stage of an international strategy.

6.3 Strategies and Integration. Lyon as a 'Sustainable Student City'

6.3.1 Relations between the Students and the Community

The *City of Lyon* is the engine behind the creation and activity of Greater Lyon. For many years the former mayor of Lyon was also the president of the *Communauté Urbaine*, and under his guidance, the urban region made a strong effort for the development of new economic activities. In the past, Greater Lyon played a key role in the development of Lyon's technopoles as a part of the economic development strategy, and this plan is due to be reactivated providing, for instance, €27m in the field of higher education. Within the metropolitan strategy, the City of Lyon is also active with its '*Plan de mandat*', a document which defines the priorities and the for the administration in charge in the period 2001–07. The plan foresees the expansion of the Gerland technopole and the establishment of a new technopole at Porte des Alpes, a new stage of reorganisation of the activity at Doua with a focus on 'engineering sciences', in close cooperation with Greater Lyon, and finally the strengthening and internationalisation of Crealys.

However, Greater Lyon is aware that a student city is not all about 'building'. Therefore it has explicitly set out the objective of making the city more attractive for students and young talents. Students are seen as a potential engine of development and an interesting user group for the city; as such, they need to be integrated into the community. The idea is to grant to students not only a pleasant stay in Lyon, but also to induce them to stay and live in Lyon after the completion of their studies. The *Plan de Mandat* sets out to 'open the different university campuses to the city' and to manage them in such a way as to enhance the occasions of integration of the students in the city. Though it is not possible to know how the city is going to implement these ideas, they are already active as far as welcoming students is concerned. The '*politique d'accueil*' of the city[3] consists of a number of actions, some of which are organised in partnership with other public and education institutions:

- a poster and information campaign;
- a *welcome day* on campus, where the city of Lyon is presented (documentation about sport, culture and tourism);
- a *cultural week*, when all the cultural institutions are open and free to students;
- a *card pass culture*, granting special discounts for cultural activities, supported by the regional government;
- the '*Espace multi-services*' offering new students support for services such as accommodation and transport, open in July, August and September;
- the '*Guide de L'Etudiant*', a handbook for the new students;
- the Internet site www.lyoncampus.org (in connection with the PUL), the only academic website in France financed by a city administration, which receives more than 1,000 hits a day.

Other initiatives aim at making more value for the community out of the students' lifestyles and their academic knowledge. The city organises a prize for young researchers, and airs a bi-weekly programme on a local TV station where selected issues of scientific interest are discussed by academic staff for a lay audience. Through the *Cycle de Grandes Conférences*, which hosts speakers with international reputation, the intellectual production of higher education institutes is given further visibility, and so is its social importance. The city also organises a 'cultural week' (soon to become a 'cultural month'), which is helps incoming students to familiarise themselves with the city.

The *Region Rhône-Alpes* is another actor with a strategic role with regard to education. This region includes other university cities like St Etienne and Grenoble – the latter being considered a truly 'student-friendly' city. The Region is not formally in charge of education, but it comes to affect the quality of the human capital in the Lyon agglomeration as it offers a large number of scholarships, mainly for students who go abroad as exchange students. In this way the city's internationalisation strategy is indirectly supported. The Region is also active in the research and development environment of Lyon, cofinancing, together with Greater Lyon, the *Crealys* organisation, which started as an initiative of INSA and has received the support of various ministries and a large number of private and public partners.

In such a student-rich area as Greater Lyon, many organisations work between the academic and the institutional world to facilitate the integration of students into the community and the full development of the human resource potential.

CROUS (*Centre Regional des Oeuvres Universitaires et Scolaires*) is responsible for participation in higher education, mainly dealing with accommodation and scholarships. Despite the name, it falls under the authority of the Ministry for Education, and as such it is rather bureaucratic, having to cater for all the 75,000 public higher education institution students in Lyon, plus a proportion of the students of *grandes écoles* and private schools. One of CROUS's missions is to achieve a certain 'democratisation' of student life, supporting students with weaker backgrounds. The 'human touch' provided by the workers who interface with students ensure it a good reputation among the student community. CROUS has established a service of psychological aid for foreign students, who might suffer problems of loneliness or with integration upon arrival. In this centre it is possible to meet people with similar problems and build social networks. Events and meetings to stimulate integration are also organised; however, the lack of resources does not allow a large number of students to profit from this valuable service. Important moments of the students' career, like their arrival (finding a house, going through the bureaucracy for resident permits) and departure from the university (finding a job), are not directly followed by the CROUS, but rather by dedicated agencies.

The *Centre d'Information International* (CII), working under the supervision of PUL and hosted in premises offered by the city, is concerned with the welcoming of exchange students. Its mission is to inform about mobility (jobs, courses, etc.), to produce information and advice about studying in Lyon and France in general, as well as about enrolment conditions, administrative obligations, problems students may face in settling down and in their daily life in Lyon, and to welcome exchange students and free-movers. The CII is also responsible for informing all students about the possibilities of continuing their studies, attending courses or work placements in a foreign country. This pilot centre is unique in France; there is an increasing interest in creating a training process to formalise it and transfer to other countries. The Com'Et space, hosted in the CII locales, is dedicated to cultural activities. It gives access to all sorts of events at reduced prices, and organises meetings with professionals from the art world and opportunities for attending rehearsals. Special discounts for transport and sport activities can also be obtained here, as well as a free guide (*Le guide de l'Etudiant*), financed by the municipality and made available to all the universities and GEs. A cafeteria and Internet service for students are also hosted at Com'Et.

Jeunes Ambassadeurs de Lyon (Young Ambassadors of Lyon) is a programme for welcoming and integration of exchange students into the economic and cultural life of Greater Lyon. JAL was created in 1999 from an idea of some CEOs of companies located in Lyon, with the aim of promoting the city and its excellent facilities internationally. JAL organises activities that focus on the good experiences of former students; when they leave after their study period in Lyon, they are expected to transfer the social network developed in Lyon to their home countries. The aim is to favour the creation of qualitative, lasting relations between the students and the companies that want to develop their international image and activities. Though highly successful and clearly a best practice to imitate, only very few students in *grandes écoles*, involved in official exchange programmes, can participate in this initiative. The project is run in collaboration with the Chamber of Commerce; the city is not a partner at this stage.

The *Central Library of Lyon*, located in Part-Dieu, should also be mentioned both as a meeting-place for students and an impressive resource for higher education and research. The library is the most important documental library in France after the National Library. It receives one million visitors every year, 56 per cent of whom are higher education students. Over the years it has developed a specialisation in research on oriental languages, and for the teaching of bibliographic studies. It was one of the first to introduce access to the Internet. This institution offers a unique opportunity to the city to market itself as the French 'city of books' and an centre of excellence for documentation (building on an existing specialisation: in Lyon there is a grand circle for librarians, the only on in France). Plans are being made for settling a 'museum' in an old building in the city centre, which may be very attractive for cultural tourists. The library is involved in partnerships with many university libraries, for instance as regards the digitalisation of works and on-line resources, and it can be seen as one of the agents behind the rationalisation of the knowledge supply in the city.

Despite this host of initiatives, one should note that a *specific* welcoming policy targeting the 11,000 international students is still insufficiently developed. The difficult integration of 'free-movers' (students who are not in any particular international exchange programme) reflects a wider issue with migrants in metropolitan areas. First of all, potential newcomers have to overcome the tiresome and excessive bureaucratic problems of registering at the city hall, obtaining a residence permit, and at the same time renting a house and applying for state grants – which are all linked in a vicious circle. Secondly, the organisation of higher education into campuses, obliging some students to undertake a daily 90 minute commuting time, reduces free time spent in the city to a minimum. It can be added that the lack of 'student life' does not favour a sense of belonging and hence attachment to the city after the completion of studies. Given the fast growth of this group in the recent years, and the future plans for expansion of international exchanges with interesting 'markets of origin' such as China and Eastern Europe, it is clear that better tools are needed to attract, welcome and manage international students.

Another issue worth raising is that the research output of the local institutions affects the local community at a surprisingly low level. HEIs do have a tangible impact in terms of jobs and expenditure, but the city is hardly seen as a 'laboratory' for research, contributing to the solution of the main urban issues and enriching the local cultural capital. Given that 80 per cent of research in applied science is financed by public and private actors, there is scope for positive synergies. The new location of Lyon III, well inside the city at the Tobacco Factory, favours the development of links with the community, for instance through events organised inside the university. This is a demonstration that an urban location for the campus is both a stimulus and a contribution to education, and that there is no prejudicial distance of the city towards the student community, as long as the opportunities for exchange are there. Indeed, students are not treated as a separate social group, and this is typical of the French tradition. A contrasting example is offered by Doua, possibly the most important campus in France, yet enjoying little integration with the surrounding urban environment, which downplays its role in the community.

More generally, the city could do more with its excellent higher education sector. *Grandes écoles* like INSA and ENS, which produce strategic knowledge for the local economy, have different interlocutors like the city of Lyon, but also Villeurbaine and the Region Rhône-Alpes, a solid financial supporter. However, none of these relations has a strategic character, and they are mainly on a unidirectional basis. It is hard for them to make contact with the city administration if they want to develop special international projects, or take the lead in local matters. Exchange programmes, networks between universities and international research agreements are developed independently of the city's aspirations. By putting together resources in this respect, the results achieved could be multiplied.

Despite the accent put by Greater Lyon on the importance of stimulating an innovative and attractive business environment, these concerns receive little attention in the new strategic document endorsed by Greater Lyon, *Millenaire 3*. Human capital is presumably assumed to be flowing to this place if there are opportunities, whatever the living and working conditions. Instead, Lyon clearly suffers from a mismatch between its 'hardware' and its attraction capacity; hence a greater effort should be done to improve the 'software' and the 'orgware' of its higher education system. Not only the content of the courses and their organisation, then, but also the way in which students and staff are welcomed in the city. Lyon's general score on the students-community link in the diagram in Figure 6.2 is therefore less than optimal.

6.3.2 Relations between the Student Community and the Private Sector

Relations between the business sector and students in Lyon are mainly geared at the institutional level. To the extent that the private sector invests in education and research, an important ingredient of the student-friendly city is stimulated. An EURICUR study on the healthcare cluster in Lyon (van den Berg et al., 2001) reported that these relations were poorly developed, and suggested how they could be improved improved. Today the situation is different to some extent. This is mainly because in recent years the higher education institutions have profited from a new regulatory situation. This has given a strong incentive to research in fields that are commercially interesting, but has also allowed students to come into closer contact with the world of enterprise. This holds especially for the *grandes écoles*, which are possibly more prepared to take full advantage of this change, and less so for the universities, where resistance to change is greater. Clearly, human science faculties have problems in marketing themselves with the private sector. In some fields of applied sciences, and especially medicine and biotechnology, the situation is better. There are plenty of examples of successful combinations of research and business, such as at Technopole Gerland. The proximity of firms and research laboratoriess and the mobility of staff within the cluster is seen as a fundamental factor of success.

However, the private sector's relation with higher education and research needs a strong intermediary organisation like the Chamber of Commerce and Industry (CCI). CCIs are relevant in French urban policy as they represent places for meeting, negotiating, exploiting the potentials of many sectors, including the producers of knowledge. The CCI of Lyon has always been at the forefront of urban management, playing a proactive role in the development of new strategic sectors. For instance, to meet the wishes of the tourism industry, a '*Groupe Tourisme*' was created within the CCI, with the aim of lobbying for the sector, stimulating and coordinating tourist activities, and finally become a full discussion partner on tourist matters. A similar role could be established in relation to the student population.

The CCI directly supports two *grandes écoles* (the Lyon School of Management EM Lyon and CPE, the Lyon School of Chemistry, Physics and

Electronics). However, it is also interested in the generalised development of links between higher education, research and development and the business sector at all levels. For this reason, CCI facilitates the transfer of knowledge, providing incubator-type facilities for companies to settle near the universities.

Other private organisations have a role in facilitating or stimulating the linkages between universities and business sector. One of them is EZUS, an independent organisation initiated by Lyon I as a private-public partnership. Its goals are to execute financial and managerial handling of projects in which industry and university cooperate, to favour technology and knowledge transfer, and to develop innovation. INSA can also be taken as a good example of how extensive the networking with the business sector can be. However, the private sector initiatives to ameliorate students' quality of life and facilitate their integration are not as developed as knowledge transfer. In the framework of existing plans like the *Schema Directeur* of the CU, where greater involvement of the private sector in the general economic development objectives of the city is stimulated, this aspect should be taken into more explicit account.

6.3.3 Influence of Higher Education in the Relations between the Private Sector and the Local Community

Lyon's economy is strong and diverse. In the last few decades, local medical and pharmaceutical firms have acquired global relevance and this sector has driven the success of a wider number of 'growth clusters'. However, today Lyon feels the need to restructure its economic development policy. There are more promising business fields, which would call for some despecialisation, but the issue remains of selecting the most strategically relevant. In the past, support to sectors without a real strategic importance, such as textiles, had little impact, and provoked a certain dispersion in industrial policy. This situation is supposed to change by giving a stronger focus to a limited number of 'core' clusters, to be actively supported at the metropolitan level, and *softer* policies towards other promising sectors, boosting, for example, creative industries. This idea is fully endorsed by the new strategic documents *Millenaire 3* of the CU and the *Plan de Mandat* of the city. To accompany this strategy, the development of a vision to attract and welcome new categories of 'creative groups', like students, is crucial.

Aderly, the economic development agency of Greater Lyon, has taken over the job of marketing the city and welcoming incomers, offering the best possible conditions. Many other organisations and associations facilitate the welcoming and integrating of newcomers to the Lyon region. The agency Rhône-Accueil, a member of the French Cities Welcoming Network. It runs 18 welcoming points in Lyon and the surrounding area. Its mission is to welcome newcomers and their families and to facilitate their integration. It is organised into different teams with a network of more than 150 volunteers.

All in all, the job market and quality of life in Lyon are attractive enough to support the current ambitions of the business community, but this favourable situation may change in the future if Lyon really wants to diversity its economy towards other sectors where different qualities of the human capital than 'technical skills' are required. In that case, Lyon should offer something more – for instance an enhanced 'cultural environment'.

6.3.4 Model Testing and Identification of Issues

Lyon has a world-class education and research sector, a very strong and diverse economic base, and a metropolitan government that at various stages has been the real engine of the development of the region. The city is attractive and the quality of life is high; the excellence in education and a culture of hospitality make of Lyon a preferred destination for many international students. In this respect, the position of Lyon is quite similar to that of Munich, another participant in this research. However, the synergies between these strong points are less than satisfactory, as synthesised in the scheme of Figure 6.1.

The first is the persistence of a 'second city' complex with respect to the French capital that undermines its ambitions. Not only investments and public projects but also human capital and creativity are drained to Paris. The city is not *charming* enough to attract young talent, and has developed a campus model that to some extent hampers the possibility of real cultural integration between students and the host community. The 'pleasure district' in Lyon is surprisingly low-scale for a city of over one million inhabitants. Lyon is an intimate city. The offer of cultural activities of the 'high-art' type is pre-eminent, with a world-class ballet, theatre and opera. Little 'underground culture' is available, except for the vibrant ethnic community, which only recently begun to to receive attention in mixed districts like St Catherine and the Croix Rousse. On the campuses, student life is indeed lively, and culturally appealing. A challenge for the city is to bring this cultural creativity into the city and connect it with other strategic sectors: high-brow culture, tourism, and entertainment.

There is also a serious shortage of land and resources destined for education and research in Lyon. Despite the massive presence of education institutions,

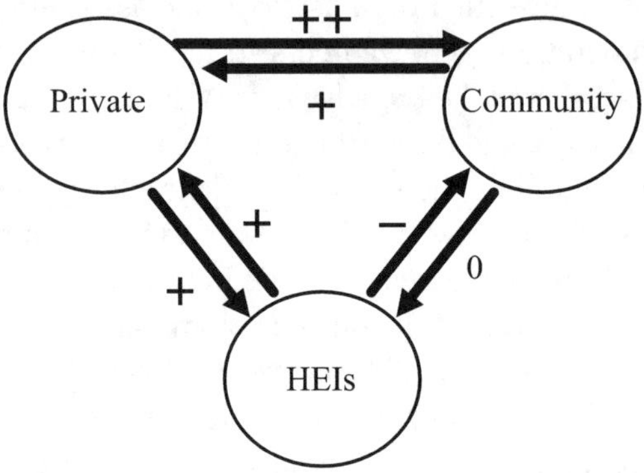

Figure 6.2 Lyon as a 'student-friendly city'

negotiation for the development of education in the city centre is problematic, to the point that the refurbishment of the tobacco factory, hosting no more than the 5 per cent of the higher education students in Lyon, is seen as a major achievement. At the time of writing, plans for further development of other specialised functions and curricula are almost impossible.

The synergies between higher education institutions, the urban region and its community can therefore be greatly enhanced. In the following section we suggest how more consistent synergies could be activated, and what role there may be for public and private players in this process.

6.4 Summary and Conclusions

In order to revive the network relations and give a more integral character to the policy towards students, some structural problems need to be overcome:

- the impossibility that the city administration is responsible for its knowledge infrastructure, which largely falls outside the municipal borders;
- the lack of an explicit mandate for the CU to deal with higher education institutions and student policies;
- the lack of channels of communication or permanent forums to elaborate common strategies between higher education institutions and local administrations.

Governance: improving cooperation and channelling resources The city government and the CU should develop an explicit, integral strategy that puts students at centre stage in the development policy of the urban region. Lyon does not have a clear vision about the proper role of the students as agents of dynamism in the city. The attitude of the city administrative bodies, both the municipality and the CU, has to become more proactive in this respect. Improved collaboration between the city and the higher education institutions is required. The city could be more receptive to 'bottom-up' initiatives from its HEIs, now that the PUL represents an interlocutor for the whole sector, and that the CU may be granted the necessary instruments to engage in this field. To achieve these objectives, the local institutions need to design a clear vision and comprehensive strategies to build a 'student-friendly city'. Lyon is a good case of excellent *hardware*, know-how or software that are rapidly improving under the pull of the private and research sector, but insufficient organising capacity.

In Lyon, everyone is committed, but there is little dialogue between the different players involved, and nobody is really 'in charge'. The city is still seen by higher education institutions as a bureaucratic decision-maker. How to overcome this mistrust is a real challenge.

Recently, attitudes seem to be changing; converging interests between higher education institutions and local bodies are demanding more cooperation in several fields. Local governments like Greater Lyon have started an integral policy programme for welcoming students. It has also given some substance to the 'return to the city' policy of the government, being the main player in the restructuring of tobacco manufacturing. However, in other fields where the city could play a role and has competencies, it remains too much in the background, for instance as regards housing policies.

Today, the links between universities and public domain are mainly taken care of by the 'representation doby' of higher education, the PUL. Other intermediary organisations like the CII and the CROUS, are backed by public bodies, but receive little help from single universities, and in this way, it is very difficult for these centres to have a 'proactive' and tailor-made policy with respect to different groups of students, anticipating possible problems and preparing a medium-term strategy for welcoming and hosting, for instance, international students; mostly they are busy dealing with day-to-day emergencies.

However, also that the higher education institutions have to commit themselves towards the environment where they are hosted, explicitly recognising Lyon as a 'laboratory' for social, economic and scientific research.

The City and the Chamber of Commerce, however, can do much to ensure that this cultural passage is accomplished.

Universities as a resource base for urban policy It is important that the city maintains the links with students who have completed their studies and left the city; they might be the visitors, the business partners or even the citizens of tomorrow. Programmes like JAL, entirely a private initiative, could be extended and become an integral part of the city's promotion policy, or even extended to the regional level with the support of the Rhone-Alpes administration which is already strongly committed to international mobility programmes.

The community could be mobilised around projects of resource-sharing with the higher education sector, and in this way accommodate potential conflicts from the 'opening' of the city to the outside. For instance, the central library should not just make room for university students, but could also be developed as a real meeting place between the host and guest communities.

Improved contacts between higher education institutions and the city bodies could be developed, starting from simple commissioned research initiatives. An obvious candidate for this is the city's tourist office. The city is trying hard to develop its hospitality function which has a promising potential in the cultural segment, but it is sometimes short of tools. It is believed that tourism researchers could provide skills and suggestions in that respect. Moreover, the tourist office, if involved at a planning stage, could help the higher education institutions to organise their international welcoming and support programmes better, for instance organising occasions to show students around the city, or involving them in cultural initiatives. In turn, student life, if well organised, makes the city more attractive for tourists who are looking for 'alternative' cultural lifestyles, thus providing an opportunity for the diversification of Lyon's cultural supply, presently only appealing to 'traditional' visitors.

There are indeed social problems in certain areas of the city and in its suburbs. Bringing students to live in such areas could reduce the social problems and favour social integration. However it remains to be seen how attractive this solution is for students, as against to the evolution in their preferences. It could certainly be successful with some of the most central areas (e.g. Croix Rousse). Such a strategy would also demand active backing from the CU as regards the improvement of the transport system.

Synergies between higher education institutions and the city in international relation programmes It is important that the city and the higher education institutions (through the PUL) get together regularly to fine tune their international policies, identifying synergies and areas of unexplored potential. Today, the *grandes écoles* have strong international ties (e.g. INSA) and well developed international programmes. In this sense, the score for public universities is still insufficient. The INSA model could be imitated and adapted to the universities, tapping resources from European funds but especially from private partners in research, that should be actively involved in 'designing' the map of international relations on the base of the market and knowledge potential. The Chamber of Commerce could be an ideal 'broker' for these processes, providing a Chamber for the Development of International Relations whose task should be to provide matching of demands and funds for staff, students and knowledge with a strategic attitude.

Increasing the number of PhD students and staff from developing countries is a challenge that involves solving basic problems with infrastructure and information; nowadays, despite the good relations between Lyon and host countries, researchers are more likely to choose more attractive locations in France or in other places (e.g. USA, UK).

Services for students and young workers in the knowledge sector The city must engage more directly in the welcoming, assistance and accommodation of students, national and international, offering particular support to those students who are not entitled to support and housing in the context of international programmes (*ca.*2,600 at the time of writing). To this end, the city and the PUL could form a partnership to set up an office, cofinanced by private partners, which would assist the students from their arrival until they obtain their residence permit in the city, and organise continuing consultation to monitor their quality of life and their level of integration into the community. Alternatively, the city could support the existing services (e.g. CII) with additional financial and human resources. Resources could be mobilised to accommodate students and young workers in those districts of the city that are in need of revitalisation strategies, and to promote modes of residence other than student residences, where contacts with the local community can improve, for instance, sharing rented apartments. This could also be a more affordable solution for free-moving students from developing countries, who are not eligible for CROUS facilities and for whom private residences are too expensive.

Opportunities for a more effective integration of foreign students into the city have been lost; the project of the municipality to set up a *Maison Internationale* reserved to researchers is redressing this situation and could be extended. The Croix Rousse is a perfect candidate for new student housing and facilities, and could target short-stayers in particular – it might turn out to be a profitable business for the city, linking other initiatives with start-ups: nowadays new firms have difficulty in finding locations because they cannot know for how long they will need the accommodation and a deposit of five months is required.

More generally, the interaction between relevant actors (CROUS, City, Region, Chamber of Commerce, real estate operators) has to be greatly improved to solve the housing problem. One principle that should be agreed upon is that CROUS cannot be the only player in this field. It will be also useful to establish links with social offices abroad in such a way that emerging problems could be anticipated.

Notes

1 The territory under the jurisdiction of PUL (*Pole Universitaire de Lyon*), uniting 13 major HEIs in greater Lyon, includes 110,000 students (information from PUL website: http://nte-serveur.univ-lyon1.fr/pul/uk/anglais1.htm). The PUL is the largest of the ten university poles in France in terms of students.

2 In partnership with other universities within the European Community, INSA has set up a European diploma called EUFORIA.

3 The City of Lyon's budget for welcoming students is of €305,000 in 2002, but the human resources in charge of developing this task are limited.

Chapter 7

Munich Case Study

7.1 General Introduction to the City of Munich

Munich is situated in the south of Germany. It is the capital of the State of Bavaria and, with a population of 1.3 million, the third largest city of Germany, after Berlin and Hamburg. The city is the centre of an urban region with about 2.4 million inhabitants. Munich is presently the only big city in Germany that is not losing population. Besides the city of Munich, the urban region comprises eight *Landkreisen* with populations varying from 97,000 to 278,000. These administrative bodies are clusters of relatively small but influential municipalities.

The city of Munich was founded in 1158, and given the status of a town (*civitas*) in 1214. A powerful and relatively free city across the years, it has always been economically dynamic and at the forefront of social innovation. After the dark years of World War II, the city regenerated quickly and it is now completely integrated into the contemporary service economy. The urban region is known for its high standard of quality of life and is regarded as one of the most economically successful in Europe. In 1998, only 5.8 per cent of the active population were on the records as unemployed, a considerably lower percentage than in other German cities (10 per cent or more).

The economic structure of the city is strong and diverse. The main points of strength are the high-tech activities and the service sector, though traditional sectors like the car industry and manufacturing are still central (cf. Table 7.1). The spread of activities across the sectors does not show up any substantial difference between city and region, although services have a distinct preference for the city. The peculiar economic structure of Munich is called the the 'Munich mix', balanced from three perspectives: the range of economic activities, the size of firms and the presence of innovative and entrepreneurial firms. Munich accommodates the headquarters of large companies such as of Siemens and BMW, but also numerous and small and medium-sized businesses, notably in the innovative ICT-branch, a growth cluster when combined with other knowledge intensive sectors with enormous development perspectives, such as the media industry and biosciences (van den Berg et al., 2001). The economic area of the city is indeed the major IT

Table 7.1 Employment in the Munich region (2000)

Sector	Region	Core	Ring
Agriculture	6,221	2,029	4,192
Energy and water supply	11,963	10,065	1,898
Manufacturing industry	236,768	137,693	99,075
Construction	44,279	20,746	23,533
Retail and wholesale trade	158,199	116,329	41,870
Transport and communication	58,413	29,300	29,113
Banking and insurance	76,385	61,817	14,568
Other services	391,479	266,007	125,472
Non-profit organisations	36,107	29,628	6,479
Regional public authorities	51,151	31,833	19,318
Total	1,070,965	676,147	394,818
Secondary sector	*27.4%*	*24.9%*	*31.5%*
Tertiary sector	*72.1%*	*79.1%*	*60.0%*

location in Germany, employing over 155,000 people in 8,000 firms engaged in ICT business and research. According to the *Financial Times* it is the main IT location in Europe, together with London. The presence of a high-quality telecom infrastructure is an increasingly important location factor for firms of various other sectors. Munich also has an excellent reputation as a city of financial services. There are over 80 insurance companies in the Bavarian capital, 12 per cent of the country's total. This makes Munich the most important centre for insurance companies in the Federal Republic by far, and, after Frankfurt, the main location for banks. More than 20 internationally-operating venture capital companies provide about €500m in risk capital, mainly channelled to start-ups in the IT and biotech sectors.

The corporate control centres, as well as the countless other high-rank firms in Munich, rely on the well-educated, highly-skilled people who are available on the local job market. Indeed, the educational level of the population of Munich is among the highest in German cities: no less than 14 per cent have a university education, against a national average of 6 per cent. This is hardly surprising, as Munich has a very strong education infrastructure, with numerous institutions for higher education hosting about 100,000 students in total.

Intimately related to the strong knowledge base of Munich is its 'cultware': generally speaking, the attitude of people and firms in Munich can be characterised as *open to innovation* (van den Berg et al., 2001). At the company level, that openness undoubtedly facilitates knowledge and information exchanges, technology transfer and the formation of innovative cooperative networks amongst firms and knowledge institutes in the Munich area, thereby contributing to the high innovative capacity of the region. This leads to a relatively fast adoption of new technologies by the public sector, such as digital broadcasting, electronic shopping, etc. Interestingly, the municipality invests in further strengthening the climate for new technologies and its adoption by citizens, cooperating on this grounds with higher education institutions and firms.

Another Munich asset is its high quality of life. The city of Munich is generally considered a pleasant place to live, albeit rather expensive. Many high-ranking cultural facilities are present, the housing stock is of excellent quality and the commercial supply is top class. The city is also close to amenable rural areas and skiing resorts in the Alps. Munich is very committed to the environment, as demonstrated by its involvement in the Local Agenda 21 programme. The programme – supported by the City Council with a subsidy of €50,000 and the promise to double any grant from the private sector – supports projects to enhance community life, social inclusion, and the use of art as a public asset. These features make Munich very attractive for interesting 'target groups' like young professionals, researchers and students, not forgetting issues of social integration.

As a pleasant, clean and classy city, Munich is also highly attractive to visitors. Munich's hospitality function can be split into two broad categories, leisure and culture, and congress and business. Fifty-five per cent of the overnight stays are for business purposes, with a main focus of fair visitors (18 per cent) and congress visitors (13 per cent). However, the strength of Munich as a tourist destination is mainly due to the traditional cultural-tourist market, enjoying an enviable palette of cultural institutions, as well as intangible elements related with its celebrated atmosphere and quality of life, and countless festivals and events, the most famous of which is the popular 'Oktoberfest'. On account of this, Munich is today among the European 'stars' for short breaks and cultural visits.

Munich is also Germany's media and film centre and a bulwark of gastronomy. However, Munich's cultural speciality surely must be music. Three large orchestras are domiciled in the Bavarian capital, with international superstars as conductors. Jazz music also has an extremely lively tradition in Munich. In the mid-1990s the city developed a programme of high calibre open-air pop and jazz events. The landmark event of 1972 Olympic Games contributed to the international exposure of Munich as a sport city. However, the city can hardly be considered a 'cutting edge' place for young cultures, as it rather projects the image of a rather traditional, 'bourgeois' cultural centre.

7.2 The Student Community and Knowledge Infrastructure in Munich

7.2.1 General Information on Higher Education in Munich

In Munich there are ten higher education institutions (three of which are universities – LMU, TU and the Army's University), with a student population of approximately 80,000 (the breakdown is illustrated in Table 7.2). Figure 7.1 shows their location throughout the city and the inner historical core. Every year, about 6,000 students obtain their degrees in local higher education. Due to this concentration, Munich established itself as a major national pole for higher education, competing with other German 'knowledge regions', like Berlin, Köln, Heidelberg. However, Munich has an advantage over some of these cities in its reputation as a very lively and connected learning environment, where scientific knowledge and the booming regional economy go hand in hand.

Table 7.2 Number of students in higher education in Munich (1997–2001)

University or higher education institution	1997	2001
Ludwigs-Maximilian University (LMU)	59,804	41,943
Technical University (TU)	17,701	18,864
State Polytechnic University (FHMM)	14,306	14,022
Others	6,391	5,399
Total	100,199	80,228

In Bavaria, there are 11 universities in nine cities. Despite their obvious points of strength, Munich's higher education institutions are increasingly challenged by smaller regional universities, which offer the comforts of 'university cities', such as a more student-friendly environment, lower housing process and better facilities as well as a less congested job market. Beautiful towns like Augsburg, Wurzburg, Nuremberg or Regensburg, which are all in the proximity of the Munich region, can therefore attract students and especially resources, offering the appeal of the 'small is beautiful'.

However, as an education and research centre Munich's catchment area is at least national and on this level in the past Munich enjoyed a certain amount of brain-drain from the north, pushed both by the quality of the university and

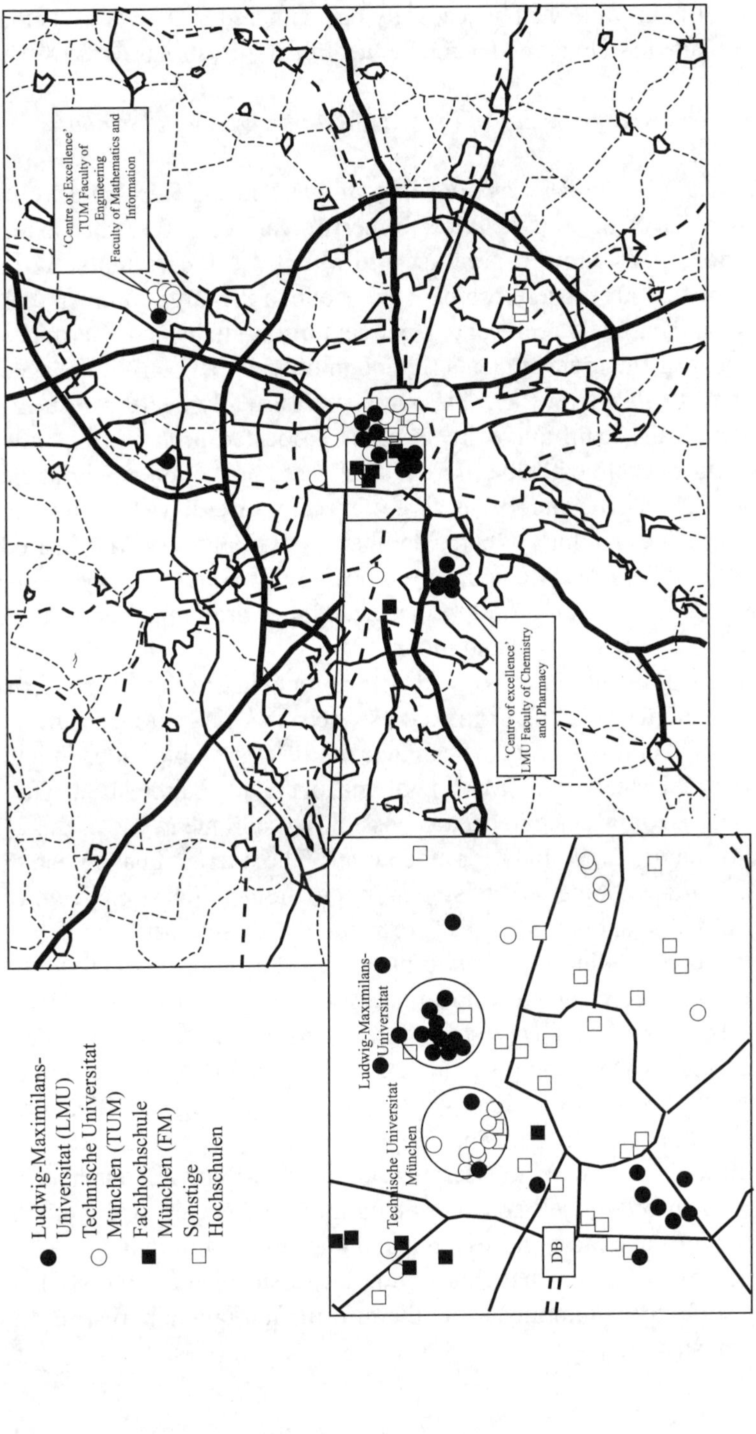

Figure 7.1 Higher education establishments in the Munich region

the quality of life in Munich. This capacity has withered in the years, as the north of Germany has developed a good educational infrastructure as well.

7.2.2 The Knowledge Sector of the City: Actors and Spatial Patterns

Universities and other higher education institutions The *Ludwigs-Maximilian University* (LMU) includes 18 faculties, formerly organised into 178 academic institutes and clinics, now reorganised to about 30 departments. With approximately 42,000 registered students, it is one of the largest universities in Germany. In fact, LMU represents 2 per cent of the students of Germany, and 30 per cent of Bavarian students (although only 25 per cent of the state budget goes to LMU). However, this number persistently decreased in the 1990s. Although the main university campus is located in the city centre, there are institutes at almost 100 different addresses, most of them within the city boundaries. Since the mid-1970s, large new complexes have been built in peripheral areas like Garching, Oberschleißheim, or Großhadern-Martinsried for the Faculties of Physics, Medicine, Veterinary Science and Biology. About 800 full-time professors and 2,500 other academics are employed, plus a workers pool of another 11,000 and more.

Of all German universities, the LMU offers the widest range of subjects and subject combinations, being ranked as one of the first in Germany in regular surveys. At present students can take 96 different subjects leading to the degrees of Magister Artium (MA), Diploma, or a State Examination. After completion of one of these degrees, it is possible to continue in one of the 24 postgraduate courses or to study for a doctorate. The international character of the LMU is demonstrated by its student population, which is made up of 13 per cent foreigners, both in regular programmes and summer schools.

Today, LMU is looking for ways to make its offer more adequate to the large numbers it attracts, and more up-to-date for the requirements of the local economy. This is done through investments in internal resources and relocation. Step by step, the LMU is merging the medical and natural science studies in a 'high-tech campus' in the southwest of Munich (Planegg-Martinsried), where 1,000 ha are currently being developed in a centre of excellence for life sciences, and adding to the existing resources of the Bio-Region Munich in the same site. A 'Stanford model' of sorts is sought for in this area, with optimal connections between education, research and business institutions.

However, LMU admits that this is not sufficient to meet the student community's quality requirements. As the information brochure distributed to students says:

> It has not been possible to increase the levels of university funding and staffing and the space available in proportion to the rise in numbers of students at LMU. For this reason the studying and working conditions are, to many respects, worse than in smaller, newer universities. This disadvantage is, however, more than compensated for by the university's favourable conditions and location.

In the future, though, these very 'favourable locations' might need to migrate outside of the city in search of more space. Changing regulations regarding the academic structure are placing additional pressure for change.

LMU's *Munich School of Management*. The school organises all sorts of undergraduate courses in business and MBAs, and has extensive links with large local and international corporations for job placements. In collaboration with the employers' associations and the Munich Labour Office, the LMU has set up a central structure to prepare students for working life, offering a system of voluntary courses in which students can acquire additional professional qualifications.

The teaching programmes of the *Technical University of Munich* (TU) comprise 25 courses in 12 faculties, leading to various German degrees, though the system is changing to suit the new international requirements. The TU hosts 19,000 students, 440 lecturers, and 8,000 other workers and researchers; this makes up 11 per cent of the Bavarian student population, though contributing a share of 18 per cent of degree examinations, which is proof of the effectiveness of TU's academic supply. The research activities at TU include a total of 15 special programmes administered by the German Research Association. Beyond the facilities offered by the faculties, TU runs special research centres of general, nationwide or international interest. TU is organised in four 'poles'. The inner city campus is the biggest, with 6,000 students, and the Faculty of Architecture is even more central. The third is in Garching, at a one-hour journey from the city, where the medical faculty is located. Then there is another campus with agriculture laboratories and faculty in Weihenstephan at some 30 km from the city.

Consistent with this university's strong focus on research quality, the TU has adopted a policy under which there is a revision of the strategic plans after a careful evaluation of the past policy every five years. An evaluation and consultative board with the task of giving advice to the TU on important matters includes important stakeholders of the private sector like Allianz and Beurs; this is a unique structure in German universities. TU has a history of flexibility, and an inclusive approach to finding solutions that fit the ambitions of students. As an example, the presidency of the university, faced with a housing shortage, has formed a task group to tackle the problem and identify solutions in cooperation with influential political interlocutors. The board also organises get-togethers and special information packages for international students. This attitude helps TU to tackle the same issues faced by the LMU with a more open, less bureaucratic approach.

The *Fachhochschule München* (FHM), with approximately 16,000 students (10 per cent foreigners) and 480 staff is one of the largest polytechnic institutes in Germany. In Bavaria, there are another 17 such schools, at least one for every region. The FHM has a strong focus on innovative methods of education and research. It is relatively concentrated in buildings in the same area close to the central station, plus a little 'dependence' in Pasing; this makes it easily accessible for commuting students, whose place of origin is indeed strongly regional. The range of topics taught includes architecture, many branches of engineering, economics and services, divided into 14 departments. Two thousand students out of the total are educated as social servants, making this the most relevant group of trainees in this career in Munich.

FHs offer a new style of higher education, with relates students' academic knowledge to practical applications. The teaching staff have to prove their ability in practical applications through a career in private industry. Recently FHM introduced some form of research obligation, beginning a process of convergence with normal universities. FHM has therefore equipped itself with state-of-the-art technical laboratories where experimental applications are being developed, for instance in space-science technologies and communications.

Students do not write final dissertations for their diplomas at FHM, but they all have to complete a period of in-company training, which explains the ease in finding jobs when they graduate; only 2 per cent of FHM students stay unemployed, scoring better than the students of 'normal' universities. FHM places a special importance on international exchange programmes and student mobility. Thirteen per cent of their students go abroad for a period, the highest mobility rate in Bavaria among FHs. It could be higher if constraints on housing did not constrain the further development of international exchange programmes. For the future, FHM wishes to strengthen relationships with UK, USA and other English-speaking countries like Singapore. Foreign students have classes in both German and in English, but FHM requires knowledge of German anyway.

Research and technology transfer The *BioRegio* company is at the heart of Munich's thriving biotech cluster, acting as an incubator and support for the start-up of new firms in the cluster, and assisting the firms as a provider of – or intermediary for – all possible services of financial and technical nature. Most organisations of the cluster are physically located in the science park of Martinsried. The location strategy was dictated some 15 years ago by the shift of some university departments in this area, notably the medical hospital of the LMU and some of the biochemical laboratories. However, a strong partner for research was the *Max Planck Institute* (MPI), also present in this area. This prestigious national institution is a milestone of medical and biology research. Though a private foundation, it is closely associated with the federal state. It is seen by many young researchers as a hierarchical and 'traditional' structure. That is possibly why many innovative research activities are born as 'off-shoots' of MPI; this is the case of the *Gene Centre*, a network of gene researchers, who were not satisfied with the cautious attitude of the old research institutions in this field and decided to go it alone, against all odds and looking at the American model. In the end they were thoroughly successful, making Munich one of the strongest European poles for breakthrough genetics and protein research.

At the moment, major building activity is happening at the Martinsried campus, with new firms (e.g. IPO, Windows) planning to relocate here. The vision of the LMU and Ministry of Science is to cluster all scientific research in one location, but at the present time there are no perspectives of expanding this site. Planegg (where the Martinsried campus is located) is a different municipality from Munich, and the community's resistance to further development is growing, whatever the beneficial impacts this may have in terms of revenues and tax base. Then, laboratories and faculties will stay in the city centre or they will have to find other suitable locations for science clusters.

Various initiatives to nurture an entrepreneurial culture have been launched. The *Association for the Promotion of New Technologies*, for instance, organises get-togethers, interactive seminars and workshops covering all aspects of business start-ups, financing and management. The Munich Business Angel Network, an informal, regional platform for business start-ups with a strong growth potential, is also located there. For the fourth consecutive year, the Munich Business Plan Competition (MBPC) took place in the Munich region, with a record participation of 386 young entrepreneurs and 126 business plans. In four years the competition has already founded 103 companies with some 750 employees financed by approximately €80m.

The *Software Campus* to be established in Munich combines elements of a business incubator with those of a research institute. It will offer approximately

4000 m^2 of floor space for start-ups, research groups and training and service facilities. Possible topics for start-ups and research groups range from software engineering and new e-commerce applications to innovative telecommunications services. Research groups are formed by experts from the industry and the universities. A seed capital fund and a funding programme will be available to support start-ups on the campus. Finally, the *Founders' Region Munich* – an initiative of the Science and Economy Region of Munich – promotes start-ups and spin-offs deriving from (and thematically close) to institutions of higher education.

Spatial patterns of higher education and research The higher education institution sector in Munich is roughly organised in four 'poles': the city centre, Großhadern-Martinsried, Freising and Garching. The universities all have their main offices downtown, but the various institutes are scattered throughout the city. This spatial spread affects the students in two ways: firstly, the underground that connects the various university locations throughout the city, the U-Bahn, has almost become a second home to many students. Secondly, much of the students' social life, especially for those not originally from Munich, takes place at huge student dormitories, some of them hosting up to 2,400 residents. The most famous dormitory complex is the 'Studentenstadt Freimann' on the outskirts of Munich. Its large apartment blocks do not make it particularly cosy. However, it remains popular among students due to its cafés, bars and sporting facilities. The small-town feel makes it easy to get to know people, especially for international students. In any case, a large share of students come from upper Bavaria. This tendency is higher for FHM than for TU and the LMU, but nevertheless the catchment area of the higher education institutions is more regional than national, and the magnitude of the student population living *in the city* is relatively small. As a consequence, the housing problem is mainly felt by international students. Moreover, rooms in residences are granted for a maximum of three years, in order to increase the turnover, and hence all the resident students sooner or later have to revert to the private market.

The recent trends in the location of the Munich universities is to transfer a higher share of their teaching and research facilities to the peripheral locations, at increasing distances from the city centre, where they find more space and can expand physically in science parks and business parks of sorts.

The decentralisation trend is a natural result of the growth of the universities. So far, the location strategy has been opportunity-driven, with some leading actors setting the main trends. *However, at a strategic level, it should be*

Figure 7.2 The campus of Garching, aerial view

Source: http://www.mpe.mpg.de/Living/garching.html.

considered that the competitiveness of the university system in Munich depends very much on the quality of life and on its association with a pleasant, safe and wealthy city. If the education sector grows disconnected from the inner city, in a rather isolated neighbourhood where no real urban facilities are to be found, this could be a serious problem for them. The combination of living and working would be an interesting solution, but there are no such projects at this stage. That is why LMU is resisting the trend to move outside; the human and social sciences studies are particularly attached to urban locations where they find a favourable cultural context.

As in other cities in this research, students are dynamic in their location attitude, responding fast to changing conditions in the housing market. 'Fashions' hardly convince a student to spend more to live in a pleasant neighbourhood. Moreover, the reaction time of students when the news of new housing spreads is very fast. Foreigners in particular, when they know that something is going to be done, move immediately and therefore the stress on one area increases, making development more difficult. Today, expansion of student settlements is possible in selected areas of the city, which has made two lots of land available at a favourable price, and with relaxed building regulations. However, local residents, immigrants and young skilled workers are also in need of similar arrangements and competition is tough. Students are seen as competitors by disadvantaged groups because of their flexibility: they can live in four in a house, pay higher prices and the turnover among this group is higher.

The lack of a genuine 'student-life atmosphere' poses a threat to the homogeneous development of the inner city, which becomes monotonous, with an excessive concentration of headquarters and office functions that keep the city 'running' during the day, but leaves it dormant at night. The traditional orientation of Munich's tourism is also scarcely linked to the creative potential of the student community.

7.2.3 The Student Community: Behaviour, Residence, Careers

Munich is quite attractive for students. The characteristics of the city play a role at least as big as the prestige of its higher education institutions in its attractiveness. In fact, faculties are not very customer-oriented, and leave it all to students to organise their own way of studying and living in the city. While most students take this for granted and instead appreciate the atmosphere of freedom and independence of life in the big city, some – especially international students, some 14 per cent of the student population – are disappointed and have a very hard time settling in this environment.

Munich's student population is not so relevant compared to the mass of the city, and it is not concentrated in areas with an atmosphere of 'Latin quarters'. This dispersion will increase with the ongoing relocation of universities on the periphery, threatening to diminish further the potential benefits that students bring to the urban environment in terms of social and cultural change. For a fairly traditional city like Munich, this loss of 'creative' input may be remarkable. In fact, despite the excellent supply of 'traditional' facilities, Munich is much less active in the cultural realm than many other cities even in its proximity. Munich still implies 'Laptop und Lederhosen': its students are more career-oriented than many other German students

Students commute easily to the peripheral campuses, so it is not much of a problem for them to keep rooms in the city centre, where they are close to all sorts of amenities, and move to the campuses during the days to attend classes and laboratories. However, in this way campuses (especially Garching) turn into real wastelands after dark. The idea of a science region clearly requires some living community to be settled there, and therefore it can be expected that some student residence will be planned in those suburbs, where a real 'student district' might develop in the future. The necessity to strike the right balance with maintaining or improving a visible presence of students in the city centre remains.[1]

Another contradiction of Munich as a student city comes from the fact of being very expensive, especially as regards accommodation: the rents paid in Munich are among the highest in Europe, up to €1,000 for a flat. Munich offers many jobs so it is easier for students to make their own living, but then it takes longer for them to finish their studies. More than the half of the student population earns part of their income through some hours of work. Indeed, students in smaller 'university cities' graduate faster than those in Munich, and though students who graduate in Munich have prestige and get better-paid jobs, the tendency for firms is to extend their attention to younger graduates of peripheral universities, and even to settle in smaller peripheral towns (e.g. Erlangen, Nurnberg, Rosenheim). There are many reasons why rents are so expensive, from the saturation of the market and the very low building activity, to the competition in the same market with users groups with a strong capacity to pay, such as young professionals and visitors, and the reluctance of relatively high-income families to rent spare rooms. After the completion of their studies, graduates would try to stay in Munich, where the job market is rich and the quality of life for young graduates is high, but the problem of housing re-emerges for young families, which might have an easier start in small, peripheral cities.

An important participant in the management of the housing issue is *Studentenwerk*. This organisation is a network of 62 local associations, present in all German territory since the end of World War I, supporting a total of 1.8m students. They are public non-profit organisations, largely autonomous as regards local strategies. They are state subsidised (20 per cent of their budget, only for restaurants and kindergarten services), but a high share of their operation budget derives from self-generated revenues and student fees (9 per cent, corresponding to a contribution of *c.*€30). The operation costs of student residences are not subsidised by the state, which only grants part of the capital investment funds (*c.*30 per cent), while the rest comes from Studentenwerk and bank credit. Studentenwerk is fairly autonomous as far as policy initiatives are concerned, though the Ministry of Science and Education maintains a right of veto. The director has ultimate strategic responsibility; for instance he decides if it is necessary to build new residence, and finds the financial partners.

Studentenwerk Munich employs a staff of 700 and is responsible for student services for all the HEIs in the Munich region, that is 17 institutions between universities and 'Fachhochschulen', covering 85,000 students, about 75,000 in Munich alone. On the whole, they provide accommodation for 10 per cent of Munich's student community. The responsibility stems from Bavarian law and includes social welfare for the cultural sector, buildings for student residence, catering, libraries, etc.

Studentenwerk, in common with some other non-profit organisations (for instance those of a religious nature) offer subsidised accommodation at an average rent of approximately €175. However, there is hardly room for everybody in this market segment. At the time of writing, it is estimated that supply falls short of demand by 1,500 units, and waiting lists can be very long, especially for accommodation in central neighbourhoods. It is not uncommon to have to wait for four semesters to get a place in one of the Studentenwerk residences. The rise in demand for subsidised housing is due especially to the increase in the number of students from abroad, who on the other hand have greater problems in finding rooms in the private market, not least because of their ethnic background. The main housing accommodation offered by Studentenwerk are in Freimann (2,400 students) and close to the Olympia Centre (1,800 students).

Today, new funds are needed to expand the accommodation capacity in student residences. In fact, the recent economic crisis ended in a shrinking of the Studentenwerk budget for students, though part of this money came back through the Army and the City Hall, thanks also to the active support of the mayor. Two thousand five hundred new apartments are planned, but the first will not be in place until 2004, whereas it is needed immediately. This means Studentenwerk has to improvise solutions to the housing shortage in the short term, and creative solutions are needed. At the time of writing there is a waiting list of 3,500 students who are staying somewhere else, and it is also an anticipation of future needs. Studentenwerk offers a home in former military barracks for 2,500 freshers on the shortest time schedule. They also offer 'first aid', with a mattress on the floor; in practice everybody gets something, but the quality of these solutions is questionable.

In Munich, these problems are more dramatic than in other cities which compete with Munich as poles of economic development. For instance, in Berlin housing is much cheaper (and the housing stock bigger), and it is easier for students with different origins to be integrated into that community, with the result that the creative input of young people is strong and very visible in the 'Berliner' atmosphere. To what extent Munich may be losing its reputation and economic strength to more 'open' and student-friendly cities, remains a long-term issue that the city has to address more proactively than it is doing.

Many *student organisations* are active in Munich. They are strongly organised along university and faculty lines, so that a certain cohesion in student action in the city to resolve the most pressing general issues of housing and university development is lost. In this respect some peculiar differences are found between universities. At TU, students have been actively involved by the board in finding solutions for the shortage of facilities, and sometimes students came up with creative and effective ideas. This 'bottom-up' approach is more likely to be accepted by students who are fully aware of the housing emergency. Some living spaces were decorated by students of art, and Studentenwerk was involved as the financial partner. In Freiham, students organised a group to discuss how to improve living conditions. At the same time, students are also concerned about making the link with the city authorities, and in this respect, they distinctively took a 'citizens'' attitude towards the problem.

At LMU, officially represented student organisations have a radical attitude, but they also address many issues of citywide relevance. Another form of association is within the student community accommodated by Studentenwerk, representing students in all matters regarding their living conditions and services in the complexes. Other student initiatives have different goals. Some are of particular interest for this research, such as MESA (Münchner Erasmus Studenten Assoziation), which helps foreign students coming to LMU within the Erasmus Programme during the first few months of their stay in Munich, and the *Entrepreneurial Group*, an interdisciplinary students' initiative promoting entrepreneurial thinking and acting in universities and colleges.

Studentenwerk also runs two open-access and two reference libraries, hosting around 24,000 regular readers, important meeting places for students from different faculties and local residents. Moreover, they manage ten dining complexes and 18 snack cafeterias with very accessible prices. Finally, they offer advisory services such as psychological support, legal advice, childcare and cultural affairs. They are also developing a network of international relations with similar institutions (among which is the CROUS in Lyon, included in this research), which serves to exchange information and know-how in student services, but also to support exchange programmes, anticipating and resolving possible problems at their source. Thanks to its independent status and inclusive attitude, Studentenwerk has come to be trusted by students and their organisations, although some repute it an ineffective, obsolete instrument. Another important facilitator is the *Munchener Uni Magazine*, a platform for information and opinions on student life in the city. Unfortunately the magazine is in German so it is almost inaccessible for most starting international students, who would nevertheless benefit from having access to this information right from the beginning of their stay. Both universities have alumni networks which organise a number of programmes and projects, especially in connection with job market opportunities.

7.2.4 Challenges for the City

The spatial conditions of Munich are on the whole favourable. First, Munich's external accessibility is excellent. The location of the city at the crossroads of the east–west and north–south corridors contributes further to its attractiveness as a business location, and this has posed serious problems on its *internal* accessibility. Due to the high share of young skilled workers who flow into the area, the demand for accommodation is high and rents grow fast – a medium-sized flat can easily take half of a normal income.

To counteract the growing sprawl, the city of Munich set up a new City Development Plan in the mid of the 1990s, the so-called Perspective München. PM consists of a set of seven overall goals and of a number of specific projects. The key words are: 'Munich: Compact – Urban – Green'. Strategic planning is today as central to policy efforts of the city as is physical planning. Munich has chosen to develop a number of strategic sectors in a participated, networked approach, placing much emphasis on shared responsibilities and mutual interests. Some of these sectors clearly owe their success to the performance of the education and research sector and to the capacity of the city to attract the best talents.

The *media cluster* in Munich is a 'cutting edge' sector. The strength of the cluster can partly be ascribed to historic factors (the division of Germany after the World War II made many media firms flee to Munich), but the strong economy of the city as a whole should also be regarded as an important element: different segments of the cluster – notably telecommunications, new media, marketing and communication activities – are strongly stimulated by the local demand. Another characteristic is the strong network culture in the cluster: strategic linkages prevail, facilitated by cultural identity and a cooperative attitude amongst actors. The technological profile of the cluster is high, due to the presence of world-class firms and top academic institutes in the fields of ICT and new media. This gives Munich a lead over other media cities, in particular in the promising new combinations of 'traditional' media and ICT.

With sales of German software companies and service providers rising by some 10 per cent to €25bn during 1998, the *software industry* has emerged as one of the driving forces behind the dynamic development of the German information technology market. In particular, the German State of Bavaria has earned a place as a leading centre of the European software industry. Bavaria hosts not only one-third of Germany's 25 largest software companies. Set up in 1998 in a common effort of industry, science and politics, *Software Initiative* is part of the Bavarian government's 'high-tech initiative'. By investing the proceeds of the privatisation of public shareholdings (amounting to approximately €1,000m) into R&D, training, and infrastructure, this projects aims at making Bavaria an even more attractive location for the high-tech industry.

Bio-Regio Munich has established itself in the last years as one of the growth clusters of the urban region. The *biotech* cluster is still in a phase of strong expansion and faces a process of consolidation. Thousands of researchers are employed in its firms, to whom a thousand between medical students and staff must be added, plus some hundred researchers at MPI, making for a total employment of almost 2,500 in the area of Martinsried alone. The recent trends of the cluster regard the passage from developing technological platforms to developing specific products; the consolidation with the increase in size of existing and new companies; and the diversification of the fields of application. Moreover, a number of non-bio firms, for instance consultants, financial brokers and incubators, have settled in the same area.

The demand generated on the job market is huge: in the next five years, 1,000 specialists will be needed to develop the cluster according to its market potential. *In a highly competitive environment with only 40 cluster locations all over the world, and the insufficiently developed academic curricula in this sector, the risks of not disposing of sufficient human capital are high.* This is the more so since overseas jobs are highly remunerative compared with German jobs, and private companies in fact compete with universities where base research is produced. The issue of establishing synergies with the local universities as regards the skills to be developed and 'rooted' in the region is therefore of prime importance. This not only calls for more coordination between universities and the private sector, but also for some degree of change in the structure and orientation of university education.

7.3 Strategies and Integration. Munich as a 'Sustainable Student City'

Munich is a strong community, illustrating perfectly how the local 'cultware' can be utilised at its best in order to achieve both ambitious growth objectives and a balanced, sustainable society. The region thrives on the delicate intertwining of large-scale industrial power, knowledge excellence and good governance. However, this virtuous circle seems to be challenged by a new form of exclusion. Those sectors of society that are at the forefront of innovation, the students and young graduates, cannot easily find room in today's Munich society, both literally and, in a broader sense, as a community. So far, the actions of different players all enjoying an enviable strength are not moving in the direction of creating more opportunities for an inflow of new creative capital in what risks to become an 'old style' metropolis.

7.3.1 Relations between Students and the Community

To what extent are local higher education institutions working for the community of Munich? Apparently, the benefits generated by education and research institutions are high, and they tend to stay attached to the place because of Munich's capacity to attract new graduates and skilled workers. The knowledge output of the local higher education institutions trickles down to the local society in the guise of a strong, diverse economy, with strong participation from local residents. Even the problem of social exclusion is limited in Munich if compared with other big German cities. The network of higher education institutions involves the main decision-makers in the city policy. Higher education institutions collaborate closely with the administration for the management and welcoming of foreign students, and in relation to international exchange programmes.

However, strategic cooperation with the city for upgrading student facilities and improving quality of life in the city is less good, and limited to physical planning. Restrictions on ownership structures and regulations make it difficult to upgrade the university infrastructure in the city centre; in such cases, the Municipal Administration did not adopt a proactive attitude. The excessive fragmentation of competencies makes things even worse, as several levels of administration with different political colours are involved. The city is the natural partner in ordinary matters regarding the spatial management of the higher education sector, but in Germany the states have sovereignty as regards education, academic careers and support, which on the other hand have strong spatial implications. Federal policy directives and state regulations orientate the strategic development of education, but the city is scarcely involved at this level.

Competition with other universities in Bavaria is probably the most pressing issue that local higher education institutions have to face. The city is mainly oriented towards traditional universities like the LMU and the

TU; it supports them with resources and promotes them consistently with the general marketing policy of the city. Less attention is paid to the FHM, which, however, is an important actor in the diffusion of applied technical knowledge. In other Bavarian regions which are relatively backward compared to Munich, FHs are more rooted in the local economy, and get more political support as they are recognised as being strategic for economic regeneration. In Munich this is not the case, despite their strategic importance, especially for the traditional manufacturing sectors like the car industry, but also for new sectors like space engineering.

However, the biggest problem for Munich is that the student community – despite its remarkable size – is hardly an agent of social and cultural change. To what extent this is a problem, and if so, whether this problem can be solved, is not easy to say. The impression is that Munich – a city that is very attached to its traditions and way of life – is losing out to other dynamic German cities as far as dynamism and creativity are concerned. Being able to attract a higher proportion of students as residents in the city is an necessary step for a more visible student community. Strategically thinking, there should be an explicit location policy to associate student settlements with the development of 'creative quarters', which currently are badly missing in Munich. At the time of writing, the city is not spending sufficient effort on this area.

The roots of this problem are in the incapacity of the city to provide room for and hospitality to students, a user group that has on average a lesser capacity to pay than other segments of city users. The issue of a subtle 'competition' for the use of the city between students and residents is reflected in a cultural 'distance' which marks the attitude of the residents towards young people and foreigners in particular. The people of Munich do not feel that they 'need' students, either economically nor socially, though they may be tolerant and benign. An example: 4,500 students live in the district of Garching, and every year they organise a festival to which local residents are invited, but the response is small. In other Bavarian cities, like Passau or Regensburg, the network of private sector, society, and students has led to challenges to the traditional way of life in these communities, making them vibrant, dynamic medium-sized cities. In conclusion, the relation of the city with the student community is not as strong and articulated as it could be, and a stronger integration has to challenge the issue of student housing. Only by becoming citizens themselves, and participating to the city life, will students become to some extent 'rooted' in local society.

Despite Studentenwerk's many efforts to tackle this problem, there is no easy way out, and the problems are bound to get worse as the general number of students is expected to increase by 18 per cent in the next five years. Most of them will need accommodation, especially students from abroad. However, the funds for these kinds of services are always short. Today solid medium-term solutions are being planned, including a student hostel, which is being extended, and the purchase of three development areas in Garching with state funds, where 200 flats can be built. The negotiation has not been without problems, but in the end the state could not afford the negative image of having students sleeping on the floor in hostels. The recent policy of the city is no longer to allow this kind of operation, although it 'tolerates' them in an emergency. In this light, Studentenwerk is trying to involve private partners in investing in student facilities, with some success. Two projects in particular appear promising: the private sector is financing a building in Bavaria for the school of architecture; and at TU, an architect from Japan will develop standard housing units. A house can be built in 24 hours, a just-in-time arrangement that reduces drastically the financial risk involved in housing development and has the potential to attract more private funding for student facilities. Three hundred and fifty of these modules are currently being completed, wholly paid for by the industry. This could be a good way of 'improvising' and could provide spin-off for other projects.

7.3.2 Relations between the Student Community and the Private Sector

Though historically the network of business and research is dense – BioRegio is an example – universities are perceived as hierarchical and therefore incapable of meeting the requirements of the information society, which are openness, networking styles of management, involvement and flexibility. As a result, the best graduate students are drawn to the USA, and this human capital is only partly recuperated in the good job market for researchers. Pioneering companies in bio-engineering and related fields had to struggle to get support and facilities. In such a competitive environment, where a limited number of centres of excellence can be found worldwide, it is easy to lose acquired advantages in the space of a few months.

Universities are hardly active players in this competition, proving to be rather static, self-perpetrating entities, without the capacity to reallocate resources among different 'chairs' to keep up with the times. The existing initiatives to match university output with local needs do not pay much attention to the 'soft' factors that add a place dimension to the job market performance. Also in this respect, the higher education institutions are not working 'for' the city. The attitude of a relatively new and more context-oriented institution like the FHM is different in this respect; they are adapting more easily to the changing requirements of society. However, this successful model does not receive the attention and the resources that it needs to become a real competitor of the traditional universities and trigger a process of change.

To a large extent, the private sector is not particularly active either. The firms that are not really rooted in Munich easily shift their attention to the human resources available elsewhere in Germany or abroad, and attract them to Munich by offering good conditions. Thus they do not have an immediate incentive to support and challenge local institutions in improving the living conditions of the student population. However, in the long term, Munich might lose in competitiveness if these central nodes are not solved and the city could lose its capacity to attract creative talent. Most of the small and medium firms are not forward-looking enough to fully realise this danger. So far, attempts by the mayor and Studentenwerk to convince the private sector to invest in students have produced poor results. Strategically thinking, companies like Siemens, whose history is closely tied to that of the city, have started investing in the quality of life and are today at the forefront of urban regeneration, but specific attention on the student community is still lacking.

To tackle this lack of dynamism, the relevant decision-makers have started to take a more proactive attitude. Through a host of creative financial arrangements and partnerships, the State of Bavaria is channelling many resources into improving the processes of knowledge dissemination and spin-offs from applied research. *Offensive Future Bavaria* and *High-tech Offensive Bavaria* are re-investment programmes of the Bavarian state government based on funds gained by privatisation of state-owned assets, totalling more than €2.8bn in the last five years. These funds are channelled into various projects and activities to foster networking between basic research, applied research in companies and support of technology-oriented start-ups, for instance through the set-up of venture capital funds. Additional public funding for start-ups comes from KfW of Frankfurt, Germany's promotional bank for the domestic economy and the official development bank for countries in transition and developing countries. Basic financing from these institutions is often topped up by private venture funds. To increase the total number of innovative founders, in addition to the mentioned start-up centres and the regional networks, the Region of Munich regularly hosts a number of fund-seeking events. Finally, the state and the city government are leading the process of rationalisation and development of the higher education institutions and other players of the knowledge sector on the territory, supporting the expansion of the poles of Großhadern-Martinsried and Freising-Weihenstephan.

The efforts of public bodies for the development of Munich as a high-class research and education location are therefore substantial, but they could be shortlived if other conditional factors are not taken into due account, as the elements that guarantee that the human-resource base for this strategy can be effectively managed. And indeed, the housing shortage problem has proved to be an important topic in the recent elections, as the inhabitants in Munich are faced with increasing competition for land and housing stock from the student and young workers' population.

7.3.3 Influence of Higher Education in the Relations between the Private Sector and the Local Community

Munich's private sector is very diverse, but nevertheless there are some large firms that played a relevant role for local development, not only economic but also social and cultural. All the leading venture investment companies, as well as numerous German and US investment companies, are situated in Munich. Research activities in the Munich area are exemplary: in 1996 more than 10,000 scientists were granted patents for around 2,000 inventions. However, under the pull of globalisation, firms which have traditionally been solidly embedded in localities are currently restructuring their relations with the local area. For instance, they care little about the social contexts of their production sites, given their increasingly cosmopolitan orientation and the flexibility of their chain of production. Could this also be the case of Siemens, one of Munich's most prestigious companies?

Though Siemens is present all over Germany, its headquarters are in Munich, along with some 30 production facilities, with a total of about 35,000 employees in the Munich area only, out of the 485,000 employees worldwide. As one of the most popular local employers, Siemens has extensive links with universities and technical schools. They target enterprising, independent and goal-oriented freshers with real marketing strategies. The marketing of Siemens at Universities of Applied Science is 30–40 years old, and is taken as a point of reference for other companies.

Along with the rest of the business community, Siemens is changing; the company is less hierarchical than it was 20 years ago, and it differentiates its product lines. That is why the focus on human resources is also changing. Nowadays the goals to be reached are set, and employees have autonomy in achieving them. This flexibility, however, is reflected in a higher volatility of the jobs; the newly employed are those who are more likely to be laid off in a recession. This gives Siemens an image problem. Students are reluctant to start a career in a company where they can be discarded just as easily; at the same time, public opinion resents the recent cuts. Owing to recent economic stagnation, Siemens operated massive employment cuts (10,000 in Munich alone), but the turnover is high: during 2002, they took on 6,500 new employees from higher education institutions.

To improve its image and be more attractive to the local human resources, Siemens has developed an extensive programme of *corporate social responsibility*. This brought them to support financially science and research, facilitating knowledge exchange and applications, and to increase the potential for innovation though the improvement of quality of life conditions for their workers. For instance, they invest in educational methods which are designed to foster creativity in primary learning. Through university projects that support internships and foster cross-border and interdisciplinary cooperation, excellence in higher education is promoted. Siemens' support however, transcends the borders of education. Investment in arts and culture should be mentioned – Siemens in this way is trying to develop a responsible image as a committed member of the society. Moreover, Siemens is active in public welfare, helping deprived groups, disabled, refugees and victims of disaster, and plays a role in prevention, helping children and supporting cultural development in foreign countries. Finally, Siemens is active in research into improving the environment and the level of safety and health in cities. The company is involved in a renovation project at an old factory site – in cooperation with the municipality – where living, working and entertainment facilities are being developed. This new housing is not specifically intended for students, though some of them, especially the top-level graduates and researchers, may certainly be interested in such developments. Again, however, Siemens pays little attention to the problem of welcoming students, and especially international students.

Apart from large and well-rooted firms like Siemens, one would argue that the real dynamism of the 'Isar Valley' is engendered by its small and medium-sized enterprise sector. Indeed, the Munich region counts some 13,000 new business start-ups each year, in sectors such as multimedia, advertising, IT-consultancy and environmental technologies. Almost 20 per cent of all German Internet start-ups are based in Munich. In a recent study by Bain and Company, the region is characterised by its 'pull as a high-tech location'. The Region of Munich actively supports this climate by offering space for business locations and by active participation in start-up networks and initiatives.

The Munich Technology Centre and four additional business areas offer 66,000m^2 for small and medium-sized enterprises. During the first four years, business start-ups pay lower rents and obtain free consulting and advice from the Fraunhofer Management GmbH. A number of large players are actively involved in the MBPW (Business-Plan Contest). This contest serves two purposes: 1) to strengthen the entrepreneurial climate by stimulating young people to set up in business; and 2) to develop new technologies and related services (for instance software). The underlying thought is to match ideas and capital.

In conclusion, the economic history of the Munich region illustrates a perfect synthesis of a highly-industrialised territory and a high-class education infrastructure, concentrated in one place and well integrated into a local network of knowledge production and application. In this context, highly sophisticated industries developed, like optics, engineering, for which the keywords are quality and innovation. As in the German tradition, the government (at many levels) is heavily involved in supporting economic development. The problem-solving approach is typical of Munich, and is deeply rooted in its cultural climate. Being from the beginning a rather liberal city, it had no need to defend itself (it was already defended naturally and was not a powerful city), and developed an open attitude, where local industries cooperate with the public sector, on one side, and the university institutions, on the other. The region is dynamic in business, but politically it is conservative and this makes the environmental context in which decisions are taken a stable, unsurprising one. But in fact this attitude does not extend to the entrepreneurial spirit. Innovation, in Bavaria, is a priority. From this point of view, the state government has a thoroughly progressive orientation.

7.3.4 Model Testing and Identification of Issues

On the whole, Munich's situation can be envied by most big European cities, and by almost all the other cities participating to this research. Munich is a wealthy, accessible and human-sized city, that offers to firms the best human capital and resources for a booming market, and to residents very high levels of quality of life. The relationship between universities and the business sector is far from disappointing, as Munich has pioneered the concept in Germany of science cluster in Europe. However, in the contemporary knowledge society, confident cities can easily lose their advantages if they do not maintain a proactive attitude and a strong governance of the local socioeconomic system. To this respect, Munich faces some dangers, which are illustrated by the negative scores in the diagram of Figure 7.3.

The first is that some of the points of attractiveness that have made the city's fortune (human size, quality of culture, traditional way of life, proximity to

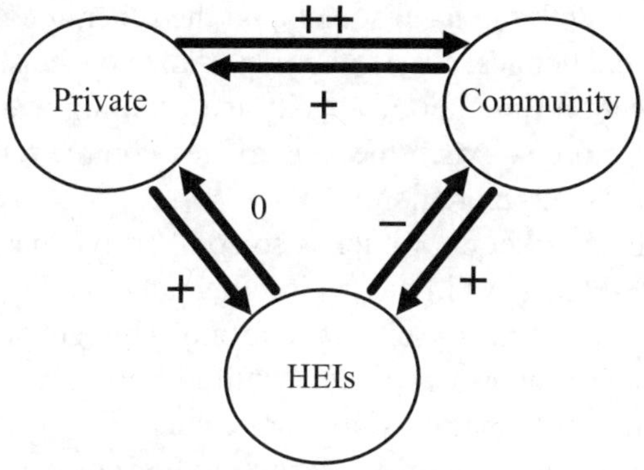

Figure 7.3 Munich a 'student-friendly city'

natural amenities) today matter less than in the past. Cities that are lively and vibrant, where fashions are created and that are architecturally innovative are more attractive, while cities that value tradition and quiet run the risk of being perceived as dull. In this respect, Munich has an image disadvantage not only in comparison with other German cities like Berlin, Hamburg or Stuttgart, but also with respect to not too-far distant European centres like Prague, Vienna and Milan. Munich is today very strong, but in the future these centres could siphon off resources and attention from Munich. At the same time the city cannot put its present attractiveness for students to good use, because there is simply not enough space to accommodate everybody. The 'creative city' environment that is nowadays a prerequisite for the success of many growth sectors, new media and culture above all, demands an integral approach to planning for a balanced, 'targeted' social mix in the city. Students in this respect have to occupy a central position in regeneration plans and economic development policies.

Secondly, the prestigious university and higher education system that is behind much of the success of Munich appears inadequate to face the new challenges of the open knowledge economy, where the tiers between research and production have to be strong and, especially, bidirectional. Today, it appears that the pull function is all on the side of business and knowledge exchange activities. The higher education institutions find it increasingly difficult to produce the resources that are needed, given their bureaucratic, hierarchical structure and the existing constraints that they face when they wish to expand physically in the city. As the scheme in Figure 7.3 suggests, the presently optimal state of the local economy could encounter a stage of crisis as the links between the education sector and the business community, on one hand, and with the local society, on the other, are not restructured to face these new challenges. Today, the importance of the student community and of the education function of the city is becoming marginal, and it risks becoming even more so when the universities complete their shift towards the periphery of the urban region.

In general, the market in Munich is so competitive that firms have hardly any resources left to invest in the environment and tend to stick to their core business. At the same time the city has enormous budget problems. However, it is felt that with a more cooperative attitude, the private sector could in fact be led to invest in quality of life and education, but they clearly need a 'guide' that is currently missing, a forum authoritative enough to convince everybody of the necessity to pool resources to this objective rather than dissipate them in sterile competition. It is also felt that this guide is hardly to be found within political circles, but it should be an initiative internal to the private and education sector.

7.4 Conclusions and Recommendations

Housing

Many people still want to come and live in Munich. This to some extent happens at the expense of an important user group of the city, higher education students and other young people, who have no easy access to the city. The private sector has no immediate incentive to change this situation. Profits in the student housing market are low, and human resources are easily attracted anyway. Therefore, in the short term it is largely up to the public sector to find solutions to the shortage of housing for students.

Today, the strategy of the city with regard to student housing consists of granting land and lowering taxes to companies building new houses, which then must offer them at low rents. The rents remain low only for a certain period (20 years), and then they increase according to the average rent level. Because the average rents in Munich are increasing rapidly, young people are progressively pushed out of this market. Clearly, a different form of housing regulation is required. However, creative solutions are also needed to diversify the market and attract investment into low-cost, good-quality housing. These are not missing, thanks to the initiatives of faculties and students at TU and the active back-up of Studentenwerk. Yet the general impression is that the municipal administration, instead of facilitating these experiments, is making things even harder with bureaucratic impediments regarding land use. This may reveal itself to be a myopic strategy. Instead, the city should have an open and integral approach to support such bottom-up initiatives; for instance, establishing a monitor office to find space in the city and being a more active 'broker' between higher education institutions and the owner institutions in the process of acquiring and regenerating these areas. Having in mind the huge works for the redevelopment of the station area of Munich, some space should be found to provide student housing at better conditions. However, an integral approach to solving the housing problem has to extend to more general issues of cultural integration between host and student community. In fact, housing opportunities are not missing, but the willingness of local residents to rent extra rooms to students at reasonable prices is rather low.

In this respect, the city administration can play an active role, for instance:

- informing the residents about the opportunities that they have to rent rooms to foreign students;
- being a warrantee in rental agreements;
- developing programmes of assistance to elderly people by students; and
- more generally marketing the student community role and potential among the local population.

Student Welcoming and International Orientation

Foreign students require support services that extend beyond just housing, for instance regarding welcoming and assistance. Finding tutors is a problem, and also making one's way to obtaining residence permits. Given that the most interesting and relevant group, but also most difficult to deal with, is that of Asian and Chinese people, a greater effort is needed in a city in which students are generally left to themselves. The attitude towards these students should improve.

In general, Munich's universities should be more internationally-oriented, developing undergraduate courses in English with more commitment. The role that associations like AISEC and the alumni networks at LMU and TU can play in the process of upgrading the international profile of the student population is precious, but it is today insufficiently developed. Until now, the city has been mostly concerned with international tourists, and little marketing of the city as a student destination has been undertaken. Today, it is acknowledged that a city marketing strategy directed at making the city attractive for young people should also focus on its educational strengths.

Student-city Environment

The effort of the city should be directed at reviving the university-private sector links. This can be done, for instance, by directing more resources towards the most 'business-oriented' schools, like the FHM Munich and some faculties and interdisciplinary groups in the two universities. At the same time, the private sector should be led to invest more in the city and in its higher education institutions, and this can be achieved by stimulating more cooperation.

The need for more space and high quality laboratories has clearly to be balanced with a care for the environmental factors that contribute to the attractiveness of an education institute. In Munich, it seems that these problems have not been given enough attention; service and business functions are 'chasing away' from the city centres other less profitable (but not less important) functions like education, and this may turn out to be a drawback. Today, instead of an overall planning of such development at a metropolitan scale, the tendency is for a new competition between core and periphery, that dissipates rather than bringing together the resources for a 'rediscovery' of the city by obliging students live their Munich experience as commuters.

As a notable exception, the project *Architektur und Kunst*, which received support from Studentenwerk, is a promising example of how knowledge produced in a university can be 'filtered' down to the society, achieving many objectives. This architecture contest focused on improving living conditions for students in the district of Oberschleissheim through original planning and design. In this way students are made responsible for the management of local resources, solving to some extent the conflicts with locals over the use of land, and at the same time leaving a valuable mark of their presence in the territory. It also gives visibility to the students, who thus have a chance to be perceived as a important group of city users. To favour such 'holistic' initiatives, which could really make a difference to the way in which the city opens up to a young segment of the population, requires soliciting open-minded people within the administration to give suggestions and implement them, pooling resources wherever they can find them.

Yet it is felt at many levels that rather than just shifting student life towards the periphery, the city should organise the conditions for a better integration of the student population in the city centre. This could be done, for instance, by providing affordable student residences – but also amenities and cultural facilities – in run-down neighbourhoods that are already popular among students, which could greatly benefit from a more visible student settlement in terms of economic regeneration and image upgrading.

It is remarkable that a 'campus' structure could be achieved 'virtually' rather than physically. LMU developed Campus^{LMU}, a platform that gives interactive access to any information about their courses and on-line contents to 40,000 students, allows the development of their own webpages, and virtually eliminates administrative paperwork. Today Campus^{LMU} is accessible through 1,000 terminals and from private homes; for the future, the university is planning to expand 24-hour access points in the campus.

Governance

The universities and higher education institutions of Munich should have a leading role in revitalisation processes, looking at future challenges. For instance, they should be more decisive in supporting the dynamism of strategic sectors like the biomedical, looking closely at the market and its requirements in terms of new knowledge and human resources. This involves the careers of the lecturing staff. The city could play an active role in offering additional research funds (e.g. an annual prize, for which sponsors have to be found) to those departments that are shown to be really working *for* the development of the business community. Also, the mobility of staff between the education and private sector could be improved with more governmental backing. However, some things are already going in the wrong direction. The Bavarian state has changed the system of property rights on innovations, which now belong to the university and not the staff, with the result that efforts in research are downplayed.

The likeliness that the university system will change and adapt to the evolving environment depends on a dynamic leadership, an active president, strong and diplomatic but with a vision, whose main task would be that of convincing the university community to work more closely with the city and the private sector, negotiating with the former the possibility and modalities of expansion, and with the latter the new specialisations and services to be offered to students. The higher education institutions in Munich are not linked through any kind of platform or organisation as happens in France with the PULs. Such an organisation – to be formalised and given real planning powers – would probably be an effective tool to negotiate development in the region with the administrations, and to collaborate in the reformulation of the internationalisation programmes of the city, which has, of course, a bearing on the housing infrastructure requirements, and on the quality of the services offered to foreign students and staff.

The creation of a cohesive 'knowledge region' demands that various administrations cooperate strategically with scientific institutions, developing networks and synergies. Plans in this sense are more likely to be found in other regions than in Munich, where everything seems to be working well already. A more focused approach is needed. The organisations of special 'scientific areas' within and outside the city boundaries where education, research, start-ups and living all coexist, should become a standard approach to urban policy, and the mayor of Munich could be the initiator of this strategy. The city and the region should support these 'scientific areas' by showing some financial commitment, and endorsing the resolution of the issue of student housing.

Note

1 At the time of writing, Studentenwerk is investing *c.*€100m to expand to 1,600 units the number of students that they can accommodate in their residences, 150 of whom will stay in Garching, while the others will be in five complexes of the 14 residences in the city centre. Another four residences are located in Freising and two in Rosenheim.

Chapter 8

Rotterdam Case Study

8.1 General Introduction to the City of Rotterdam

Rotterdam is located in the southern part of the province of Zuid-Holland, and it is the heart of the Rijnmond Region, including smaller municipalities and the port (Figure 8.1). Together with Amsterdam, Utrecht, the Hague and a few smaller cities, the Rotterdam metropolitan region (1.4 million inhabitants) belongs to the Randstad, one of the largest urban areas in Europe. This atypical ring-shaped urban region houses about two-fifths of the Dutch population (corresponding to six million people) and holds a dominant position in the economy, in particular in the tertiary sector.

In the wake of the flourishing of port activities in the nineteenth century, Rotterdam's population grew to 732,000 residents in 1965. From the 1970s onwards, social, political and economic factors favouring spatial

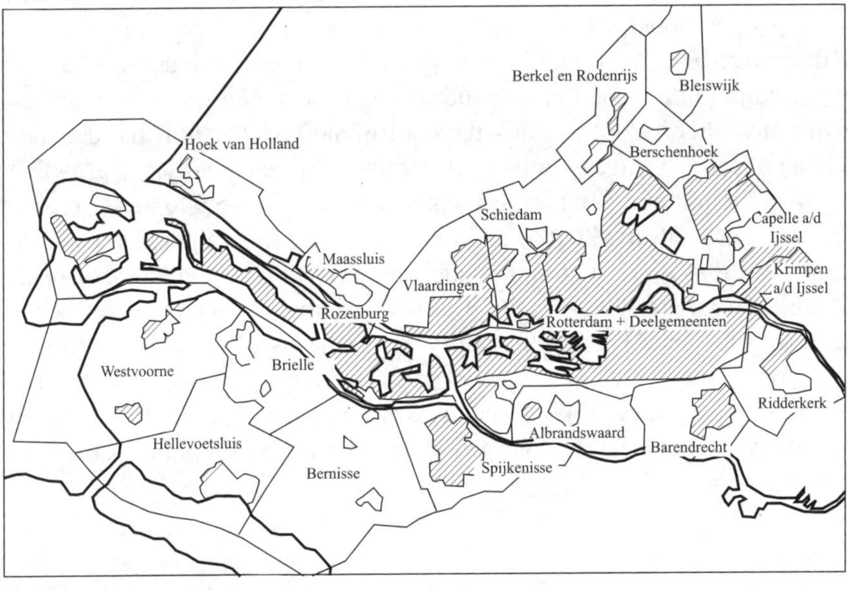

Figure 8.1 The Rijnmond region

deconcentration caused a net population decrease of some 160,000. This, in combination with the effects of the energy crises of 1973 and 1978 and the decline of the port, resulted in serious social-economic problems. Drastic policy changes on the national and local levels helped to turn the tide, and stabilised the population to the present 660,000, in a trend of slight growth. Today, Rotterdam is the second largest city of The Netherlands.

The population of Rotterdam is relatively young compared to the rest of The Netherlands. About 200,000 people are either studying (approx. 180,000) or teaching (20,000). Rotterdam is a truly multicultural city, which is one of its points of distinction. In a number of Rotterdam's municipalities, more than 80 per cent of the residents aged between 0 and 23 years are of immigrant origin. Overall, some 43 per cent of the city's population are non-indigenous, and of these, 65 per cent are ethnic minorities (EUA, 2001). Most people speak two or sometimes three languages.

The number of job opportunities in Rotterdam amounts to 284,000. Strikingly, about half the jobs in Rotterdam are held by people from outside Rotterdam, mostly from the Rijnmond region or the rest of Southern Holland. In Table 8.1 the economic structure of the Rotterdam economy is shown. Traditionally, the Rotterdam economy was dominated by the port-related industries. The transport and distribution sector are particularly strong, as are the financial and business services sectors that are often associated with maritime operations.

In the last 20 years, important changes have occurred: industrial activities such as shipbuilding and the food industry have lost their dominant position. As in many other European cities, the share of industrial activity has declined, and that of service activity grown substantially. More recently, the growth of employment in the region has been considerable, with 46,940 newly created jobs between 1996 and 2000. The impressive increase in the financial sector lies behind this score, but significant improvements are noted also in the culture and tourism sector. In the Rijnmond region as a whole, employment increased substantially as well, mainly due to the growth of existing firms in the region. The city itself is slowly but steadily changing from a grey industrial hub to a world-class commercial centre, whose mixed population and cultural eccentricity is increasingly regarded as an asset for development rather than a disadvantage.

Table 8.1 Economic structure of Rotterdam, 1996–2000

Sector	Employees 1996	Employees 2000	% change 1996–2000
Transport, distribution/communication	40,512	40,902	1.0
Financial services	16,590	19,047	14.8
Business services	33,260	60,162	80.9
Public administration	12,806	15,442	20.6
Education	17,712	20,220	14.2
Health and welfare	35,360	39,585	11.9
Culture, recreation and other services	9,190	12,567	36.7
Horeca	7,174	8,553	19.2
Agriculture and fishery	501	675	34.7
Energy extraction	9	–	–
Industry	31,157	30,105	–3.4
Public utilities	1,853	1,761	–5.0
Construction	15,020	16,228	8.0
Wholesale	15,523	16,151	4.0
Retail	22,627	24,836	9.8
Total	259,294	306,234	18.1

Source: COS, 2000.

8.2 The Student Community and Knowledge Infrastructure in Rotterdam

8.2.1 General Information on Higher Education in Rotterdam

The knowledge infrastructure of the Rotterdam region, including the important academic pole of Delft, consists of a large number of institutions. We discern five types: 1) universities; 2) higher education institutes (HBO); 3) research institutes; 4) large firms; and 5) others (including consultancy firms, engineer's agencies, etc).

8.2.2 The Knowledge Sector of the City: Actors and Spatial Patterns

Universities and other higher education institutions The region of Rotterdam hosts two universities: the Erasmus University of Rotterdam (EUR) and the Technical University of the neighbouring city of Delft (TUD), some 15km to the northwest of Rotterdam. Both universities offer a wide variety of education facilities and programs, and are heavily engaged in research, both fundamental and applied, in many different fields. The EUR offers economics and business, law, philosophy, history and art, social sciences and medical and health sciences courses. In Table 8.2, the number of Erasmus students and graduates are indicated. All faculties except medicine are located in the Woudenstein complex in the east of Rotterdam. Medicine students are hosted in Hoboken, near the Academic Hospital in Rotterdam-West. TUD offers only technical education, and is very popular among architecture students from Europe and beyond.

Table 8.2 Erasmus University, number of students in the academic year 2000–01, first year, total and graduates

Programmes	1st year students	Total students	Graduated (in academic year 1999–2000)
Economics	1,005	4,122	437
Law	757	3,271	265
Business administration	941	4,309	388
Medicine	299	1681	*435
Public administration	168	671	87
Econometrics	62	376	54
Health policy and management	120	460	36
History of society	75	331	30
Social Sciences	83	303	25
Art and culture studies	78	274	26
Philosophy	89	201	9
Modern Japanese studies	3192	9	
Total EUR	3,708	16,091	1,366

Source: 'EUR at a Glance', http://www.EUR.NL/eur-uk/eur/index.html.

Both institutes organise courses in English for both undergraduate and postgraduate students, participating extensively to the Socrates exchange programme. A number of Masters and specialisation courses at the EUR are directly relevant to matters of urban management, like the Masters in Management of the European Metropolitan Region, the Masters in Port Economics and Logistics, and several specialised courses with in the Masters in Business Administration. The IHS (Institute for Social Development, associated to both the EUR and the TUD and located in the Woudestein campus) also organises Masters courses in urban topics, which attract a large number of foreign students and government officers every year.

In the rest of this study, due to its relevance as an institution 'of' the city, we focus our research on the Erasmus University. However, the reader should be aware that TU Delft is becoming increasingly integrated in the Rotterdam economy through the intense links that many companies and departments have with the laboratories and research teams. For an 'architecture city' like Rotterdam, an intense knowledge exchange with the Delft School of Architecture is more than obvious.

Rotterdam also counts several institutes for higher education, so-called HBO in the Dutch education system. The data show that this education sector is the most dynamic: Figure 8.2 compares the growth of HBO students with those in university education. 3,300 students have obtained a HBO diploma in Rotterdam in 2000. The HES is specialised in economic-related education and hosts about 5,000 students, 1,200 of whom graduated in 1999. The HR&O (High School of Rotterdam and Environment) is by far the largest institute, and offers a very broad range of education, including technical, economical, medical and design.

Possibly the most interesting HBO for our study is Ichthus. The School, which has recently moved to a fancy new building in the Kop van Zuid area

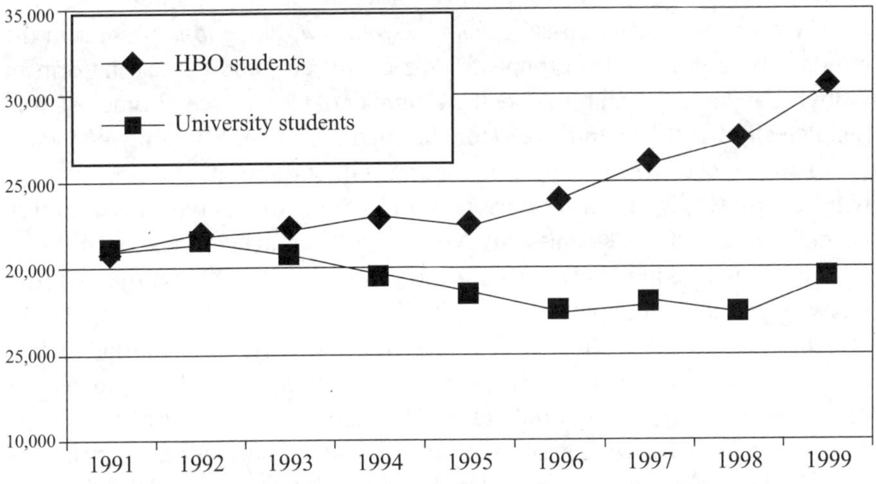

Figure 8.2 Students at university level and HBO, evolution 1991–99

Source: COS, 2000.

opposite to the new Luxor theatre, is particularly strong in business economics and communication. Ichthus in Rotterdam has some 8,500 students, 400 of which are international students, eight times as many as three years ago. Ichthus is becoming more and more market oriented. However, the school offers a learning environment that emphasises the social component. The original Christian tradition was left behind consciously to broaden the student base, and now the school is extremely careful to support the cultural development of strong minority groups, who are invited to share ideas and pastimes with Dutch pupils. Recently, Ichthus and Hogeschool Rotterdam (including the HES School of Economics) merged in the *Hogeschool Inholland* conglomerate.

Research and technology transfer With the exception of the universities and The Netherlands Economic Institute (NEI), there are no major research institutes located in the Rotterdam region. However, the large firms in the region are suppliers and producers of knowledge as important as the HEIs. Transnational companies like Shell, Unilever and Robeco have their own research and development departments, and transfer knowledge and technology to the network of their suppliers. With 24,131 employees in 6,514 firms in 1995 (De Langen, 1998), knowledge-intensive business services have grown substantially over the past decade and they are still in full swing. These play a very important role in the region as a source of knowledge for other firms, possibly larger than that of universities or research institutes.

There are also many so-called *intermediary organisations* present in the region. Some represent a group of firms or a sector and act as platform or lobbying apparatus. Examples are the Chamber of Commerce, Prometal (from metalelectronics), Het Initiatief (for the media sector), Stichting Waterstad (tourism), and the Recycling Foundation. Others aim at bridging the gap between knowledge infrastructure and firms, or at stimulating innovation in general (Syntens; Delft Kennisstad). A number of units at the education institutes that act as intermediary between education and business. Other projects aim at knowledge and technology transfer, in which regional actors participate. An initiative particularly worth mentioning is the foundation 'Kennisinfrastructuur Port Rotterdam' (Knowledge Infrastructure Port Rotterdam). The objective of the programme was twofold: 1) to stimulate concrete improvements and renewal in the knowledge system of the city and the port; and 2) to stimulate the coherence, pairing and cooperation between existing knowledge-initiatives in the region, aimed at the strengthening of the port.

The private sector is also variously involved in projects that stimulate in-company training of technical students and start-ups. To name a recent one, the Technocenter in the suburban district of Krimpen aan der Lek received a €18m subsidy from the Rotterdam business community.

Spatial patterns of higher education and research Knowledge functions are only partially an 'inner city' function in Rotterdam: the main establishment of the Erasmus University, as well as the main research institutes and knowledge transfer facilities, are at the city's edge. However, Erasmus University's medical school and some of the most important HBO schools are in central locations, as is the bulk of student residences.

Because of the ongoing concentration of higher education studies in Rotterdam, more and more students, and especially international students, are looking for accommodation in the city. This puts pressure on the housing market. In addition, large companies seek accommodation for their trainees. In this context, the usual tools of state-subsidised housing are no longer so easily available. However, international students would pay up to €800 a month for a room if they were not subsidised. Hence the necessity to be creative in the organisation of projects that put together student facilities to be supplied at convenient rates with other, in-the-market initiatives, that may exploit the high capacity to pay of firms and young professionals. One example is the new combination of offices, student housing and penthouse flats in the Kop van Zuid area, in close proximity to the new Ichthus building. Another mixed-use development pertains to the growth of the University Hospital.

The question remains of whether these uses mix well together; for instance, student housing has proved to be not completely compatible with residential functions in other areas of the city. However, the Rotterdam Development Corporation believes that mixed-use development manages to promote social development even in a market context, and the results achieved so far in Rotterdam's development policy have been mostly satisfactory.

The city government has two places in mind to develop as 'student areas'. One is in the vibrant western part of the city, with an eye to young, single tenants; and the other is the cultural cluster of Witte de Withstraat, already a popular meeting place for students and young professionals attracted by the cultural dynamism of the city.

An important actor in the development of the education function of the city is Stadswonen. This private organisation delivers services such as housing and catering facilities to students. In recent years Stadswonen has evolved from a university-subsidised, semi-public organisation to a private company with a vision of housing opportunities for students and young people. Like similar organisations in this study (e.g. ESU in Venice, Studentenwerk in Munich) it is diversifying its mission, supplying non-students target segments. This step becomes necessary as private investors reduce the funds for student residences, and changing patterns in the students' housing market (more heterogeneity, attention for quality, purchase power) demand differentiated responses. Stadswonen enjoys good networking with private investors and local building corporations, and on this account they could enlarge and upgrade the housing supply for students in the city, for instance by restoring historical buildings.

The key for Stadswonen is customer orientation – hence the tendency to endorse the students' preferences, which revolve around location and quality rather than price. They are also convinced that students are a strategic market; once they have finished their studies, they could become 'up-market' tenants. Stadswonen is therefore explicitly building in the perspective of 'seducing' fresh graduates into the city, and some projects, like the renewal of the central area of the city as a 'Latin Quarter', point in this direction. Stadswonen is expanding its accommodation supply (at a planned rate of 10 per cent a year), but this hardly satisfies the demand for apartments in central locations. Though they are not the only suppliers (the private sector and other organisations are also active in different segments of the market), there are waiting lists of nine to 12 months for a room. Some exemplary student apartments like 'Het Pakhuis' (on the Kop van Zuid) represent a dilemma for the public sector; the apartments are so nice, that postgraduate students do not wish to move out after the completion of their studies. However, this is proof that policies that enhance students' quality of life have a durable impact on the city's human resources. In many respects, therefore, Stadswonen has developed an autonomous vision of a student-friendly city, which calls for attention of the city planners.

8.2.3 The Student Community: Behaviour and Careers

When faced with the issue of profiling the student population of a city, we should first define it precisely. This is by no means an easy task, due to the ambiguity about the 'relevant area' (should the education institutions of Delft and Den Haag be included in this analysis?), and the increasing 'transversality' of learning people with respect to age and status. Indeed, official accounts of the higher education sector seldom take into account 'continuing education' and 'in-company' activities, which bring many adults and professionals into the ranks of students.

However, in-depth information regarding the university student population is not diffused either, and this is of prime importance for an explicit city policy which addresses the Rotterdam students. To this aim, a survey has been carried out with a sample of more than 300 students at the Erasmus University in order to identify the most important tracts of a substantial segment of the student population of Rotterdam.[1] The information derived from this survey helps address the main issues regarding the student population's spatial behaviour, students' perception of the city, and the quality of their life – hence the need for specific policies tackling these factors.

Housing Many Rotterdam students (42 per cent) live at home with their parents, and commute to the university during the day from their place of residence, that is in the majority of the cases either the city itself or its immediate surroundings (Rjinmond region). Those who do so because they cannot find a room or afford the rents charged are a minority in this group. In most cases, they feel they have no reason to live on their own, and the university is reasonably close to their family house.

The students who live on their own come from Rotterdam in less than 10 per cent of cases, 38 per cent from within the region and the rest of Holland or abroad in the majority of cases. A large share prefers the neighbourhood of Kralingen, traditionally a student area, close enough to Erasmus University to be reached by bicycle or even on foot, and apparently far safer and quieter than the rest of the city centre. However, this district is also one that presents the best features as regards housing and quality of life, so it is not surprising that it has become difficult to find accommodation there, and that rents, especially compared to the quality offered, may be quite high. At the same price, students can obtain better quality houses in the new peripheral districts; in the the lively districts of Croswijk or Oude-Noorden, they can save €30 a month. Most people are happy with their accommodation, whatever price they pay. Accommodation in Kralingen, the Centre and Oude Noorden is the most appreciated. Kralingen also hosts most of the foreign students in the popular 'International Student House', located there. There is a negative side to this: since this location is extremely close to the university, foreign students have little incentive to find their way to the city centre or other interesting districts, and are not integrated into the local community during their short stay in Rotterdam.

Most students (38.7 per cent) obtained their room via informal channels. Few obtained their rooms via the dedicated organisation. Ten point four per cent of the respondents obtained their room via an advertisement, and in this way they managed to obtain cheaper rents (on average €185, compared with €240, which is the average rent paid for Stadswonen-rooms), even though the quality obtained is lower on average. Foreign students pay on average much more than Dutch students do, probably because they are more likely to go through formal channels like Stadswonen in search of accommodation, but also because on average they demand higher standards (and are less satisfied than their Dutch counterparts).

Study and living habits The preferred meeting place for students seems to be the 'club' or 'disco-bar'. There are many such facilities in the central and western districts of the city, as well as other popular places such as restaurants, bars and cinemas. Few students make regular use of cultural facilities, despite the good supply of such assets in the city. In one year, a typical student goes fewer than eight times to the cinema and twice to visit an exhibition or a concert. There is in this respect a remarkable difference with foreign students, who – possibly because their time in the city is more limited and they enjoy a one-off experience – pay more frequent visits to cultural events, and in general demonstrate a more positive attitude towards the arts. The average number of weekly hours spent on sport courts is almost four. Belonging to sport or student associations is also a popular social opportunity, and the membership rate increases with the year of study. Students' unions can be a way to meet other people, especially when you move to a new city and have no acquaintances or family in that city. A quarter of the students of the Erasmus University are members of a student union.

A substantial share of the students are in paid employment, either full- or part-time, and in this way they have a source of income that covers part of their living costs. This attitude increases with the years of study. Students who live with their parents, however, also have jobs. The average workload per week is low, at almost 12.5 hours, rewarded at an average of €8 per hour, yet slightly more that the national average (Studentenmonitor, 2000). This may suggest that Rotterdam is an expensive city: students have to work more in order to pay for their life styles; on the other hand, the salary per hour is slightly higher, compared with the €6.6 on the national level. A recent enquiry among a small sample of university students (*Erasmus Magazine*, 2002) affords a better insight into the budgets of university students. They can count on a budget of approximately €670, half of which, on average, comes from part-time jobs. Students, on average, seem to live above their incomes: they spend approximately €830. Housing costs do not account for more than 30 per cent, whereas 'leisure' expenditure accounts for 16 per cent.

Students use trains and buses to travel to Rotterdam (54%), and trams, the metro and bicycles to get to the university. Students living on their own use the bicycle more often. From the districts of Kralingen, Crooswijk and Prins-Alexander, all close to the university, 78 per cent of students come by bicycle.

Choice of city Students have many options regarding the choice of their study location. Economics can also be studied for instance in Tilburg, Groningen, Amsterdam, Maastricht and in other foreign cities. A research project by the Ministry of Education (VVAA, 2001), regarding the factors influencing the first-year students' choice of a university, shows that the image of the university and of the city, the proximity of the university and the attractiveness of the student life are the most important. But what were the reasons for students of Erasmus University to study in Rotterdam?

As Table 8.4 shows, the attractiveness and the proximity of the university are the most relevant factors for the students to study at Erasmus University. The image/attractiveness of Rotterdam count in this choice for only 11 per cent of students, against a 17.4 per cent national average (VVAA, 2001). Students living with their parents stayed closer to their university of origin, so the proximity of the university was more important in the choice of Erasmus University than for those who live on their own. Foreign students who choose to come to Rotterdam do so primarily because they value the university's reputation (20 per cent), and very few because of the attractiveness of the city (less than 4 per cent). Therefore, there are gains to be made by promoting the city better among potential incomers.

Table 8.3 Reasons for studying in Rotterdam

Reason for studying in Rotterdam	Living situation On my own (%)	With my parents (%)	Total (%)
Attractiveness of city	12	10	11
The university	49	38	44
Friends/family	11	7	10
Proximity	26	61	41
Numerus fixus	9	2	6
Other reason	15	7	12

After their period of study, students choose a place to settle and work. Factors influencing such choices are the location of future job opportunities and the attractiveness of the potential residence. Only 8 per cent of the students thought that Rotterdam would play an important role in this choice. Apparently, the city as such is less important. It appears that age, the study orientation and the year of study are not related to this choice.

Perceptions about the quality of life When they are asked to grade different aspects of Rotterdam's life, it can be seen that most students are satisfied with the shopping facilities in Rotterdam. Opportunities for going out and meeting places are also very much appreciated especially by open-minded students who may come to Rotterdam from a very wide region for the reputation of its clubs and cafés (Rotterdams Dagblad, 2002). The worst aspects in the perception of the students are safety and the cleanliness of the city. Male students feel safer than female students, but the difference is very small. In particular, the perception of the city from people living in 'good' neighbourhoods is low.

The average grade for housing in Rotterdam is 6.7. Kralingen, Oude Noorden, Prins-Alexander and the centre appear to be the best appreciated areas to live. The average grade given for public transport in Rotterdam was 7.1. On the other hand, Rotterdam is not considered a clean city (a score of 5.6).

The perceptions of foreign students differ remarkably from those of Dutch students. For instance, the average grade for 'safety' is rather high (6.9), as well as the one for the cleanness of the city (6.7). Evidently, they are used to lower standards in their country of origin. On the other hand, they do not think that working in Rotterdam scores well (an average grade of 6 compared with 7.2 by Dutch students).

Integration with the local community To derive information about this particular issue, some 'proxies' have been used. The fact of being officially registered as citizens (and thus the possibility of voting for the local representation bodies), the use of facilities in Rotterdam and membership of associations and unions are considered a sign of integration into the local community by students of Erasmus University. In this respect, two-thirds of the students living on their own are registered at the City Hall as residents. The use of the city's resources like libraries, theatres, cinemas, etc., is also widespread, but it is higher among registered students, giving the impression of a fundamental difference in attitude towards the city between these two groups. Even if integration in the student scene says little about integration into the community, it is observed that also being members of students associations is more frequent among registered students, while the contrary holds as regards sports' associations.

Synthesis Rotterdam does not lack entertainment and cultural opportunities when compared with other cities, as in the last decade Rotterdam has developed pleasant attractions like festivals and cultural activities. However, Rotterdam lacks accommodation capacity in those areas that are the best appreciated by students. Moreover, basic facilities like transport, safety and cleanliness are judged to be below average, with repercussions on the quality of life perceived by students. This also extends to the facilities supplied by Erasmus University, like free study places, a cosy atmosphere and on-campus facilities like pubs and PC laboratories. Many students who did not know the city before, especially the international ones, are much more positive after they have lived in Rotterdam for some time. After a year they like the city and they become informal ambassadors in promoting the city to others; it is therefore advisable that the city actively supports this 'networking' function together with the university.

8.2.4 Challenges for the City

The seaport complex is a major pillar of both the Rotterdam and Dutch economies. In spite of the recession that hit hard in the 1970s, a wise policy of support for investment and socioeconomic regeneration allowed Rotterdam to maintain its position as the main world port. In the mid-1980s major changes occurred. The economy was once again thriving, and Rotterdam profited from the renewed interest in major cities as locations for business services and as places of residence for higher income groups. Non port-related Rotterdam companies began to invest on a large scale in new office buildings, in particular in the town centre, which thus acquired an increasingly international status.

In total, nearly 1,350 internationally active companies are established in the region. Most of them belong to small-scale enterprise. Compared with the other three major Dutch cities – The Hague, Utrecht and Amsterdam – Rotterdam boasts a strong representation of the internationally operating building, transport and distribution sectors.

A comparison of the economic development with that of The Netherlands as a whole reveals that Rotterdam lags behind on several other points. That is due in part to its sector structure: the city has a generous representation of sectors that are not fast growers nationally either, such as industry and transport. But even within the sectors, growth is lower in Rotterdam than in The Netherlands as a whole. In terms of returns, turnover and profitability, too, ventures in Rotterdam are weaker than the national average. From figures measuring R&D input and output, the innovative capacity in the Rijnmond region seems rather low in comparison with other industrial regions in The Netherlands. Moreover, most R&D seems to be accounted for by a few large companies. The innovative capacity of the dominating small-scale enterprise is disappointing. More effort should therefore be put into the existing 'knowledge potential' in the Rotterdam region.

Economic regeneration in the 'post-industrial' city The economic benefits of the expansion of the port of Rotterdam accrue mainly to inland regions, where transport and distribution activities increasingly locate (van den Berg et al., 1997). With this in mind, the economic strategy of the municipality is directed to the broadening of the economic base of the city. In particular, the municipality focuses on a few sectors that are regarded as potential generators of economic and employment growth. A prerequisite for Rotterdam to develop whatever innovative functions is to host excellent higher education institutions, but this is not sufficient. They also have to produce a capital of knowledge and human skills that is as rooted in the local environment as possible. One of these sectors is the audiovisual industry; others are medical technology and the recycling industry; a particularly interesting development for this study is the leisure sector.

In recent years, the city developed more and more explicitly into a city of art and culture, within a strategy pushing the soft, 'knowledge' functions in contrast with the 'heavy' investments of the past. The city's appointment as Cultural Capital of Europe for the year 2001 can be seen as the best momentum for this strategy. The development of a strong cultural tourism sector, perceived as an engine to improve radically the quality of the living environment, is now an objective which matches in importance – and strategically integrates – the physical redevelopment. The policy makers recognised that the enormous potential of the city stands on the development of innovative and modern tourist products – on the verge of being 'futuristic' (van den Berg, van der Borg and Russo, 2003).

The flood of new buildings, including the highest office building of The Netherlands, the towering 150m-tall headquarters of ING, a new casino, a new theatre, some exclusive high-rise apartment buildings, etc., have given the city core an aspect of modernity. The metamorphosis is mostly due to private investment. The shopping centre has been given new impetus thanks to a partially low-level shopping street.

Corresponding with this outstanding hardware is a vibrant, on-the-edge atmosphere, centred on the emerging cultural cluster of Witte de Wittestraat. Several festivals have been added to the already renowned Film Festival,

Heineken Jazz Festival and Poetry International, the Summer Carnival and the Rotterdam Marathon. The main consumers of these new urban products are the young, students and outsiders, and the visitors who choose Rotterdam as their destination on the basis of special interests. There seems to be little integration between the new face of Rotterdam and its residents, as the low levels of participation of the indigenous population in the events held on the occasion of the ECC 2001 event show. The result is that the whole process gives the impression of being artificial and shortlived (Richards and Hitters, 2002).

Social revitalisation, education and redevelopment Rotterdam has low education levels (cf. Table 8.4), and a more serious situation with low-income groups. University graduates and those with professional education diplomas are less well represented than in the rest of The Netherlands or the other big cities in the Randstad, both in relative and absolute terms. Though social policies in the last decades tended to avoid the emergence of 'problem neighbourhoods', some areas of the city are in need of regeneration. They have been targeted by various revitalisation policies, aiming at functional mix and at the involvement of private investors in social projects.

Table 8.4 Composition of the population per education level, in Rotterdam and in The Netherlands, 1996

Education level	Rotterdam	Netherlands	Other cities in Randstad
Base education	26	16	19
Secondary schools	25	25	18
High schools	32	39	31
Schools for professional education	10	14	16
Higher education/university	7	6	15

Source: CBS (EBB'96) SCP-bewerking.

The master plan for the Wilhelmina Pier, the economic gravity centre of the Kop van Zuid, is based on an integrated residential and working area with a number of leisure and other urban facilities. The plan put particular emphasis on the relationship between the project area and the surrounding neighbourhoods, among the socially, economically and aesthetically weakest in the whole conurbation. The new buildings are real eye-catchers on the Rotterdam skyline. Other residences, port-related offices and a theatre have settled in the area. Among them is a student house, which demonstrates how policies providing for the student population are perfectly compatible with general regeneration objectives.

Thanks to these developments, Rotterdam is slowly but surely enriching its status as a city offering high quality opportunities for shopping, entertainment, sports, culture and art. The inner city has regained its position as regional shopping centre. The Erasmus Bridge is the symbolic milestone of the new face of the city. Rotterdam is associated with a dull image, hence the importance of adding colour and atmosphere. From this point of view, the synergies between the vocation of the city as a research and education centre and the development of a vibrant cultural scene are high. However, most of the new projects seem to be having difficulty in defining a user mix that is really consistent with the aspirations of the city. Tourists are not flocking to Rotterdam, despite the investments in culture and hotels; and the Kop van Zuid, despite its renovated image, is practically deserted at night.

In the process of redefining the socioeconomic profile of the city, the student population should be a key factor, both as a user group, and as a capital in terms of knowledge. This resource base is an increasingly important factor in determining the location behaviour of firms, and therefore competitive cities have to make the best use of it. From this point of view, students may be beneficial to the city in the long term. Also, students can contribute to solving the social problems that still plague Rotterdam: getting students to live in run-down districts may upgrade them, as is shown by the attractiveness of Rotterdam West as a student residence area. It should also be noted that because of their flexible behaviour, students represent an ideal human resource pool for many industries connected to the tourism, leisure and entertainment sector, which relies heavily on low-skilled, part-time jobs.

8.3 Strategies and Integration. Rotterdam as a 'Sustainable Student City'

Rotterdam faces the challenge of upgrading its human capital and knowledge base to a level that is consistent with the ambitions of the city in accomplishing its transition to a post-industrial, 'light' economic structure. This objective is not easily achieved, first of all because of the persisting cultural legacy of the 'port city' in its very DNA, and secondly because of the relative isolation in which the three ends of the 'sustainable city triangle', community, students and private companies have lived the city in the past.

8.3.1 Relations between the Higher Education Community and the City

The community of Rotterdam can be variously described as parochial, conservative, and working class. The strong suburbanisation of the higher-income families to the peripheral district following the urban crisis of the 1970s and the consequent social imbalance in the city centre are barely matched by an inflow of young professionals attracted in the city by cheap house prices and a stimulating cultural environment. For many years the city has lacked a real centre, that is, a meeting place where people spend their free time and nights out. In latter years, the opening of commercial malls in the centre and, more importantly, the revitalisation of important districts in the city around the Museum Park, offer more opportunity for contact, but still this seems to apply more to tourists and students that to the real 'Rotterdamers'. However, students (especially foreign students, who, according to the survey presented earlier, are more eager to 'experiment' with all that the city has to offer) may prefigure a tendency, that of the reconcentration of the socio-cultural fabric of the city towards the centre. It is interesting to note that the main group of attendees of the cultural events organised on the occasion of the Cultural Capital Year were young people, not necessarily from the city itself, but temporarily residing here. The attitude of the residents towards the student population and Rotterdam as a 'cultural city' has also changed over the years. From an initial lack of enthusiasm, there is now a certain awareness of the potential that the development offers for the solution of some of the city's social problems, such as the difficult integration with the ethnic communities or the economic transformation of the city. The extraordinary participation of the citizens in events such as the Dunya Festival each May or the Harbour Days in September seems to prove this assertion.

Student settlements generate benefits for the local community, but also costs, as is argued in the introductory chapter of this study. In some parts of Rotterdam student housing is not an issue. Moreover, the impact of students on the Rotterdam's economy is quite small compared to the impact of student populations in some other cities. Even so, the high concentration of students in some neighbourhoods is problematic. This explains the halt put by local organisations to the expansion of student settlements in Kralingen. It has been necessary to draft a 'covenant' between the student associations located there and the citizens, in order to regulate issues like the frequency and organisation of parties, or information on such events. However, other experiences in the city show that a good balance should be sought, and that 'mixing' is still the key word when planning for student settlements. One should not forget how once run-down districts in the west and north have improved considerably when the student presence started to be consistent, as a result of the lack of opportunities in other parts of the city.

Rotterdam has an image problem compared with other university cities: it is mostly seen as a big, dirty harbour city. Yet an ongoing change is happening; world-class and 'trendy' institutions like the Dutch Architecture Institute and the Berlage Institute of Architecture are now located in Rotterdam, and five large architecture firms have moved here. This clearly supports some synergy between development of a specific vocation for the city and its knowledge infrastructure. The ideal setting for a city-university synergy is that of an 'open environment' where knowledge circulates and network relations can be freely shared. At this moment the network is too small. The faculty and school directors are regularly summoned to the Education Department of the city, but issues concerned with physical and strategic development are barely touched in these meetings. Ichthus has developed a good network with companies, (municipal) institutions and schools, and plans to do more with the harbour organisations. The Faculty of Economics has recently decided to develop a long-term strategic relationship with the Rotterdam Development Corporation. This promises to improve the coordination of city and universities as regards important aspects such as development plans, support to the development of strategic sectors, and city marketing.

However, today the students are not incorporated enough into the human capital and customer base of the city, and this means that the city is not making the best out of the potential resources available, nor is it preparing the ground for an expansion of this potential. On the contrary, many bottlenecks in this respect are already visible, for example in the housing market, where the serious shortage for international, short-stay students risks affecting the success of new courses and programmes. About 30 per cent of the students have a non-native background. International students are supposed to play an integrating role, but currently this is not happening. It could perhaps be achieved if a higher proportion of local people attended the university, but it is a long-term process in which the city is already playing a part.

To be significant actors for local development, higher education institutions should also be relevant from the social point of view. They should put their own expertise and physical presence at the service of the community, even in the neighbourhood where they are located, balancing in this way the tendency to be agents of 'gentrification' or cultural conflict. On the other hand, as the theoretical framework for a sustainable city-university relationship suggests, the university can 'use' the local context as a laboratory for research. Obviously the social and human sciences have most promising links with the direct environment, but scientific faculties may also be important. A good example is offered by the medical faculty, which manages the most central of the city's hospitals. The students are exposed to urban medical problems: they are faced on a daily basis with poverty and addiction-induced diseases. However, this challenging arena is somewhat neglected by the faculty, who concentrates on international issues. The network of the medical faculty clearly exceeds the urban level, but local links should be better developed and the initiative should come from the university itself. An opportunity in this sense comes from the new development planned in the next 20 years, which will bring in the Hoboken campus area plenty of physical space for spin-offs. A significant impact is instead achieved by the Ichthus school, located on the edge of one of the most deprived neighbourhoods of Rotterdam. In this district, Ichthus facilitates neighbourhood relations between the students and the community, acting as a link between these peripheral districts and the rest of the city, not only from the physical, but also the cultural point of view.

The City Council is active in promoting the city, either directly or through its 'development arm', the Rotterdam Development Corporation, a powerful actor in many redevelopment and housing projects. The city is variously influencing the education function of the city and its interrelations with other social and economic spheres, for instance through economic development, physical planning and social policies. All these domains require a certain cooperation right from the planning stages with a host of other actors. The departments in the city administrations responsible for carrying out the programmes above are numerous (education, physical planning, culture, and social affairs, for instance, all come under different departments); as a result, policies for the development of higher education sometimes lack cohesion.

Important changes, however, are emerging. The City Administration is now convinced that different policy areas have to be combined to tackle the problems of the city, with its insufficiently developed knowledge infrastructure and human capital. It has therefore adopted a 'chain approach' which focuses on the links between opportunities for economic development – especially targeting the 'new' growth sectors like health sciences and media – and the educational and research requirements. This vision has resulted in a working team 'Education-chain Group', which includes representatives of the private sector and education, and which is steering a number of operational projects to enhance the strategic linkages between education and entrepreneurship, pushing the development of new specialisations and supporting start-ups in these sectors through an 'incubator' policy. The SME sector is especially targeted.

The problem remains that much more attention has been dedicated so far to the 'physical' aspects – e.g. the development of higher education institutions – and less to the *soft* elements (quality and diversity of housing, cultural amenities for the student population, safety, etc.) that may influence the behaviour of student settlement with respect to the city, and in general may improve the city's capacity to attract new settlers. It is recognised these days that even investors value the factors related to the environment and quality of life in a place highly, in addition to traditional location factors such as the quality of the infrastructure and of the human capital.

To sum up, though everybody would agree that students are important to a city, up to now the city has lacked a consistent strategy towards the student population, to make sure that this important asset is fully integrated into the capital available to the city for its development. This is reflected by the relatively low position that the issues listed have in the policy agenda of the city and on the insufficient development of networks to attack the main problems. The difficulty of hooking up with initiatives coming from private parties such as Stadswonen bears witness to the fact that, when confronted with policy priorities, the city does not always consider students an important resource for its socioeconomic development, contenting itself with the actual situation, which can to be considered 'constrained' under many respects. The city does not think of itself in terms of a 'student city'. Problems such as social security, drugs, ethnic conflicts are much higher on the political agenda. *However, there is now more awareness that the 'knowledge infrastructure' is essential for dealing with these problems at source.*

On the whole, the relationship between the city and its education sector is rather disappointing (Figure 8.3). The university is barely seen as an asset of the city, and vice versa. As a result, students receive little attention in city policy as a city users' group. Therefore the overall score in this relation is negative. However, new initiatives by the city offer an important opportunity for improvement; they will be successful to the extent that other stakeholders like the university and the private sector will also realise the importance of this new approach.

8.3.2 Relations between the Higher Education Community and the Private Sector

There are natural ties between private firms and the education and research sector, but they lack an underlying strategy that also considers their durability and their social impacts. The Faculty of Health Sciences is a 'natural candidate'

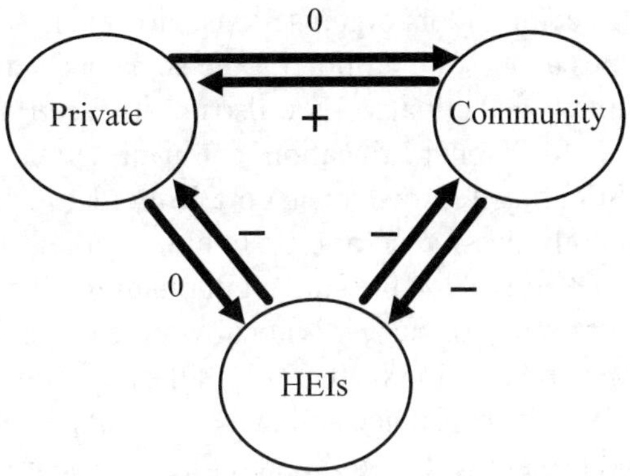

Figure 8.3 State-of-the-art: Rotterdam as a student city

for supporting R&D in booming fields such as pharmacy and bioengineering, but links with the industry are surprisingly underdeveloped. The planned expansion of the medical campus will reserve more space for incubators. At lower levels, formal relations and networking are present, mostly concerning large companies, as well as links between the higher education facilities, the city, the port and the Chamber of Commerce. To some extent, the synergies seem to be more promising between the private sector and the HBOs. Perhaps this depends on the fact that, increasingly, firms are looking for non-specialised diploma holders who they can train 'in-house'. Compared with other cities in this study, Rotterdam also lacks physical places where firms and HEIs get together to exchange knowledge and share research projects. Rotterdam Transfer and Kennishaven are positive experiences, but lack visibility. The business area Brainpark is also a success, but links with Erasmus University are poor.

The city has so far not been overly effective in involving the private sector in developing a sustainable relationship between the city and its education infrastructure, but the situation is gradually changing as the vision of Rotterdam as a 'knowledge centre' is gaining ground in political circles. However, to implement this vision, the city needs committed and proactive partners. Instead, neither the higher education institutes nor the business community seems to appreciate the urgency of the situation. This may be due in part to the lack of 'social embeddedness' of the actors who could make a difference in building the 'student city', Eramsus University, on one side, and the big Rotterdam companies on the other. The EUR is just 30 years old. Though there is an interest in stimulating the links, the rooting of the network is not yet there. A few 'networking initiatives', like Rotterdam Morgen, are all from part of the private sector and its organisations. However, the composition of this club determines a lack of attention on the social aspects involved in the process. Other publicly-led initiatives are also unsatisfactory in this respect; in the 'Strategic Table for Higher Education' companies are almost completely absent, and in 'Stichting Kennishaven Port Rotterdam', higher education institutions are merely passive actors. It should be noted that when private partners have plans that include the student population, like Stadswonen and the city housing corporations, the city is not always ready or willing to include their plans in the general development policy of the city, preferring a top-down approach that encounters resistance and lacks market perspective. The links between HEIs and private sector could therefore also be greatly improved.

8.3.3 Influence of Higher Education in the Relations between the Private Sector and the Local Community

Even if the economy of Rotterdam is recovering from the dire years of the recession, becoming more diverse and competitive, the strategic attitude of the business community with respect to the city is problematic. First of all, a strong commitment by the private sector towards the long-term perspectives of the city as a business location seems to be missing. This can be a sign that in the global economy, firms increasingly have a 'free hand' as they may relocate part of their production chains whenever they find more convenient conditions elsewhere. In this way they tend to ignore the fact that social objectives and local issues can be fruitfully incorporated in their business strategies, as recent research shows (van den Berg, Braun and Otgaar 2001). They believe that they can reap the full benefits of globalisation, but they do not seem ready to support the associated costs. For instance, they clearly profit fron the human capital that is made available by local institutions, but in the majority of cases they would not financially support the development of facilities like student housing.

The private sector in Rotterdam can be divided into three broad categories:

1 companies strictly related to port activity, in the sectors of trade, logistics and processing;
2 banking, financial and other 'command functions' who have important sites or headquarters in Rotterdam;
3 small and medium-sized enterprises located in the Rotterdam region (e.g. food processing, software, media, etc.).

These three broad groups exhibit specific attitudes regarding their 'embeddedness' in the Rotterdam socioeconomic environment, and consequently in their capacity to connect with Rotterdam's main educational and research institutions. Despite a host of initiatives to build a concrete link in terms of knowledge transfer between research and economic activity (e.g. Transfer Rotterdam and Kennishaven), only the firms in category 2 seem to have developed structural links with the main education institutions. This may depend on the fact that these companies have to build on their location advantages and they can take profit from the excellence of studies in the faculties of economics and business and administrations. Several Masters and specialisation courses are co-organised and sponsored by giant firms such as Shell, ING, Robeco, etc. It remains to be seen to what extent such links lead to richer opportunities in the job market for Eramsus graduates. Though exact figures are lacking, there is an impression that skills acquired in Rotterdam eventually leave the city towards more attractive locations. This is probably due to the high mobility in the sector and the fact that many such firms are also present in other sites in The Netherlands and neighbouring Belgium and Germany. However, one should question why Rotterdam, though convenient (housing is certainly easier to find and cheaper here than in Amsterdam or Brussels) and lively, is not an attractive place for highly-skilled workers who want to start a career, for instance, in finance.

As regards the port companies, there seems to be a certain 'cultural distance' from the city, possibly the result of a difficult relationship and years of mutual mistrust, that led the port economy to be independent from the city's 'software'. Port companies participate very little in the formulation of development plans in the regions, merely acting as a passive but powerful guest. Links with the educational sector are no exception: it is only recently that the economic faculties at the EUR have started a number of specialisation courses in maritime economics, management and logistics.

Finally, the small and medium-sized enterprises, especially those associated with the 'new economy' and the development of growth clusters, are still too young as a sector to have rooted structural relations with the city's higher education institutions and other strategic services for knowledge circulation, as the ICT sector. However, the flexibility and openness of this sector is promising. New graduates are more likely to start their own businesses and settle in the area than workers who look for employment with the big firms. This, of course, is on condition that the business climate is favourable and the land at the edge of the city remains cheap for the development of new business areas.

On the whole, there seems to be lack of leadership in the private sector. The Chamber of Commerce, whose role should be that of making the sector aware of the risks of uneven development, has difficulty in keeping such a heterogeneous sector together, and has few resources with which to operate. It is expected that this will change soon. In fact, an organisational reform of the Chamber is under way, which should lead to a rationalisation of its structure according to an 'area approach'. This promises to improve the relationship with the territory and the networks (for instance, through financial backing to development projects, as happens in other countries such as the United Kingdom).

8.3.4 Model Testing and Identification of Issues

The analysis suggests a situation in which the involvement of the education function in the general strategies of economic and social development of the city is not optimal. On one hand, the higher education sector contributes only partially to the knowledge capital available to local firms and institutions. As a result, the upgrading of local skills is slower than necessary to catch up with national levels, and the social impact of local business is relatively limited. On the other, the imperfect integration of the student population in the local community is taking away opportunities for Rotterdam to achieve a more balanced social base. Students do not enter into contact with the city, leaving aside the money spent in bars and discos, and they do not appear to be particularly interested in remaining Rotterdamers after their period of study. International students, who are more eager to discover the city, do not always have the material opportunity to do so. The relative lack of safety and the unfriendliness of the city as regards basic services may influence these choices. The resident population is not particularly willing to improve this aspect, as they mainly perceive the 'dangers' of noisy student communities and the possible adverse impact on housing opportunities for lower income groups. The positive impacts of the student population in the city are strictly 'indirect', at the level of image creation; Rotterdam is increasingly known as a 'trendy', on-the-edge city, with a good university and plenty of education opportunities. But this is not enough to activate a positive cycle of development.

Of the three bidirectional links of the analytic framework (Figure 8.3), none seems sufficiently developed. The relation between higher education and private sector has to be improved, especially as regards the 'market orientation' of the local HEIs – and Erasmus University in particular – and their consistency with city strategy for the development of growth clusters. This is also reflected in the imbalance between the dynamic business environment and the human resource base of the city; and in the awkward relation between HEIs and the community, which, with some exceptions, is barely visible, and may become conflictual to the extent that a satisfactory solution to the housing problem is not found in the medium term. In this context, the city faces a series of threats to sustainable development. The first is that businesses which are currently firmly embedded in the local environment would leave in search of better job market conditions. The second is an erosion of the good reputation of the Erasmus University. The third is the incapacity of the city to finally bridge the economic, social and cultural gap that still divides it from the rest of the cities in the Randstad.

The initiatives of the city concern mainly knowledge diffusion and technology transfer, which is now more focused towards the development of new growth sectors. The city is also careful to accommodate the development plans of the HEIs, but did not develop a consistent vision as regards the accessory services and the improvement of the living climate for students. As a result, physical development risks being accompanied by a bottleneck on the demand side; and the knowledge produced in the local HEIs and transferred to the private sector leaving the area for cities that offer better conditions for new graduates. There remains the need for commitment to a *cultural change* in the business sector, especially for universities and SMEs, which should move closer together to profit from each other.

8.4 Summary and Conclusions

The education function is an important asset for Rotterdam's society and economy, both qualitatively and quantitatively. But how well is this potential utilised at the present time? An excellent education and knowledge infrastructure is a fundamental prerequisite for the successful transformation of the Rotterdam economy. But there are other factors that put pressure on the higher education institutions, in the context of a rapidly transforming city, both on the physical and on the socio-cultural level. These can be summarised as:

- *changes in the educational structure*. The current study curriculum in The Netherlands, which comprises a number of degree programmes to be completed in four or five years, will be changed under the terms of the Bologna Declaration, which harmonises the structures at the European

level. All Dutch universities will have to transform their educational structures in order to comply with the new regulation;
- *the on-going internationalisation of the university and the city*. The number of foreign students is increasing; graduate courses especially see a higher share of international participants every year. At the time of writing, international students number 550 at the EUR (this includes post-doctoral students), and total 2,000 to 3,000 every year when all the schools and courses are considered. The changing student profile leads to the need for adequate teaching and learning approaches in the international classroom, which may demand a more careful consideration of foreign recruitment of staff;
- *the need for a proactive attitude* as regards communication and promotion in an increasingly competitive environment. EUR competes with many universities in a small country, where mobility for young people is particularly high. Therefore, to maintain its eminent position the Erasmus University, as well as other higher education institutions in the city, must promote its strong points more effectively;
- *the challenges posed by the development of strategic clusters*. This is may be the most crucial topic. If Rotterdam wants to pursue a development policy based on innovation clusters, it needs a strong support and interaction with research and education available locally. Among the clusters selected by the city and the region as potential 'growth poles' for development, there are some that deserve particular attention as far as the education provision is concerned.

Education, Knowledge Generation and Transfer

To redress the imbalances in the schema of Figure 8.3, the university should work more clearly *for* the city. In this way they can expect the private sector's and community's awareness of the potential offered by knowledge institutes to increase, and with it, their willingness to *invest in a new cycle of development based on the attraction and integration of the 'knowledge workers' into the local society*.

Building on the city's strength, a *centre of excellence in transport and logistics*, could be implemented in cooperation between the city, its port-related industries, and the universities of the region, as it is clear that both the Erasmus University and the Technical University of Delft are competent in different aspects. This new centre could be strategically located in an attractive new building in the city centre, to mark the centrality of education as a characteristic of the city, and to make the integration of the student community into city life easier and more visible. However, a complementary objective for Rotterdam's economy policy would be to diversity its knowledge base. Because Rotterdam has so decidedly pointed at building a 'culture city' environment, this strategy should be supported by a clearer focus on education in art and design. Some of the schools for professional education today offer these courses, but a 'selling point', like a multifunctional education and exhibition centre, is missing. Architecture has the NAI and the Berlage Institute, creative industries should have something similar, possibly strategically located in the proximity of the new cultural cluster of the city 'Witte de With'.

Finally, we noted in our study that despite the intentions, Rotterdam does not have sufficient capacity for knowledge transfer and business incubation. It is therefore recommendable that the university assumes the lead in involving departments of the Technical University of Delft and private firms in a project for a world-class knowledge transfer centre in sectors of excellence.

Housing Policies and Student-friendliness

The student housing market has become increasingly segmented in the last decade. Students want to live only in the city and not in the suburbs. This could be changed by promoting other neighbouring cities as student locations: the regional dimension should not be neglected. So Rotterdam could make the best out of its close relations with regional centres as Dordrecht and Gouda, provided the mobility to and from the higher education institutions is well organised. However, to assess the validity of such initiatives, better knowledge about the habits and situation of the student population is needed. More generally, it is apparent that housing development for the new 'knowledge class' of Rotterdam (students and new graduates) is a matter for city marketing: fine-tuning demand and supply of these facilities and increasing their customer-orientation by keeping in mind the different ambitions of various user groups. In this respect, it is paramount to appreciate and facilitate existing bottom-up initiatives by concerned stakeholders, which, out of self-interest, may contribute to a more sustainable, student-friendly Rotterdam. It must be remarked again that these students are very heterogeneous, as they range from young freshers who want to party as much as possible, to hard-working post-doctoral students and learning company managers. From a planning point of view, student housing initiatives should recognise this conflict.

Stadswonen has developed a vision that puts students at centre stage in development strategies. Students, young graduates and workers in the new economic sectors of the city are not distinct groups, but phases of life projects that need to be rooted to the place, with integral housing and marketing policies. Nowadays, these groups are the most promising users of the city's various redevelopment projects, as they provide the necessary socio-cultural features that allow these projects to become truly regenerative. This, however, needs a proactive policy of communication and support. Students should be informed right from the start of their study career of the opportunities that they have to become citizens of the city, so that they can make their choices in due time. This must hold for the lower-level, technical students of HBOs, who are crucial for the reinforcement of the education capacity of the region; and for high-level, international students, who may supply Rotterdam's business sector with fresh inputs as far as the restructuring and internationalising strategies are concerned. To these groups, the city should offer a wide range of opportunities. *The communication policy of the city should stress the role and the dynamism of the young, student section of the community, informing potential incomers of the existing opportunities to build a life career in Rotterdam.*

Because the first critical step is the capacity of the city to attract students (many of whom now prefer to stay in their family home and commute to Rotterdam from regional towns), the Rotterdam universities, in collaboration with the city, should support this 'weak' segment with information and help desks. In the medium term, if the initiative finds the support of private sector sponsors, first-year accommodation capacity should be guaranteed for all those who apply, as in British colleges.

Socio-cultural Integration

As our survey reveals, a first step towards a good integration of students into the community is the quality of the welcome and services that they receive at their own education institute. Recently the EUR has tried to widen the range of services aimed at a closer integration between different groups of students. The activities of Erasmus Cultuur and Studium Generale are tools to help the student population get in touch with Rotterdam's cultural life. Another important initiative for acquainting students with the city, the university and all sorts of institutions is the EUREKA Week. At this event, new students are presented with a booklet which gives information on many things to see, do or visit in Rotterdam.

The city also has the problem of building a network to bring students and the community closer. Opportunities for interaction are extraordinarily limited for a city of the mass and dynamism of Rotterdam. As the students themselves suggest, more 'meeting places' are needed. These come naturally in a city with a diffused university settlement, like Amsterdam or Venice, but needs some planning in a situation in which students are concentrated in peripheral campuses or in 'islands' like the International Student House. It is not clear, though, which authority should assert leadership concerning the relationship between students and the city. In some respects, universities handle certain developments earlier, and anticipate emerging problems thanks to their broader vision. The only best practice comes from the Ichthus and their 'neighbourhood policy'; they maintain a legal advice centre in the Afrikaander District of the South of Rotterdam, and a centre for youths and their problems in Rotterdam North. In Kralingen, the student associations try to give a positive impulse to the neighbourhood by organising days for the neighbourhood children. This develops goodwill amongst their neighbours in a 'delicate' environment. Similar initiatives should be encouraged.

If the answers provided by students reveal a 'conservative' attitude towards the city, this may be attributed to a lack of capacity of the city to let itself 'be discovered' in its good points. A more open attitude by foreign students to this respect is revealing. *Therefore, the city should also play a proactive role by promoting its cultural richness better among students.* This may be the way to turn a point of weakness (the perceived lack of safety) into one of strength, as it is nowadays accepted that cultural thrills and curiosity are boosted by an 'unfamiliar' environment. The city should also play an educational role in this respect, targeting the student more explicitly in its cultural strategies, with an eye to personal preferences. This does not necessarily call for a 'commercialisation' of culture, but rather, for a strategy to bring the valuable cultural output of Rotterdam within the bounds of the city campuses. The cultural potential of the city to achieve this level of interaction is enormous, but underutilised. For example, for various events and gatherings only Dutch students are entitled to discounts and communication is almost exclusively in Dutch. The city should address students as a target group of cultural programmes and attractions more explicitly, for instance, creating a friendly environment for international students around the many events (e.g. Dunya festival, Filmfestival), of which the student population presently is hardly aware.

Students also ask for better study facilities and more student-only living areas in the city, as well as more possibilities to travel after midnight. This contrasts remarkably with the tendency of students in other cities in Europe to 'disperse' themselves in the city centres (also among cities included in this research, like Lyon or Munich), but it is understandable given the diffused structure of the city of Rotterdam which inhibits interaction. Still, the city could pursue such interaction in subtler ways that just putting up 'student districts'. This may be done at the neighbourhood level, bringing students closer to the local population, but also pushing foreign students to share houses with Dutch students, instead than reserving for them luxurious but sterile (for the city) 'international houses'.

Internationalisation and Promotion: Rotterdam as an International Knowledge Centre

The most important actor in the field of education and research, the Erasmus University, does not, at the time of writing, seem completely ready to meet the challenges from the ongoing internationalisation of its student base, failing to become a real stimulus for the creation of a world-city environment. Despite the ongoing acceleration of the process towards the adoption of English as the primary teaching language, more has to be done in terms of international courses and curricula. The medical faculty will start international Masters only in two years time, and at this time the structure is not yet clear. A Peer Review Team in charge of evaluating the threats and opportunities for the EUR indicated that the two communities of Dutch and foreign students seem to stand apart. More international and intercultural teamwork may help the process of integration. Also, careful planning in the location and structure of student residence and meeting places might achieve good results in this respect. The Peer Review Team also recommends stimulating further the mobility of Dutch students abroad, in order to familiarise them with a foreign environment; in fact at the present time the balance of incoming and outgoing students is negative.

Universities are not used to being oriented towards the market and utilising promotional tools, but this will become standard practice in the future. Some universities learn faster than others, and in this respect, the EUR lags behind, showing little creativity in its communication policy. Moreover, once at the EUR, students, and especially international students, are insufficiently assisted compared to students' experience at other universities in The Netherlands (e.g. Tilburg, Utrecht, etc.). However, the university alone is not to blame for this. Marketing the 'student friendliness' of the city should be the result of a concerted effort by the main urban players. Today, higher education opportunities receive little exposure in the promotional material of the city aimed at visitors, business partners and potential investors, but this will also have to change in the future, as the successful experience of other cities in this study suggest.

Governance

An important question is: what do the university and the city mean for each other? It is felt at various levels that the distance between Rotterdam and the knowledge economy is still great, and not necessarily reducing with time. The OBR, the development arm of the city, has identified as a strategic objective 'to bind the knowledge to the place', but attention to students as the main target of these initiatives is still missing so that until now it has been difficult to implement this vision. The 'education chain' approach recently adopted by the city in its socioeconomic development policy instead puts a natural focus on students, and especially on the continuity between different stages of a person's education career, from the choice of the university-city to the choice of a job. It is expected that new initiatives will follow consistently, addressing students as an important group of city users as the target of an integral strategy.

Even within the private sector, despite the effective network of personal contacts and ties, there is little cohesion between the big companies. There is also an issue of lack of clear leadership in this respect. This could be provided by an actor with 'free hands'. Whereas there are actors in the city who are aware of the situation and have a proactive attitude towards it (Stadswonen, OBR and Rotterdam Marketing, to name a few), they do not seem to make contact at higher levels of strategic planning, nor is the city doing what it should to bring them together and involve them in the elaboration of a vision of Rotterdam as a student city. Moreover, these actors need investors for their initiatives, that can only be convinced if they see personal advantages, which is not always the case, especially as regards student housing.

This work could be started by building some sense of 'pride of the city', that could come out of informal meetings and a permanent 'platform', where important stakeholders in the private sector are informed of the city's strategy and needs with respect to higher education, and have the opportunity to circulate ideas and proposals.

Note

1 This survey has been carried out designed and carried out by the students Ralf Bosman, Koen Fasten, Marc van der Linde. The following section is largely based on their findings as drafted in the report 'The Profile of Rotterdam Students'.

Chapter 9

Utrecht Case Study

9.1 General Introduction to the City of Utrecht

The city of Utrecht counts 265,000 inhabitants, and is thus the fourth largest city of The Netherlands after Amsterdam, Rotterdam and the Hague. Utrecht is at the centre of a conurbation (Utrecht's urban regions) of approximately half a million inhabitants, while the larger Province of Utrecht territory includes more than 1.1 million inhabitants. The population of the city reached its height at the end of the 1960s, at 280,000. After that, it fell to some 230,000 by the end of the 1980s, and has increasing since then. Approximately 70,000 Utrecht residents work and live in the city, whereas the rest of the city's workforce is employed elsewhere. More than 120,000 workers come to the city every day as commuters. However, Utrecht is a meeting place for different population groups. As in the other major Dutch cities, the number of residents of foreign descent is increasing. The largest ethnic groups are Moroccans (nearly 8.7 per cent of the total population), Turks (4.7 per cent), and the Surinamese (2.8 per cent). The city hosts the largest concentration of students in The Netherlands, about 50,000 students, a large part of whom lives temporarily or permanently in the city.

Utrecht lies in a central position within The Netherlands, at short distances from the centres of Amsterdam and Rotterdam. This is one of the factors behind the success of the Utrecht region as a business location. Utrecht ranks second after Amsterdam in terms of employment growth, and comes well before the Dutch average. Utrecht's job opportunities in 2000 amounted at approximately 194,000, with a 31 per cent increase in the last ten years (Gemeente Utrecht, 2002). The Province of Utrecht scored even better, with a 43 per cent increase over the same period. The local economy holds some strong trump cards in financial and business services, in the ICT sector (Utrecht tops in The Netherlands as an ICT region), and in the biomedical and biotechnical clusters, which include environmental technologies. Table 9.1 illustrates the economic structure of Utrecht and its region. Financial and administrative services are clearly dominant especially in the inner city economy. Finally, Utrecht is an important distribution centre. In 2000, services showed the largest yearly increment in jobs in the last years.

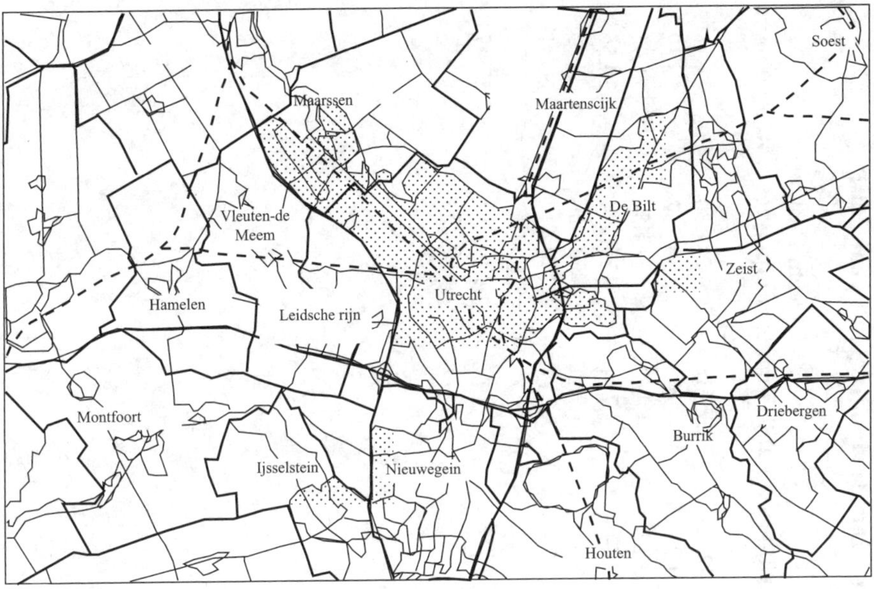

Figure 9.1 The urban region of Utrecht

In comparison with the other three major Dutch cities, fewer branches of international companies are housed here: Utrecht is first and foremost a national centre. However, in recent years, the situation has changed. The congested Amsterdam and the unattractive Rotterdam are losing out to Utrecht as favourite representation centres. By the end of 2000, there were 1,050 foreign-based companies (mainly American, German and British) in the province of Utrecht, including giants such as IBM, Oracle, Mercedes Benz, and Bayer.

Utrecht is very much the place for meetings, as is demonstrated by the success of the Jaarbeurs, one of the most popular venues in Europe. In 2000, this establishment organised 95 fairs and meetings, attracting more than 2 million visitors. In the creation of office space, too, Utrecht scores highly, third after the capital Amsterdam and the government seat The Hague. As a result of these favourable factors, the yearly growth rate of the gross regional product has been constantly exceeding the national figure (Figure 9.2). In 2001, Utrecht registered the highest number of business start-ups among all Dutch cities, with 60 new firms for every 10,000 inhabitants, providing some 28,000 new job opportunities in the last five years (Gemeente Utrecht, 2002: 15–16).

However, economic and geographic reasons alone would not explain the competitive strength of Utrecht; the city and its surroundings are particularly

Table 9.1 Economic sector breakdown, City of Utrecht and Province (urban region), year 2001

	City of Utrecht			Utrecht Province		
	Establishments	**Employees**	**% workforce**	**Establishments**	**Employees**	**% workforce**
Industry	614	10,524	5.42%	3,365	51,180	8.38
Construction	773	8,183	4.21	4,765	39,832	6.52
Agrobusiness	83	421	0.22	4,418	11,654	1.91
Trade, catering, repairs	3,910	29,977	15.44	17,764	125,594	20.56
Transport, storage, comm. companies	383	11,047	5.69	1,764	32,060	5.25
Banks, insurance, services	4,254	69,404	35.74	18,514	179,612	29.41
Administration, education, health and other services	3,187	64,617	33.28	11,409	170,817	27.97
Total	13,204	194,173		62,000	610,750	

Source: Province Utrecht (http://www-test.prvutr.nl/utrecht-biz-locations/).

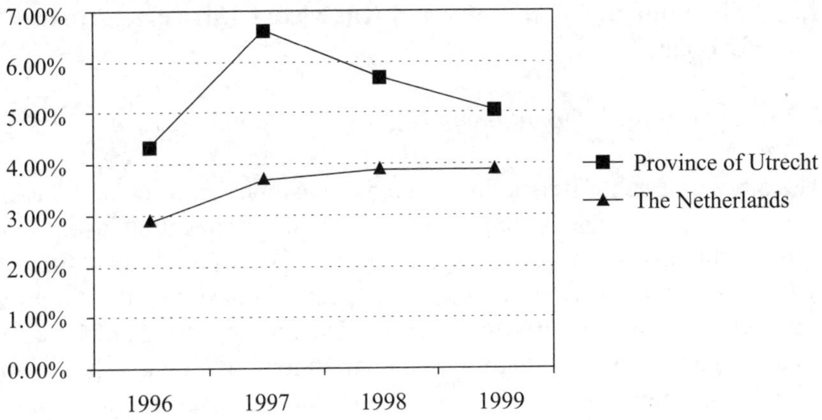

Figure 9.2 Gross regional product, yearly growth rates

Source: Province of Utrecht.

attractive as living and working locations. The historical centre provides the perfect ambience for a citywide shopping and entertainment area. The pedestrian canalside wharves hosting waterside footpaths giving access to the cellars of the canal houses (many of which now have been turned into restaurants, boutiques and craft shops) contribute to a unique open-air atmosphere, which is the perfect stage for Utrecht's rich cultural offerings. A wide range of museums and cultural institutions is available, side by side with interesting contemporary buildings. In the evenings, the music centre 'Vredenburg', the municipal theatre, the smaller cinemas, theatres and podia offer high-quality cultural entertainment. Utrecht is also the city of major festivals, including the Springdance Festival of Modern Dance, the Dutch Film Festival, the Festival *aan de Werf* and the Holland Festival of Early Music. The rural areas surrounding the city centre are among the best preserved natural environments in The Netherlands.

Quality of life is therefore considered high. Firms can attract all the most highly-skilled workers to the area, and count on a large and wealthy local market to sell their products. In the city of Utrecht alone, there are 40,800 residents with a higher education level. The availability of motivated and skilled personnel is clearly a success factor for a region which holds the highest number of highly-educated people in the country.

9.2 The Student Community and Knowledge Infrastructure in Utrecht

9.2.1 General Information on Higher Education in Utrecht

The city of Utrecht offers a fair range of educational and research facilities. It is home to the country's largest and oldest higher education institution, the Utrecht University, one of the largest Universities for Professional Education, the *Hogeschool van Utrecht*. and the largest school of art, the *Hogeschool voor de Kunsten Utrecht*. Moreover, the University College of Utrecht has recently joined this club, an international honour college that is part of the Utrecht University. A total of some 55,000 students receive higher education in Utrecht (Table 9.2).

Table 9.2 Higher education in Utrecht

HEIs	Approx. no. of students	Main characteristics
Utrecht University (UU)	22,000	Strong emphasis on variety and quality of education and research
University College Utrecht (UCU)	600	A unique educational experience, international environment
Hogeschool van Utrecht (HVU)	30,000	Close relation with the industries and employers' organisations
Utrecht School of the Arts (HKU)	2,700	Selective university of international status; well-developed relations with the private sector

All kinds of specialisations can be found in Utrecht: arts and humanities, healthcare, social and economics sciences, engineering, journalism and communication, law, languages, life sciences. The newest addition is a business school offering an MBA. Faculties and institutes are used to cooperating in international networks and as partners of prestigious national institutions. The 'Utrecht Centre of Knowledge' (*Kennis- en Innovatiekring regio Utrecht* KIRU) is a cooperative network of educational and research institutes, aiming to interface knowledge suppliers and consumers, and thus contribute to creating an even more attractive business environment in the region. The presence of high quality knowledge institutes cooperating in the KIRU has turned the region into the 'knowledge centre' for the whole country.

9.2.2 *The Knowledge Sector of the City: Actors and Spatial Patterns*

Universities and other higher education institutions With more than 22,000 students, 14 Faculties and nearly 7,200 staff members, the *University of Utrecht* (UU) is the largest in The Netherlands. It is a prestigious, classical university, which, over its 365-year history, has developed into one of Europe's leading research and educational institutes. The university was founded in 1636 to replace the medieval monastery school, founded in 695 by the Irish monk Willibrord. The university reached its peak in the first part of the twentieth century. The UU offers a wide range of Bachelor's degree programmes and Master's degree programmes taught in Dutch and in English. Together, these cover virtually all disciplines in the humanities, social sciences, exact sciences and the biomedical sector. UU makes a point of maintaining an open attitude with respect to the social issues of the contemporary world. UU has earned a reputation for placing strong emphasis on the variety and quality of its research, which takes place in several prominent research institutes, and in the departments of the various faculties. The Academic Biomedical Centre, which is unique in The Netherlands, has recently been created by combining five biomedical faculties. UU coordinates 20 national research schools, is a partner in four of the six national research schools, and participates in more than 100 European and internationally-financed research projects. Two university holdings administer 13 research-based organisations, with the aim of stimulating knowledge spin-offs.

Over the years, UU has acquired an international dimension. Every year, hundreds of foreign students come to Utrecht to take one of the hundreds of courses taught in English. International staff and foreign short-stayers are offered attractive accommodation facilities. Moreover, as the costs of living and studying may be prohibitive to international students, Utrecht University has created various funding opportunities. Under the 'Bologna Agreement' framework student exchange programmes with other top universities are expected to intensify, with mutual recognition of courses, coursework and research, as will exchange opportunities for lecturers and researchers.

An international honour college is part of the Utrecht University, a joint venture between different departments of the University. The *University College of Utrecht* (UCU) is the university's most internationally-oriented institution, and the only one of its kind in continental Europe. It is remarkable that the initiative to found this college was taken by the city and notably by the mayor, who mobilised his personal network around the idea of a centre of excellence meant to extend further the visibility and prestige of Utrecht as a world-class education centre. UCU is a small and highly selective college. It hosts approximately 600 students, 200 of whom are foreigners from all over the world. The UCU's campus stands halfway between the two main UU locations, on the site of a former military plot, easy to reach by bus and by bicycle. All students at UCU live on campus. Faculty, tutors and students form a unique community of diverse talents and interests, skills, and backgrounds, which is reminiscent of the special intellectual atmosphere that made the fortune of Britain's top colleges or of the 'Ivy League' universities of the US.

The UCU is specialised in the undergraduate teaching of liberal arts and sciences; students choose a programme in sciences, social sciences, or arts, which leads to a BA or BSc Honours degree. These degrees enable them to apply for graduate programmes in all related scholarly and professional directions. Among the characteristics of the college are its small classes and individual attention. This encourages intellectual exchange among students and professors in a friendly and informal atmosphere. The college faculty strives to connect the fields of study, thereby challenging students to integrate different disciplines in their curricula. The college community offers all the support students might need, from high quality teaching and career services to companionship, recreational facilities and accommodation. UCU students also have access to all of the academic, social and recreational facilities that Utrecht University has to offer. Moreover, in the second or third year of study, students have the option of studying abroad in an affiliated or approved institution of higher education for a semester.

Dutch Universities for Professional Education (HBO) prepare students for a variety of professional careers. With a registration of 30,000 students and 2,500 members of staff, the *Hogeschool van Utrecht* (HvU) is one of the largest and most comprehensive HBOs in the country. The HvU was founded in January 1995, yet it is not at all new, as it merged three pre-existing well-established and respected institutes whose history goes back to the nineteenth century. Today, HvU includes six faculties. A wide range of training programmes and courses (more than 70) is available, leading to BA as well as MA degrees. The six faculties have a large degree of autonomy. Each faculty is responsible for student enrolment, and for developing and conducting its programmes. This allows the faculties to respond directly to the training needs of the professions and to the preferences of their students. Three hundred and twenty-five students participate in exchange programmes with EU countries and 75–100 with non-EU countries. Over 200 exchange students are hosted every year. Special courses are given in English, which attract 700 students every year. The school also cooperates with universities and institutes of higher education in many countries around the world.

The degree courses at the HvU are always practically-oriented. This accounts for the fairly tight course structure, for the emphasis placed on applications, and for the external training periods that are an integral part of every programme in the third and fourth (last) year. Furthermore, the school offers excellent facilities in the field of study counselling, international mobility and support. As a rule, HvU has close relations with the industry associations and the organisations that employ the graduates. The board of the school is completely made up of private sector representatives, who have a stakeholder interest in the school. Their role is not to finance the school, but to control and advise. The marketing of the school is done through an excellent web presentation and the participation at fairs and meetings.

Another important institution is the *Utrecht School of the Arts* (*Hogeschool voor de Kunsten* – HKU). HKU offers a wide range of courses within its 24 schools, which are grouped into five faculties: Visual Arts and Design, Music, Theatre, Art, Media and Technology, Interfaculty of Art, and Economics.[1] The HKU is one of the largest Schools of the Arts in The Netherlands and the third largest in Europe, with about 2,700 students and over 500 members of staff. Yet the school prides itself on being very selective to keep high its reputation. Out of 3,000 applications, only 600 students are admitted every year. Besides internationally-acknowledged first degree and MA courses, HKU offers introductory courses, postgraduate courses and contract education. In artistic and educational terms, the school's approach is both traditional and innovative. The school stimulates students to learn at an individual level, working in ateliers and studios, where they are supervised by lecturers and tutors, who cultivate their projects, in teamwork and through an interdisciplinary approach.

The HKU has an international reach that is quite unusual for an HBO. Indeed, the school was conceived not as regional facility, but as an international school according to the *British* legislation, an anomalous status within Dutch higher education. In The Netherlands, HKU is considered a professional school, but internationally it is a university. The MA courses are indeed held in English. HKU has cooperation agreements with more than 100 institutions around the world. This consists mainly of student exchanges, for example in the context of the Socrates programme. One out of ten students at the Utrecht

School of the Arts is a foreigner. Most of them are enrolled in one of the regular full-time courses, but an increasing number of international students are enrolled as exchange or Masters students. The school also cooperates with many institutions in the area of internationalising the education programme or helping set up an education infrastructure.

The relations with the private and non-profit sectors and with government are well developed. The biggest faculty is the faculty of *Visual Arts and Design*, which pays a great deal of attention to the world of art and media business. This is demonstrated by the fact that all the practical teachers are not just teachers but also artists, designers or technologists and are therefore fully aware of the latest developments in their fields. The faculty maintains proper contacts with the professional associations and with companies and institutions in the field of work. Teaching programmes are adjusted and new programmes developed together with these players. Internships for students are compulsory, and plenty of opportunities are offered to travel and come in contact with different cultural environments. Moreover, the faculty is regularly visited by firms seeking to have an assignment carried out by students. These assignments are regularly implemented as part of an education project. To coordinate external projects and competitions, the faculty has set up its own project bureau, the *Utrecht Art and Design Centre*. All in all, the HKU gives students the opportunity to practice everything that they learn. Students have also the opportunity to display their work using the School's exhibition venues. Another interesting faculty is the *Interfaculty – Arts and Economics*, which is a collective effort by the four faculties at the HKU to promote cooperation and to perform a number of communal tasks regarding the market profile of various artistic expressions.

Most of the HKU faculties are located around the historical centre of Utrecht.[2] The HKU is thus central to the cultural and student life of Utrecht. For several years students at the HKU have treated art and culture lovers to a series of events, during which they exhibit their graduation work. In addition, concerts are regularly given in the concert hall of the Art and Science Building, at various points in the city and outside of it. The Academy Theatre and the Academy Gallery offer good locations to admire the work of HKU students.

In Utrecht there is no 'university pole' as in Lyon or Lille; however, the local higher education institutions have joined forces in an interesting project called *Utrecht Studiestad* (*Student City Utrecht*). This project aims at coordination in promotion between the city and the higher education institutions. It sprang from the recognition that the character of the city is very important for the first choice of the university by first year students, even more than the reputation of the higher education institutions. Similar concept images and logos are utilised in the different brochures and websites mentioning the attractiveness of the city. A cultural festival targeting students interested in studying in Utrecht is being planned, which could result in a greater awareness by the local residents about the student community.

Research and technology transfer Most of the research in The Netherlands is conducted in *national research schools*, which are assessed every five years by the Royal Academy of Sciences and Humanities. Groups of researchers from various Dutch universities and non-academic institutes collaborate in these schools. They also train the next generation of researchers to PhD level. In addition to these research schools, the Ministry of Education, Culture and Science has created six top technological institutes, bringing together the country's most brilliant minds in key scientific and technological fields.

The *University of Utrecht* participates in more than 60 research centres, grouped in institutes and holdings, in fields of activity like biomedical sciences, natural and exact sciences, human and social sciences, humanities, information and communication technology. In this context, the UU and the its Medical Centre support researchers and starting entrepreneurs in their commercial ventures by offering them guidance, legal advice and expertise. In 1998, two holding companies were set up (*Utrecht University Holding* and *UMC Utrecht Holding BV*). UU Holding advises new businesses in their first five years of operations, helping them to establish an independent market position. Since its foundation, UU Holding has played a key role in the creation of ten new enterprises (the research company of the Utrecht University) all of them closely connected to Utrecht University. Two-thirds of these companies are active in the field of biotechnology and cooperate intensively with the Academic Biomedical Centre. UU Holding has been instrumental in developing an 'Incubator Group' at the faculty of Life Sciences. UMC Utrecht Holding focuses on technology transfer, patenting, and intellectual property rights. These two holding companies are planning to merge in the near future.

The *Academic Biomedical Centre* integrates biomedical research and education at UU and the University Medical Centre. The centre provides facilities for biomedical research and education, offers a variety of courses in the biomedical field – from short-term courses to a three cycles of higher education – combines research and investigates newly developed areas across the traditional disciplinary borders, and supports the entrepreneurial capacity of researchers by assisting them to translate their scientific findings into market-oriented applications. The ABC participates in the Genomics Centre

Utrecht-Wageningen, thereby putting together its expertise in the biomedical and veterinary fields with the expertise of the Wageningen University and its Research Centre on agronomics and environmental issues. Next to the Uithof campus lies the *business park of Rijnsweerd*, where large international firms have set up their offices. However, the level of interaction between most of the UU faculties and this cluster of mostly financial and consultancy companies is rather low.

Spatial patterns of higher education and research The education and research function of Utrecht is spread all over its urban region. Higher education stayed entirely within the city boundaries until recently, an essential part of the fabric and history of the city. The Academy Building still forms the heart of the university quarter in the old centre of the city. Two large faculties, the faculty of law and the Faculty of Arts, are located there. Today, however, all other faculties as well as parts of the Hogeschool van Utrecht have their buildings in the campus *de Uithof*. The campus is located on the eastern ring of the city, in a nicely-landscaped green area not too far from the city centre, whose land use and conservation plan were drawn up by the famous architect R. Koolhaas. The territory involved in higher education, however, is even wider, on account of the fact that the neighbouring towns of Hilversum and Amersfoort host parts of the HKU, and that the students live all over the region, including suburban centres like Zeist and Houten.

All the higher education institutions of Utrecht are easily accessible by bicycle and bus, though there is concern about the safety of walkways and cycle routes at night. Utrecht is highly accessible for regional commuters. However, there is a high level of congestion on the ring roads, and it is not so easy to reach the campus from suburban towns and other cities, where the main part of the academic population stays. As a result, driving remains the most popular means to reach the Uithof, which on the other hand suffers from insufficient parking availability. For these reasons, improvements of the rail links between Utrecht and the small cities in its region are being planned.

In any case, the good national accessibility combined with a shortage of housing in the city contributes to keep the number of commuters among the student population of Utrecht high, not to mention local workers and shoppers. According to the monitoring system set up by the university, the number of students who commute is increasing from year to year. There are differences in this respect between students coming from university and other higher education institutes. While the UU has a more national reach (of 22,000 students at the University of Utrecht, 52 per cent live within the

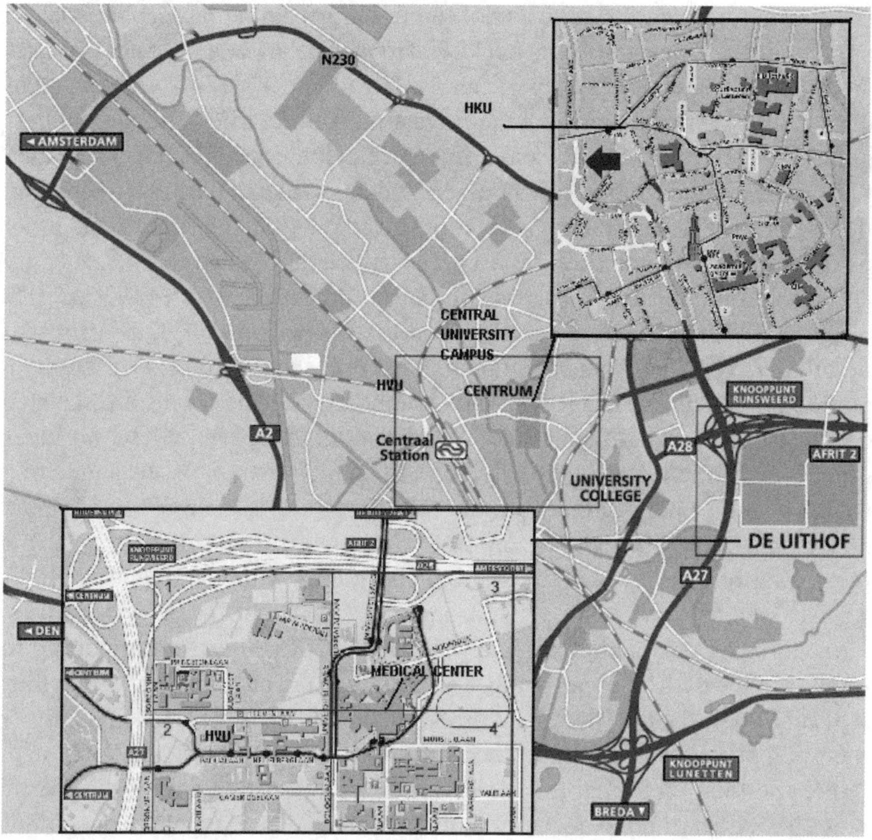

Figure 9.3 Higher education in Utrecht. Detail of city centre campus and de Uithof campus

boundaries of the city of Utrecht, the rest commuting), the HVU has a more regional scale: of 32,000 students, only 17 per cent of the students live within the city boundaries. Given its selectivity and international focus, the HKU is not constrained by the lack of housing opportunities. Because of its central location, however, people can easily study or work in Utrecht while living somewhere else, or alternatively, they can live in Utrecht, while working in somewhere else. However, at the time of writing the housing shortage emerges as the main reason for commuting. Indeed, the solid growth of the economy and of the 'user population' of Utrecht require some expansion of residential capacity. With more housing opportunities, it could be expected that some of the present commuters would prefer to become residents in Utrecht again.

Both the city centre and the Uithof campus offer substantial accommodation opportunities in student residences (11,000 rooms in the city, thousands of others on the Uithof campus), but this is clearly not sufficient, and 60–65 per cent of the students rent their rooms in the private market. Renting a private room is quite expensive, at €250–400 monthly rent. Finding a room is not easy either. Various commercial organisations assist students in their search. An interesting initiative is proposed by *Vastgoed Belang*, a small group of brokers and other partners who commit themselves to guarantee a certain balance between quality and price for student houses. The brochure 'Finding a Room in Utrecht' is available by the student counsellors and the *Foundation for Youth Housing in Utrecht* (SJHU). This is a non-commercial agency dealing with rooms in private houses and residences in the Utrecht region. The rooms are selected on the basis of a reasonable quality/price ratio. Moreover, SJHU provides practical information and legal advice on matters related to renting and letting rooms. Another important institution in this respect is the *Stichting Sociale Huisvesting* (*SSH*), a housing corporation that owns reasonable, inexpensive residential properties. It offers some 6,000 rooms with communal facilities such as kitchen, shower and toilet. Contrary to SJHU, SSH uses waiting lists: the waiting period is approximately 23 months. It is therefore important to register at the earliest opportunity. Students register, and they know that they will get a good-quality room sometime in the future. In a SSH student house or flat, students may legally sublet their room for up to six months, but only with written permission from the SSH. SSH has recently acquired rooms in nearby cities like Amersfoort to develop some student residence, but these operations are notably small-scale and disconnected from regional planning.

The UU itself offers a limited number of own furnished rooms, flats and houses. The housing officer at the International Relations Office of the UU has the task of finding (temporary) accommodation for foreign students and staff who visit Utrecht for a short period. Despite the housing shortage, the International Relations Office presently guarantees that all requests are satisfied, on condition that the request for housing is submitted two months before the arrival. The housing officer also coordinates the allocation of accommodation in rooms or houses offered to the university by private house owners or by students who are willing to sublet their room.

In synthesis, the shortage of housing facilities for students is still a major problem. The Utrecht University is the biggest university in The Netherlands, and apparently the housing problem in Utrecht is the harshest in the country. This is in part due to the small size of the city, and in part to the peculiar structure of the housing stock;[3] however, other factors are relevant. For instance, graduates tend to remain in student residences for some years after graduation, because of the sky-high prices in the normal market and of the good average quality that they get in the market for students. The University is now trying to enforce a rule that allows them to stay for maximum of six months after they have finished studying, but this goes against the national laws. Only University College students are compelled to leave their residence after graduation. This problem threatens to hamper the higher education institutions' internationalisation strategies, and ultimately their capacity to attract the best talents.

Not surprisingly, the housing problem is at the centre of intense discussion at many levels, and notably in meetings between the different higher education institutions and the City Council. The result of studies and discussions is a report ('Student Rooms in Utrecht') that gives a broad picture of the situation and of the plans for the future, but is also used as a marketing tool to convince all influential decision-makers, including the boards of the higher education institutions, of the importance of this issue. The long-term plan is to accommodate all the excess demand by 2008 (Table 9.3). One should welcome the fact that student associations, the higher education institutions managers, SSA, Vastgoed Belang and the city planners meet and search for solutions. However, the situation has been ignored for so long that now there seems to be very little room for a structural solution. It may take decades to change the structure of the housing market, and there are few private companies willing to risk housing investment. Important actors are more or less willing to accept that in the future the great majority of students or graduates will be a travelling community which 'uses' Utrecht rather than lives in it; and this may create imbalances and distortions in urban development patterns.

9.2.3 The Student Community: Behaviour and Careers

Students come to Utrecht attracted by the prestigious university, but obviously also by the special atmosphere and the central location of the city, that allows them to commute to it in a short time from all the most densely populated areas of the country. It has been estimated that almost 20 per cent of the students manage to find accommodation in the city, and in this way they are integrated into city life; the rest return from student residences to the parental house for the weekend, or commute daily. That percentage decreases to 17 per cent for the regionally-oriented HVU. For students at the HKU, it is exceptionally stimulating to be in a city with many museums, concert halls, cinemas, theatres and galleries. There are plenty of opportunities to exchange ideas with people

Table 9.3 New demand and new (structural) supply in units

	2002–05	2005–08	Total 2002–08
New demand			
Regular students			1,600
Short stay			1,000
Total new demand			2,600
New supply			
Acquisition	180	90	270
New housing development	400	1,200	1,600
Private supply	200	230	430
Turn-over	75	225	300
Total new (structural) supply			2,600

Source: Broere et al. (2002).

from other disciplines, and in the past this has already stimulated many in their creative work.

Only around 5 per cent of the students finish their studies within the normal period of four years. The long duration of study curricula is determined to a large extent by the fact that they need to work in order to cover Utrecht's high living costs (estimated at around €700 per month). After their studies, most Utrecht graduates look for accommodation in Utrecht and start a professional career there, but only a few manage to find suitable and affordable opportunities. At the Utrecht University, an *Alumni Bureau* coordinates, stimulates and supports students wishing to find a job in the city, acting as a liaison between the faculties and the business community. The alumni network database includes a large number of entries (almost 6,000), 25 per cent of whom stayed in the region for their first job.

Student life in a sociable city like Utrecht is vibrant. Student associations are particularly active, even in local politics (almost 20 per cent of students are affiliated to political organisations), and this enables them to influence the city council agenda. Active participation in the local social and cultural organisations is typical. For example, the student federation *USF*, a union of 50,000 students from the university and the universities for professional education, is heavily involved in the housing issue. Talking to both the university and the city government, it raised the problem and succeeded in putting housing policy on the university's agenda. Thanks to USF, the university is now fully aware of the problem and has started a new policy of cooperation with the city. In general, student organisations in Utrecht seem more active than in other cities. The reason may lay in the fact that the Utrecht student population is more diverse, both disciplinary and socially. This greater mix generates a higher feel for cohesion initiatives and the establishment of communities. *International students* are not so organised, with the exception of those at the UCU. The 600 students at the College have different cultural, social and ethnic backgrounds. What they have in common is their international orientation, their intellectual interest, and their ambition to make a difference. Since everybody lives on campus, the students form a community of their own in which they learn to respect, understand and accept each other.

The *Cultural Centre Parnassos* offers a full range of cultural opportunities, such as elementary and advanced courses in music, dance, and theatre, to students and staff, as well as a meeting place for foreign guests and organisations and an information centre. Parnassos was opened in 1994 by the university, in a central location. Its main goal is to bring about a lively cultural atmosphere inside the university world. Almost 2,000 students visit the centre throughout the year. Together with HKU, Parnassos works as a catalyst for the cultural activity of the city. The interest in theatre and in the many festivals that now take place in Utrecht originated there, enhancing once and for all the cultural climate in Utrecht, which tended to be rather dull some 30 years ago. Today, Parnassos also aims at stimulating horizontal cooperation in cultural policies, and co-organises many events with other venues of the city. A recent meeting between the board of the university and the City Council has underlined that Parnassos may play an important role in Utrecht's candidature to host the European Cultural Capital event in 2013. The building of Parnassos is divided into two parts: a guesthouse for international students, and the cultural centre itself, for all students. At present, the centre in some way serves students who are unwilling to discover the city's other cultural institutions. On the other hand, foreign students tend not to make use of the centre's rich offerings, which are still targeted predominantly at the Dutch population.

Among other initiatives involving student life in Utrecht, two are worth mentioning. The *Uitwijk* is the social cultural centre of the university and is a meeting place for students and staff. The Uitwijk offers a number of courses for students and staff at very reasonable prices. Twice a year the Uitwijk offers a movie programme and a second-hand book market, free of charge. A weekly newspaper, *U-blad*, 20 pages long, is available for students and staff of Utrecht University. The articles deal with student life in Utrecht (mostly housing and leisure) and higher education issues. *U-blad* is an independent foundation subsidised by the university.

9.2.4 *Challenges for the City*

The factors that make Utrecht a thriving city in the Dutch and international context are many. First, it is a pleasant, well-preserved historical city. As such, it is attractive to highly-skilled citizens. The historical inter-relation between university and city has generated a liberal, innovative, culture-rich milieu that is believed to be the ideal setting for high-ranking functions in the new knowledge society. As such, Utrecht competes with bigger but disadvantaged centres in the Randstad – the congested, tourism-flooded Amsterdam and the uncomfortable, hectic Rotterdam. Utrecht is clearly a leisure centre, which would be very attractive to tourists were it not for its proximity to Amsterdam.[4] It is in any case very successful in the congress and fair market.

The city enjoys optimal accessibility, lying at the junction of main traffic axes that connect Amsterdam and Rotterdam/The Hague with the east and south of The Netherlands and with Belgium and Germany. Utrecht is at the border between the highly urbanised, metropolitan west and the middle and eastern parts which are full of natural beauty spots. Utrecht Central Station, the nexus of the national rail infrastructure, is the busiest railway junction in The Netherlands. It is also the most important railway station in The Netherlands.[5] The expectations are that, by 2015, the station will handle around 205,000 passengers a day, almost double the present amount. It will be connected by a high-speed train service from Amsterdam to the Ruhr area in Germany, and, via Rotterdam, to Brussels, Paris and London. In the near future, Schiphol Airport will be accessible within 20 minutes.

The pressure on the city's real estate market and on other attraction factors is therefore likely to increase even more. The development requirements for incoming economic activities are analysed by the city in its *Real Estate Monitor* (Gemeente Utrecht, 2002), which produces forecasts of the business location demand for the next ten years. The emerging picture reflects the increasing pressure on the city centre and on the westward-shifting CBD of the city, where yearly rents peak at €200 per square metre.

To realise its potential as a growth location, a city has to provide incoming actors with human capital, good accessibility and top-quality housing and business facilities. Utrecht has a proactive attitude with respect to the latter issue; it is responding to unfavourable developments concerning its internal accessibility; and it enjoys a natural advantage with respect to external accessibility and the quality of the human capital. Yet the danger is that the Utrecht economy will face a human resource constraint especially in the medium segment of the job market (secondary–technical education). This is typical of medium-sized cities with many university students: there are not enough jobs for all the graduates in this area, and it is assumed that universities produce human resources for a larger regional or national territory. Today, 66 per cent of the local job positions are occupied by non-residents. So, for Utrecht to retain the highest possible share of the economic welfare created in its firms, it needs to develop more education opportunities at a lower level.

In the next ten to 15 years, the whole city will receive huge investments in infrastructure, internal and external accessibility to meet the growing demand for new tertiary activities in the city centre. Among these, the new district of *Leidsche Rijn* (see Figure 9.1), promoted as the motor of a wide process of urban regeneration, and the redevelopment of the station area. Both projects fit neatly into the national policy of consolidating municipal competitive positions by encouraging the construction of large-scale residential areas and massive multifunctional projects on the outskirts of the main cities.

The development in *Leidsche Rijn* will be finished in 2015. It involves the construction of 30,000 new houses. The *Leidsche Rijn* project is expected to raise Utrecht's population again to more than 300,000 inhabitants by 2010. In parallel with the development of this expansion scheme, the area of the municipality of Utrecht has been enlarged with the territory of the municipality of Vleuten-de Meern, also located to the west of Utrecht. An inflow of residents can be expected from the centre but also from other parts of the city to the west side. This promises to improve the dynamism of the housing market and give new opportunities to 'starting citizens' to buy a house in Utrecht. Also the housing market for students could improve provided there is some strategic planning behind it. The *National Strategic Plan* (Fifth Memorandum, not yet formalised by the national government), which gives guidelines for physical development, is expected to assign to Utrecht the task of building 5,000 dwellings on the city's edge, plus 10,000 dwellings within the built-up area, where it is still possible to expand the stock. However, the Memorandum does not say much about the social groups targeted by the housing programme. It is believed that urban professionals will appreciate family houses with gardens, but there is hardly any mention of the demands posed by the 'knowledge functions' of the city, and in particular by the university and its community. This is revealing of a tendency in city planning: the student community is considered a given and not a resource to be utilised strategically. More attention needs to be given to the 'soft' planning elements which are going to be crucial in the definition of the identity of Utrecht in the future.

9.3 Strategies and Integration. Utrecht as a 'Sustainable Student City'

9.3.1 Relations between the Students and the Community

Hosting the largest concentration of students in The Netherlands, Utrecht is the case of a real 'student city', characterised by an historical co-evolution of the city and its academic community, which has brought welfare and dynamism to the local community. Indeed, compared with Amsterdam or Rotterdam, Utrecht stands out as a middle-sized city where the elements of the quality of life play a relevant role. For economical, social, and cultural reasons students are the lifeblood of the city of Utrecht. In Utrecht, the inter-relation between the higher education institutions – and in particular its historical university – and the city are dictated by the relative size of the academic community. The development of the university has largely coincided with the development of the city; students have contributed to the dynamism and vibrancy of a middle-sized city, bringing to it prestige and international exposure. However, this symbiosis has hardly been the result of a concerted planning effort. Until recently, when the housing problem has become explosive, although the City Council has seen the presence of the higher education institutions and the students as a positive factor in economic, social, and cultural life, there was no explicit city policy towards education nor any effort to plan a balanced and sustainable development of the city's infrastructure for students. The lack of cooperation between higher education institutions and the city became visible when there started to be some competition for the use of the urban land between the university and other user groups.

At that point, it was clear to the city that new tools were needed, and a new cooperation effort was required. There is today more interaction between the board of the university, the city and the other higher education institutions to see how far they can help each other to reach their goals. From regular meetings, several projects have been developed. For example, actors of the higher education institutions and the city have begun to cooperate to find a structural resolution of the housing problem. Studies have been realised and discussions have taken place; the awareness of the problem at many levels is certainly higher though there is still no decision on how to tackle the problem once and for all. In the 'Utrecht Studiestad' project, the city has taken a leading role in the recruitment and welcoming of students for the first time, which could have strategic importance in terms of planning the quality of their stay, whereas the single universities were only interested in promoting their fields of study.

At the time of writing, the municipality of Utrecht is trying to improve the service to its citizens, and understands that the supply of municipal services should match the demand better than it does at present. In the programme 'Utrecht under Construction, 2001–2006' that intention is translated in a slogan, 'De Betrokken Stad' ('The Involved City'). This stands for a programme of activities intended to strengthen the relationship between residents, entrepreneurs and social organisations with the city and its administration. One of the 17 points of the programme is to establish District Service Centres, as 'desks' where the local residents are welcome with a range of questions about housing, working and living in the district.

However, the impression remains that the community at large is not willing to consider its academic community as a real asset for the city, that has to be managed and used for the better. In that very programme no specific service for students is foreseen in the programme, which instead could greatly enhance the 'sense of belonging' and the quality of life, especially for foreign students and newcomers. It could have been a very good opportunity for coordinating and jointly marketing local initiatives in the fields of students housing and job market. The lack of a strong cohesion between different actors is also highlighted by examples of carelessness, as in the case of the student houses that remain empty during the summer – a more strategic attitude by the city towards the higher education institutions would lead to a more efficient use of these precious spaces, for instance to accommodate groups and visiting students. It is also quite remarkable that in the large planning developments of the city, such as the *Leidsche Rijn* project, students or young starters are hardly targeted.

The city is now putting a new focus on 'the talents of Utrecht' (well-educated students of all levels including higher education), which is the new keyword for the development strategy of the city in the new policy document 'Economic Profile Utrecht 2010'.

A promising example the attitude of the city towards higher education is the creation of the University College of Utrecht, in which the city's support – and that of its mayor in particular – was critical to obtain the necessary land and infrastructure from the Ministry of Defence, to boost the project and to overcome a certain initial scepticism on the part of the UU. In this case, the operation of the city government was remarkably far-sighted, because very little short-term return could have been expected: the UCU is small and the economic impacts of its student community are limited. In spite of the fact that its students come from all over the country, UCU's graduates are definitely not expected to stay in Utrecht after their studies. However, UCU does contribute significantly to the international reputation of the city as a centre of excellence in studies and constitutes a link with similar poles (Oxford, Stanford) which could be exploited to build a solid network of cooperation in parallel fields, like R&D and technology commercialisation.

At a lower (but no less important) level, the education and research sector of Utrecht is so strong and diverse that it could be expected to contribute substantially to urban policy in terms of elaboration and intellectual inputs. However, even in this respect the occasions of interaction could be improved. A good first step has been taken by the Incubator at the Uithof, which was built through cooperation between the city, the university and the Medical Hospital yet, the universities could make better use of the city as a laboratory. The good working relations of the city's planning Department with the Department of Geography at the University of Utrecht should be mentioned among the positive experiences. This collaboration has been useful in many planning activities that prepare the ground for the future of Utrecht.

Even a locally-rooted institute like HVU has but little strategic links with the city, though there are regular meetings (every two or three months) with the City Council and the mayor concerning issues as housing, infrastructure and transport. These meetings are formal and problem-oriented; they barely touch on long-term planning issues. A promising exception is the start of cooperation between the academic hospital, the city and HVU's Faculty of Health Care to organise a Masters degree in this field. Housing, sport and cultural facilities are already shared with the university. The school participates also to the management and financing of the Parnassos Centre.

Finally, the Utrecht School of Arts (HKU) offers a noteworthy contribution to the community, as it enriches the cultural life of the city and the surrounding region. The banks of *the Vecht* are regularly enlivened with initiatives like the HKU Festival and other cultural events. The opening by HKU of a Faculty of Theatre Studies in 1990 can be seen as the real starting point of a process of bringing important theatre productions to a city that thus far has hosted few initiatives in this field. Moreover HKU's Art Management course will be helpful in future development of the city in the creative industries. This is an example of a good practice in strategic cooperation between the city and one of its higher education institutions. On the other hand, the HKU, given its international projection and status, is not involved in the planning of events or in the many festivals organised by the city. Students participate as performers, and some members of the school participate in the festival's board committee, but outside of any formal agreement. However, over the last ten years the City Council has shown increasing interest in the social and cultural education of students. This change in attitude should be a sufficient motive to establish new working links with the HKU in terms of capacity building.

9.3.2 Relations between the Student Community and the Private Sector

Utrecht is a highly attractive place for companies, because of its excellent location and because of the presence of high-class higher education institutions and human capital in this region. Many students train through stage periods in the companies of the region, thus participating to knowledge transfer. Banks are keen to have good relations with the students associations, stimulating and financing some small local activities. They know that they will become their future potential labour market.

At the level of the single institutes, the network of relations with the private sector is quite well articulated. However, the strategic links between the university and the private sector are rather underdeveloped. The lack of a technological faculty or, until recently, of a business faculty, makes it difficult for UU to have operational links with the business sector, but the general attitude of the university is not particularly prone to market orientation, probably a legacy of the national public sector rather than local focus that the UU is used to have. It is significant that one of the biggest Utrecht employers, Douwe Egberts, does not have any kind of relationship with the UU as far as stages or joint research project are concerned, in spite of the fact that this firm employs all kinds of skills. Companies recruit employees from other cities and universities, and this is possible because of Utrecht's relative centrality. The model by which local workers are 'defranchised' from the citizenship of Utrecht is therefore exacerbated, and generates problems such as a high level of communing traffic. The creation of a Faculty of Management has marked a turning point in this attitude. However, the UU has good research relations with the pharmaceutical industries, which extensively support R&D programmes. A growing relationship between the scientific faculties and the ICT sector in the region has also been noted. The University College is strongly oriented towards pure sciences and human studies. However, paradoxically it is the most visible part of the university education. The corporate world is interested in talent, irrespective of which field of study they may come from. Many private firms are interested in the school, presenting themselves in career weeks for summer jobs.

In The Netherlands, institutes for professional education (HBOs) are perhaps the segment of higher education with the best connections with the private sector, and Utrecht is no exception. For the HVU, the focus is especially on the small and medium-sized enterprises of the region, like ICT and the local branches of banks, while large companies with headquarters in the city prefer to link up with the university for research purposes. These institutional relations bring graduates close to the private sector employers during their study years. The links between the HKU and the professional field are also well developed. Special arrangements ensure a permanent contact between the HKU and the business community, the government, the cultural sector and the non-profit sector. Partly because the HKU works actively with other organisations, it is aware of the requirements in the job market, and modifies curricula or parts of courses to meet the wishes of the employers. In recent years, HKU alumni have been doing well; there is a clear trend towards recognition by the business community and the government of the added value of people with a creative background.

The development of the private sector in Utrecht is mainly geared to the service industry. Knowledge from research and education can contribute more and better to the development of the local economy. Yet nowadays it is not so easy for private companies to find the knowledge and information that they need in the higher education institutions. The Chamber of Commerce plays a significant role in putting together the knowledge generated at the university with the interests of the business sector. Small and medium-sized enterprises, in particular, take advantage of the expertise offered in many fields. In Utrecht, the Chamber of Commerce is quite active in stimulating cooperation among the city, the higher education institutions and the private sector to enhance the technology transfer that is quite low in the city with respect to the rest of the country. This is achieved by facilitating the contacts and cooperation efforts between the private companies and the key people at the university, HBOs and research institutes. The Chamber of Commerce acts also as advisor for students, offering training to starting entrepreneurs, mainly HVU students. Moreover, the Chamber of Commerce tries to help schools and students to become more entrepreneurial and offers them services and advice. It is now developing a strategy to reach the HBOs better.

Three years ago the Chamber of Commerce, the province, city, university, Hogeschool van Utrecht and Syntens, a company whose task is to inform small and medium-sized enterprises about technological possibilities, initiated a project, cofunded under the EU RITTS programme. The project produced a strategic document for cooperation. At that point, the province expressed an interest in being more involved in the higher education institutions, and the Utrecht Chamber of Commerce took the leadership in the development of UNITé (Utrecht Network for Innovation and Technology), an umbrella organisation grouping the main administrative bodies like city and province, which is developing a range of initiatives connected with skill development and matching in the region. The aim of UNITé for the next 3–4 years is to stimulate projects for cooperation between the knowledge sector and private companies, to build a network of knowledge transfer. One of the recent projects is the CUBE, a multitenant building for technological start-ups in the west of Utrecht, initiated by the HVU.

Despite these good links, there is room for improvement. The main issue to tackle has to do with connecting and transferring the high-quality knowledge produced in the local higher education institutions. The human capital formed in Utrecht is very volatile for the characteristics of the city, which cannot retain as many graduates as it produces. At the same time, knowledge transfer mechanisms are not quite as developed as in the rest of the country and in particular as they are in other knowledge-intensive urban areas which compete with Utrecht, like Eindhoven and Amsterdam. The university is active in incubating new firms, but the results are not comparable to those that could be obtained in a science park environment. In spite of the extensive network of liaisons between the private sector (as well as other public or non-profit employers) and the higher education institutions, these links seem to have little strategic orientation, at least until very recently. This is reflected in the fact that the private sector has no commitment whatsoever towards student housing and infrastructure, nor this is considered an area for investment in development. All this is left to the public sector and to spontaneous initiatives like Vastgoed Belang. The efforts of the Chamber of Commerce and their joint initiatives with the region are bringing a real improvement to these conditions.

9.3.3 Influence of Higher Education in the Relations between the Private Sector and the Local Community

Given the geographic characteristics of Utrecht, it is not surprising to note certain peculiarities in the local economic structure. The business sector of Utrecht is strongly nationally- or internationally-oriented, and chooses Utrecht as a favourable location for the quality of its connections as well as for the advantages of the local environment, from the high availability of human capital to the amenities of the city and its region. The transnational companies can easily recruit their staff nationally or internationally; this human capital adds to the resident population, making the local community wealthier. However, the city has to care about not losing its competitiveness in the context of a rapid evolution of these very favourable conditions. The 'unrootedness' of the private sector, especially when talking of services, financial institutions and large firms' headquarters, means that they can easily decide to switch location if the advantages decrease: the high qualified labour force becomes more and more important for enterprises to stay in Utrecht. Furthermore, much of the human capital formed in the excellent Utrecht's higher education institution will have to flee to other locations.

Utrecht has a surplus of university graduates, which is to be expected given the national importance of the university. In this way, it seems that the city does not exploit the advantages offered by the universities of its region to the full. It was also highlighted that the private sector faces a certain difficulty in finding workers in the low-to-medium sectors of the job market. Whereas there is no shortage of higher-educated staff, it is very difficult to attract lower-educated workers to the city, as that would involve having to discount high travelling or living costs. However, it is not necessarily a problem of Utrecht's education, even if there is a certain pressure towards lower-level education. It only highlights that the potential for expanding the city's high-profile businesses is all in place: the university produces many talented people who nowadays find jobs elsewhere, but were there more opportunities in Utrecht (and new housing solutions) they could be easily filled by the local workforce, adding to the economic circuit of the local economy.

These considerations extend from human resources to the industrial structure of the region. In the University Hospital, departments test prototypes of technical equipment and systems used in everyday hospital operations and are also patented to a larger market. However, one could not speak of a real 'health cluster' centred on the medical sector. The local suppliers for these laboratories are mostly low added-value component retailers, who settle in the proximity of the hospital and in nearby villages. Strategic interaction between them is rather limited, as is their innovative 'pull'. Instead, the customers of the healthcare-system developers and integrators are global. Recently, the City and the Province of Utrecht have adopted a proactive attitude to redress this situation and attract other high added-value business sectors which would complement the existing chains. This approach is based on the vision expressed in the document 'Economic Profile Utrecht 2010', the key strategic document for the future economic development of Utrecht. In the document, four leading sectors for the future are identified, namely (inter)national commercial services, the medical/health cluster, the hospitality sector and small and medium-sized enterprises.

Therefore, a number of challenges will be met in the future, which will affect the private sector's behaviour. The first is to link the good external accessibility – to be further reinforced in the wake of the HST connection – with a satisfactory internal accessibility at the regional, metropolitan and inner-city scales. When confronted with a high traffic level (as the Dutch main railway hub), the city had to endow itself with a station and commercial centre – the *Hoog Katherine* mall. Although probably necessary at the time, this is now regarded as excessive because it presents to the world an image of Utrecht that does not reflect its true points of strength; it will therefore be 'restyled' in the new station redevelopment project. In this way, both the internal accessibility and the image of the city are expected to improve. The second is to provide the necessary land and infrastructure for the further development of specialised service functions on the edge of the city centre, as well as to education and research functions that are required to follow this development.

9.3.4 Model Testing and Identification of Issues

Utrecht is a case of a 'student city', in the traditional sense of the expression, possibly the only such one in the set of cities that participated in this research. The co-evolution of the city and its academic community permeates the very fabric and culture of the local community, bringing to it welfare and prestige, as well as social dynamism and quality of life. It is not by chance that Utrecht is considered one of the most pleasant places in The Netherlands to live, and a favourite choice of freshers who increasingly make decisions about their study career looking at the *places* where they are going to spend a substantial part of their life. One should not forget, though, the excellence of Utrecht's educational and research institutions.

Problems in this relation have been limited, and the biggest issue – the scarcity and high prices of housing – is considered a natural feature of the limited size of the city, which is in any case being extended with the new development projects like *Leidsche Rijn*. The social composition of the city has changed so much in recent years that a certain 'natural selection' for incomers who wish to become citizens themselves is in the order of things.

This favourable situation should not hide the threats inherent to this model of development (negative scores in Figure 9.3), like the possible loss of competitive advantage of a city which can hardly find room for further development, both in the private and in the education sectors. Tackling that constraint demands a strategic, cohesive attitude by all the actors involved.

On this basis, Utrecht has lagged behind in the past, but the situation is today improving. The UU, HvU, UMC, the City and the Province are discussing the future development of the Uithof, especially the development of a life

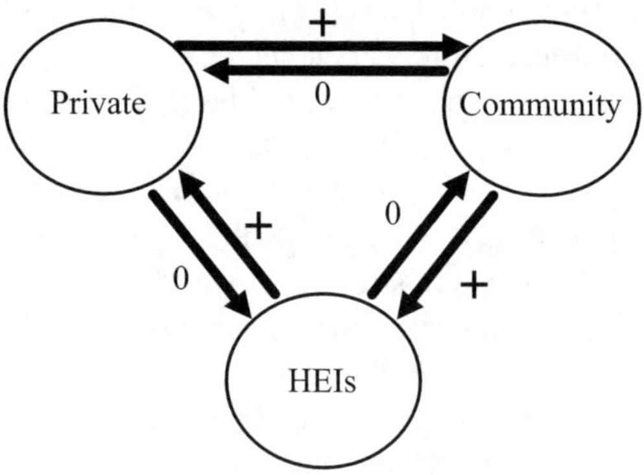

Figure 9.4 Utrecht: a 'student-friendly city'

sciences business park on the Uithof which should complement the existing incubator activities. Utrecht Studiestad gave different administrative bodies the opportunity of cooperating in the marketing of Utrecht as a student city, even if the results could be better.

9.4 Summary and Conclusions

A general priority for Utrecht's student-city policy is to define the city's *identity*. Is Utrecht a historical city, a traffic hub, a workers' and students' location for a few hours a day, or something else? The debate on this issue has to be as wide-ranging and participatory as possible, and is likely to have decisive consequences for the attitudes towards planning. The City Administration has given their economic development strategy a new slogan: 'Utrecht, a meeting place for talent'. This slogan hints at an enlarged user base for the city, but also at a more inclusive decision-making process. Attacking the conditions that now make Utrecht a 'commuters' community' demands first coming to terms with its *internal accessibility problem*. The development of the 'Stationsgebied Utrecht' area may contribute to a change in the modal split in favour of the train. It has, therefore, to contribute to a decreasing growth in car traffic on the Utrecht ring road and within the city.

An *improvement of the connection between the city centre and the campus* is required. Today there are frequent and fast bus links to the campus, but a visible, enjoyable and safe pathway connecting the city to the campus is lacking. This leaves the perception of the Uithof as 'far out' on the city edge and increases the sense of the university's seclusion from the community.

Creating an 'urban corridor' that links the university establishment, the business district, the residence and the entertainment area of Utrecht may give some consistency to the cityscape.

After all, if Utrecht cannot retain its 'knowledge community' within its municipal borders, it can plan for *satellites*, which would represent a better solution than a huge city user group made up of commuters. Currently the regional management of the city' development problems is being tackled with blunt tools. Important partners in future planning efforts might turn out to be the Region or the Province, because the planning scale for higher education functions is obviously larger than the municipal. Coordination among the municipalities around Utrecht will become a key element. Today, this issue is taken care of by the regional development agency BRU, but a specific concern for students is lacking. A platform for collaboration in this specific domain could be organised around existing networks such as the Unité initiative.

The city and the university cannot solve the *housing problem* by themselves – though they have shown that, with specific initiatives, they can ease it, and today, the university is proud to be able to offer accommodation to all its international students.

The achievement of structural solutions to this problem will depend much on the funds raised at the local and national government, but it also requires creative solutions and improved coordination among all the higher education institutions and the city, based on the recognition that it might be a constraint to attracting or retaining young workers. A promising solution might be to increase the presence of students – especially in their first years – in neighbourhoods of the city where there are minority groups, stimulating in that way integration and social, cultural and economic vitality. To be successful, this initiative must also be carefully planned in terms of 'critical mass' of students as against the resident population. However, this may not be an optimal solution from the point of view of the performance of students. It has been suggested that academic activities such as classes, laboratories, and libraries may have a more sustainable impact on city centres than just residence. In this respect, Utrecht may regret that the presence of the university in the city centre is getting less. The necessary move of space-intensive education facilities to the Uithof could be recomposed by leaving some important educational functions in the city centre, and creating the 'urban corridor' mentioned above.

Another important point to keep in consideration is that the *Leidsche Rijn* development will attract residents out of the city centre and liberate houses in the centre of Utrecht for students who prefer to live near the cultural facilities and meeting places. The danger is that the social composition will come to be unbalanced, with an excessive number of students (and a consequent decline in housing values and quality as has been observed in other student-intensive neighbourhoods in this study). Again, this calls for a solution to the housing problem that is also *spatially balanced*, possibly involving other levels of government. The development of the education function in Utrecht has to be harmonious: excessive concentrations of students could lead to problems of a different nature.

Utrecht has to learn to use the prestige and variety of its high-quality educational and research capacity more strategically, to market the city's assets to visitors and firms better. The city could be promoted as an alternative destination to Amsterdam, offering a different range of amenities in a safe, human-sized atmosphere where the local lifestyle occupies a central position, whereas – increasingly – Amsterdam is seen as an unsafe, inaccessible city. This possibility has been reinforced in recent years by the cooperation achieved between Utrecht's museums, for example in marketing strategies, which make it easier to act consistently to attract more visitors. Utrecht is first and foremost attractive as a meeting place and a cultural hub, but these qualities are connected with its student community, and they should be given some prominence in the marketing strategy of the city. Moreover, there are many student facilities that are left empty during the summer months when they could be utilised to host visitors. When students leave, in the summer months, the city tends to be considerably quieter; indeed, for the hospitality sector, students are more important than tourists as a consumer group. The challenge for the city is to offer visitors a 'package' which includes not only monuments and shopping facilities, but also the 'atmospheric' elements underlining the dynamism and variety of its young population. Similarly, these qualities may be attractive for those young, beginning entrepreneurs in the creative industries (fashion, art, and design) who could reinforce and diversify the economic structure of the inner city. Today these operators prefer the ultra-congested and expensive Amsterdam, where they enjoy a 'cluster environment', but the proximity of Utrecht should be better taken advantage of with a targeted campaign to attract these new functions and start a local creative cluster, located midway between the prestigious Amsterdam market and the Hilversum media hub.

More generally, the city could improve the *communication strategy* regarding its assets and the many opportunities available to all city users. Students should be a specific target: they need to be better assisted, welcomed and made to feel at home. To this end, the city and its higher education institutions could collaborate in the project of a city web-portal for education and student life, in various languages, where the assets of the city are presented in connection with cultural events and educational opportunities.

As far as the governance in the process of building a sustainable, student-friendly Utrecht is concerned, planning initiatives should be extended to the general economic development problems of Utrecht. The lack of focus of higher education institutions in the local businesses – and especially the SME sector – as an educational and research partner hampers the function of *generating and transferring knowledge*, recreating the dynamics of the innovative business district. Looking only at large international companies, as the university does, might have adverse effects as far as spin-offs for the local environment are concerned. The Unité initiative – where networks are stimulated among government, university, HBOs and SMEs – is just one year old at the time of writing. However, it is recommendable that Unité becomes a leading axis for economic development policy. In this framework, there would be the opportunity to involve the university more closely in local development strategies, enhancing the practical orientation of their study and research activities.

Notes

1 In Amersfoort, HKU runs a unique Carillon School.

2 A notable exception is provided by the Faculty of Media and Technology, which is in Hilversum, where it enjoys the proximity of the local media cluster. Remarkably, this relocation did not receive any public support despite its strategic relevance, apart from some EC funding.

3 The housing stock in Utrecht consists of 101,000 dwellings, of which 39 per cent are owner-occupied, 16 per cent private tenement houses, and 55 per cent corporation tenement house. Some two-fifths of the dwellings are one-family houses; the remainder are apartment buildings (high or medium-high).

4 Nowadays, the capacity of Utrecht to attract overnight staying visitors is moderate: with 4.1 million stays (2000) it comes after Amsterdam and Rotterdam as the most visited Dutch city.

5 Utrecht was the first Dutch city to develop a station as a high-density multifunctional area. This includes the *Hoog Catharijne* complex, which stretches to the very centre of the city, containing a commercial mall, many offices and apartment buildings, officially opened in 1973.

Chapter 10

Venice Case Study

10.1 General Introduction to the City of Venice

Venice is the capital city and administrative centre of the Veneto region, in the heart of the Italian northeast. The Municipality of Venice extends for 189.4 km^2, partly on islands located in the middle of the Venice lagoon, with the historical centre occupying a major cluster of islands. The rest of the municipal territory includes a section facing the sea, and a main inland section including a middle-sized city in its own right, Mestre – 150,000 inhabitants – and Marghera, a smaller town but a major industrial and port area.

The metropolitan area nicknamed 'Pa-Tre-Ve' covers the municipal territory of Venice, and extends to the neighbouring provincial cities of Padua and Treviso. The former is an art city and main service centre, site of one of the most prestigious Italian universities. The latter is a smaller historical city, mostly famous for being at the heart of a fashion district with a very distinct brand image.

Venice is a well-known international attraction, possibly the most famous tourist city in the world; yet its historical centre in the heart of the lagoon is a 'problem area'. With young households pushed out of the centre by inaccessible housing prices and lack of specialised jobs, the population in the old city fell from 170,000 to 70,000 within about 50 years (Figure 10.1). The recurring floods are a source of economic uncertainty and additional costs for entrepreneurs, who find it more convenient to relocate inland. The difficult accessibility of the old city explains why many residents have moved out of town following the jobs. However, the very survival of the city is at stake. Unfortunately, 30 years of discussion on how to defend the city from the increase in the sea level did not sort any agreement on the remedies, so that many important investors are scared away because of excessive risks.

Whereas the historical centre and the smaller islands surrounding it concentrate on the beauty and richness of Venice's millenary wealth as the Mediterranean centre of commerce, artistic production and military power, the economic heart of the city is more and more shifting towards its 'back garden'. At the provincial level, the City of Venice is losing population and

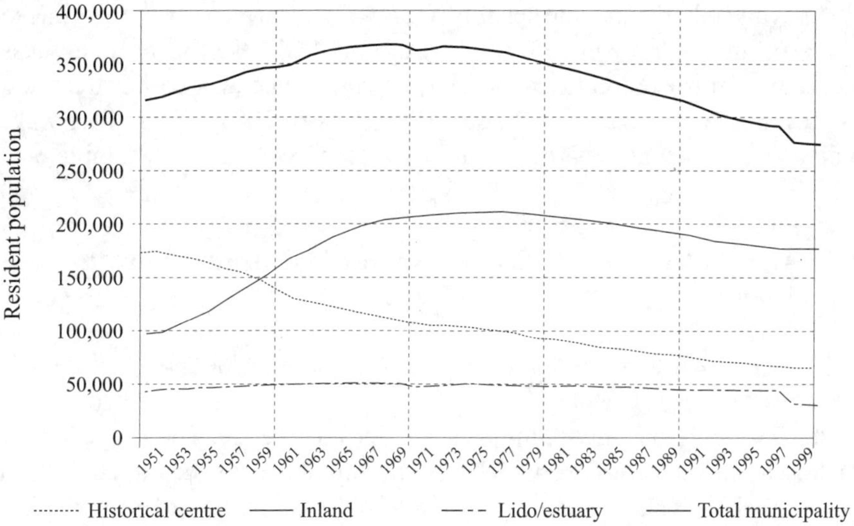

Figure 10.1 Population structure in different sections of Venice Municipality, 1951–2000

jobs in favour of the rest of the province, both to neighbouring and non-neighbouring municipalities.

Since the 1930s, the city has been able to boast an industrial area of prime importance. However, the restructuring of the Italian heavy industry, the loss of competitiveness of the industrial port and the huge environmental problems caused by petrochemical productions have caused a downward drift in the importance of Porto Marghera. Fortunately, rather than provoking a real crisis in employment, such decline has gone hand-in-hand with a structural change in economic activity, from the polluting and state-supported port industry to a highly competitive and spatially dispersed district economy, driven by small and medium-sized enterprises. Hence the unemployment rate in the area continues to be one of the lowest in Italy, reaching full employment in some municipalities, and averaging 5.1 per cent at the provincial level.

The metropolitan region of Venice is fully embedded in the so-called 'third Italy' economy, though with distinctive characteristics, and with emerging problems that could cause a loss of competitiveness in the long term – the most important of which is the poor quality of the regional infrastructure. Also, it is acknowledged that the quality of the human capital needs to be improved structurally.

In this context, the historical core of Venice is experiencing problems in finding a role. Tourism is a very strong sector and provides the city with jobs, revenues and other intangible benefits (the international atmosphere, the image of a beautiful and romantic place). Yet, the city finds itself increasingly less at ease with this economic specialisation, which generates costs, stress and displacement of all sorts (cf. Costa, 1996; Russo, 2001). Therefore, ways to diversify the economy are sought, but the context is not a favourable one.

10.2 The Student Community and Knowledge Infrastructure in Venice

10.2.1 General Information on Higher Education in Venice

Venice has a long-lasting tradition as an education centre. The University of Ca' Foscari in Venice was born in the second half of the nineteenth century as a Superior School for Commerce, with the goal of forming a class of entrepreneurs and clerical staff which would live and operate in the city. The model at which the new school was looking at was an international one right from the beginning; the study of foreign languages was seen as a natural complement to that of economic disciplines. The strong community ties have been present since its foundation. Higher education was conceived as an economic 'function' of the city. After the Second World War, and especially in the 1960s, the school again began to forge a visible role in the community life.

However, further university development resulted in a substantial lack of support for the university development plans, on account of the rising 'conflict' over the use of the urban space, with the massive growth of the student settlements going hand-in-hand with the ongoing exodus of residents. The outcome of this attitude was a low-profile, incoherent development in 'bits and pieces', rather than a growth process based on strong options and an overall vision of higher education as *the* strategic activity in postmodern, tourism-submerged Venice. Things started to change in the 1990s, when the city began to plan university settlements according to a vision. Disused manufacturing sites in the old city and in the inland district of Mestre were chosen to host faculties and services of the two main universities, regrouping academic activities in three main 'campuses'. This decision was much debated, especially in the light of the existing alternative possibilities and of the increasing demands placed by national regulations. However, this reorganisation process is now in progress.

In the last decade, the institutional ties between the city and the university have grown stronger. The Venetian academies are acknowledged as, possibly, the only Venetian 'large enterprises' still growing, one of the city's most vital assets, producing revenues, jobs and all sorts of intangible benefits. The 'marketability' of faculties such as Economics, Architecture and Urban Planning in a context such as that of Venice and its region should not overshadow the importance of the humanities for a city that has cultural excellence deeply rooted in its very existence.

Today, there are four higher education institutions in Venice Ca' Foscari University, IUAV, VIU (Venice International University), and the Academy of Fine Arts. Moreover, there are a number of smaller educational and research centres such as the Music Conservatory. The first two are scattered among a number of locations throughout the historical centre (Figure 10.2), and more recently in the inland area of the city and in neighbouring towns as Treviso, Mira and Portogruaro. The two bigger Venetian academies support a number of collateral activities, such as student housing (both private and in dedicated housing structures), cultural facilities, specialised shops and commerce, sport facilities. These are also concentrated in the 'university area' (the Dorsoduro quarter, in southwestern part of the historical centre).

10.2.2 The Knowledge Sector of the City: Actors and Spatial Patterns

Universities and other higher education institutions In total more than 25,000 students are registered on courses offered by the two main higher education institutions of the city (Table 10.1). The number of foreign students registered in regular courses is rather high, at 290 in 2002, and on the rise. Many other exchange and part-time foreign students spend some months in the city.

The principal higher education institution of the city has its headquarters in the historical palace of *Ca' Foscari*, from which it takes its name. The university occupies 18 buildings in the historical centre, two locations in the inland conurbation and three locations in other cities. Ca' Foscari is mainly oriented towards social sciences and humanities, although in the last two decades it has managed to diversify its supply. Today, it includes four faculties: Economics, Sciences, Foreign Languages, and Literature and Philosophy, offering a total of 13 first degree courses plus 13 three-year diplomas (or 'short degrees'). The dropout rate is rather high; despite the new national regulations on study curricula, with only 8 per cent of students obtaining their degree within the regular period of four years. The university also offers 12 Masters courses (MAs), some organised in cooperation with other universities, and ten specialisation courses, plus many doctoral programmes. In total, more than 16,000 students are enrolled in first degree courses, 6,000 of which are

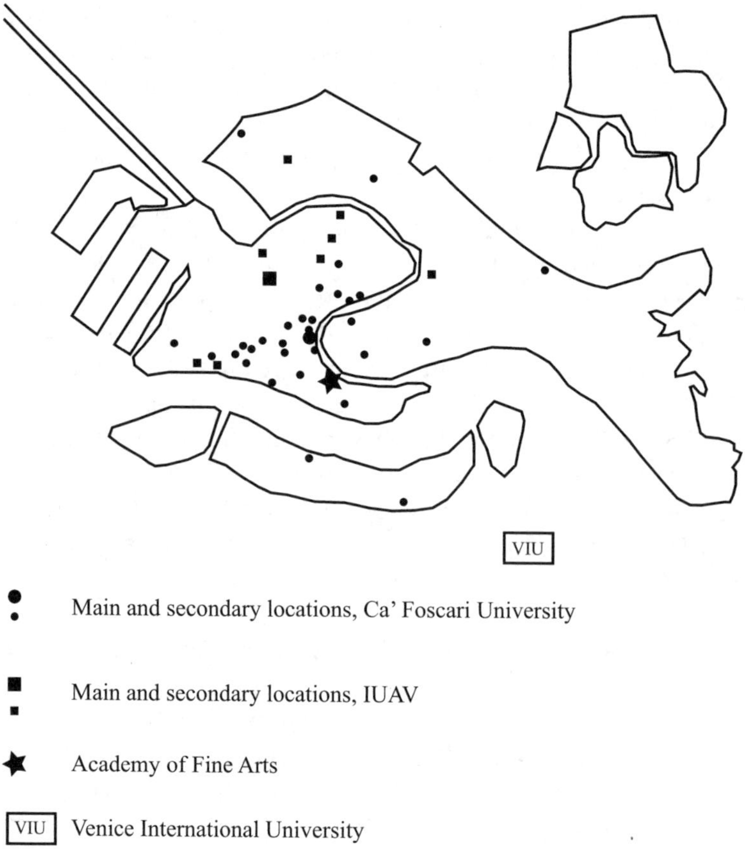

Figure 10.2 University settlements in the historical centre of Venice, 2001

in the Faculty of Economics. This number has declined since 1993, due to the demographic decline and to the opening of competing faculties in the proximity which intercept a remarkable share of newly-enrolled students. The number of first-year students is also declining (from 2,800 in 1999 to 2,300 in 2001). On the other hand, the emergence of short diploma courses offers an alternative to full degree careers; in 2002, more than 1,800 students were registered in these diploma courses.

The bulk of educational activities is organised around two main poles: the S. Marta-Zattere complex, a former maritime-industrial site in the southwestern part of Venice, and the new S. Giobbe complex on the north side of the city, where the Faculty of Economics is almost entirely housed. Some lecture halls

Table 10.1 Registered and first year students in various faculties at Venice universities, academic year 2001/02

First year students

Faculties	Total registered students	Of which, % long degree courses	% change over academic year 2000/01	Residents in Province of Venice (%)	Residents in other provinces of Venice region (%)	Foreigners* (%)
Economics	1,146	72.8	9.9	35.1	60.4	2.6
Literature and Philosophy	753	84.2	–4.7	34.0	52.5	0.9
Foreign Languages and Literature	640	94.4	–11.7	25.9	42.7	2.3
Sciences	369	63.1	35.7	50.4	45.0	0.3
IUAV, Achitecture	987	80.4	11.5	19.5	54.6	4.6
Total Venice university faculties	3,895	79.6	4.8	30.9	53.0	2.5

Total registered students

Faculties	Total registered students	Of which, % long degree courses	Residents in Province of Venice (%)	Residents in other provinces of Venice region (%)	Foreigners* (%)
Economics	5,955	82.6	34.8	59.4	0.9
Literature and Philosophy	4,874	90.5	38.5	47.1	0.9
Foreign Languages and Literature	3.875	97.2	28.4	37.8	1.9
Sciences	1,522	85.2	54.0	42.4	0.7
IUAV, Achitecture	8,915	92.8	17.3	52.1	3.3
Total Venice university faculties	25,142	90.1	29.5	50.1	1.9

* Regular students, excluding exchange students.

are decentralised to locations around town. The Information Science Faculty is almost entirely located in the inland industrial area of Mestre.

The headquarters of the *Istituto Universitario di Architettura di Venezia* (*IUAV*), one of the oldest and most prestigious architecture schools in Europe, are in the Tolentini Complex, by one of the main entrances to the historical city. Similarly to Ca' Foscari, IUAV has been forced to spread its activities over a number of historical buildings throughout the centre. Recently, disused industrial sites have been taken over by IUAV for more space-consuming activities. IUAV was born as an athenaeum of architecture; over the last 30 years education and research in the field of urban planning has also been developed. After the general reform of academic studies, starting from the academic year 2001–02, the three study courses of IUAV (Architecture, Design and Spatial Planning) offer five first degree triennial courses and 11 specialised degree biennial courses. Eight thousand nine hundred students were enrolled in 2001/02, a decrease over 1998–99 when 10,554 students were registered. Six hundred and forty-three students were registered on three year diploma courses. Five main doctoral programmes are also offered, plus ten other as shared programmes. Like Ca' Foscari, IUAV is deeply embedded in a large educational network that serves as a tool for mobility for students and faculty staff. Inter-university agreements and shared programmes are held with many countries all over the world, ensuring the international mobility of students and staff; arguably, Venice is a privileged destination of exchange programmes. IUAV has endorsed a number of continuing education programmes for teachers and administrative/technical staff.

The *Venice International University* (*VIU*) is the most recent addition to the higher education club in Venice. Born as a result of a partnership between the two Venetian athenaeums and four foreign universities, an important banking foundation and the Province of Venice, it provides higher education and a number of research centres, through undergraduate, graduate and continuing education courses, and scientific research. The format is that of selective, targeted educational supply, backed up by centres of excellence in research and by strong links with the rest of the local academic community. The location of VIU is an important trademark of its activity and represents a unique example of a 'campus' in the Venetian higher education panorama.[1] VIU's educational activities are organised as a joint undergraduate programme in cooperation with the partner universities. Each associated university offers teaching staff and a number of courses in English – the official teaching language of the VIU. Currently, two Masters courses (MAs) are offered, one in Economics and Finance and one in Cultural Mediation towards Investment and Integration. Moreover, VIU regularly organises summer schools. The total number of students enrolled in VIU courses each term is around 100, many of them coming from abroad. The contributions of the partners are not limited to the organisation of educational programmes, but also extend to research activities, through which VIU has established an important link with the local socioeconomic context. A number of research centres are based at the VIU. Among these, two can be singled out for their potentially fruitful contribution toward local issues. TeDIS (Centre for Studies on Technologies in Distributed Intelligence Systems) is concerned with the impact of ICT on Italian industrial districts as a departure point for general considerations about the evolution of new technologies in an economic scenario, where SME networks play a dominant role. The rooting of TeDIS and its experts in the northeastern economic community ensures the practical relevance of its high-profile research output. The TEN Centre (Thematic Environmental Networks) coordinates and supports research on environmental issues at an international level. The VIU stands out as the result of a cooperative effort between a network of educational and research partners and important sectors of the local community; its embeddedness in local and international networks is therefore a basic and constitutive element. However, it is relatively young (it was founded in 1997) and its role within the local networks can be expected to develop and become more complex in years to come.

Ca' Foscari, IUAV, the Academy of Fine Arts and – more recently – VIU interact and influence the local socioeconomic environment in many ways and at different levels. The role of the universities as education and knowledge production centres, and their rooting in the society, leads them to be partners in or members of many local institutions and projects. Ca' Foscari and IUAV actively supported the birth of VIU, the Science Park VEGA, the THETIS consortium of Maritime Technology, and *Veneto Innovazione*, an incubator of innovative projects and processes. Moreover, they play a crucial role in the management of many Venetian cultural institutions. Finally, they participate in physical redevelopment and regeneration processes in the city, providing precious inputs to city planners.

The teaching class itself is a prominent sector on the Venetian society. Most faculty staff – especially within IUAV – set up their professional firms and offices in the city, became citizens of Venice, and are actively engaged in the socio-political life of the city. For instance, they are regularly elected to administrative bodies such as the city or region councils, they participate on the boards of directors of public companies, they manage important cultural foundations, and so forth. It is worth recalling that the two most recent mayors of Venice come from the academy. There are, however, differences between the universities. Whereas IUAV staff and departments are widely involved – through projecting activity – in the process of the physical transformation of the city, and certain faculties at Ca' Foscari contribute with their research and educational output (e.g. the Environmental Science Program), the synergies with the city could be increased with respect to other faculties such as Economics, Business, Information Sciences and Design. These have clear strategic importance in the continuous process of redefining of the city's functions and infrastructure.

Research and technology transfer Venice's scientific park is VE.GA, a cooperative effort between the city, the universities, the port, and sectors of industry. The park, which is still at an early stage (a second building hosting state-of-the art facilities has recently been opened), accommodates small firms and research groups engaged in knowledge development and transfer, in particular in fields such as information technology, environmental management and biochemical sciences. The centre is slowly gaining status in the difficult context of the dismantling of Venice's industrial port, where it is located, facing the main corridor that links the city with its hinterland.

Venice is the seat of many research organisations, some of which are linked to the universities, while others were set up by city government institutions, such as *Consorzio Venezia Nuova* (dealing with the lagoon hydraulics), COSES, and *Consorzio Venezia Ricerche*. Moreover, private research groups at an international level, like *Fondazione ENI Enrico Mattei* and UNESCO Venice Office, have chosen Venice, despite its peculiar problems, because of its prestige and exposure. In short, hosting a working meeting in Venice is much more attractive (though more expensive!) than doing so in many other places in the world. In the field of human sciences and history, Venice also hosts important private institutes and 'fondazioni', a network of not-for-profit institutes dealing with the promotion of local culture, its conservation and promulgation.

Spatial patterns of higher education and research Though the prestige of Ca' Foscari and IUAV goes well beyond the regional context, the availability of numerous alternatives to economic studies and the relative difficulty of finding a house in Venice limit its 'reach'. Therefore Venice's universities appeal to a student base that is mainly regional (approximately 30 per cent of students are resident in the province of Venice, and one-half of them come from the rest of the Veneto Region). The number of commuters is between 18,000 and 20,000 (with a daily mobility pattern of around 3,500 students on busy weekdays). These add to the 1,500–2,000 students who are resident in the city or in its immediate surroundings. About 500 *fuori sede* are foreigners. The majority of students coming from the hinterland consist of 'commuters' and they only visit the city occasionally. The student population of IUAV is more 'outward oriented' than that of Ca' Foscari, due to the relative lack of competing institutions nationally, and to the institution's international prestige. In total, the number of students who stay temporarily in the city ('*fuori sede*' students) is estimated at 7,700 (4,150 at Ca' Foscari and 3,575 at IUAV (Bonifacio et al., 1994)) out of a total of almost 30,000 university students. *Fuori sede* students are accommodated in private rooms and flats (*c*.6,000), or in 'institutional', subsidised residences (1,300). *ESU*, an organisation affiliated to the regional administration, has the task of enhancing participation in higher education. It manages a number of student residences in the historical centre (five) and in inland areas and distributes grants with which students may rent private flats and rooms. Moreover, ESU manages a number of catering services at special prices, and other facilities for students. Other student housing is provided by private (usually religious) organisations.

Private accommodation is available on the market at standard prices, but finding it is difficult and in many cases accommodation is offered with informal arrangements on the black market. Neither of the two major Venetian higher education institutions has a real 'campus' structure, though there are concentrations throughout the city as regards educational facilities and student residences. Student housing is concentrated in the university area, but inland areas where rents are cheaper are also popular among students. A research project of some years ago (Bonifacio et al., 1994) contributed to shedding more light on issues such as the housing conditions of students and on the degree of 'overlapping' between the housing market for students and for residents. It is estimated that around 6,500 bed-places were offered in 1,650 private flats (the figures for the historical city were respectively 4,800 and 1,280). Students as tenants have little protection (40 per cent of tenancy agreements are 'informal', that is, not stipulated by a formal contact). Consequently, the quality of housing and maintenance tends to be rather sloppy – 17 per cent of bed-places are at ground floor level, a serious problem in a city like Venice where flooding is recurrent and the housing stock is in a bad condition. The total estimated expenditure by students living in the city amounts to €25m, plus €10m in housing rents, though the estimated per capita rent of €167 for a room in the city centre is likely to have increased considerably since the report was published. Recent trends, however, are not favourable to student housing.

In fact, whereas in past years lodging students was an attractive method for landlords to earn revenue, putting to value a remarkably underutilised housing stock, there is a risk that tourism will take many houses away from the student market in the future. This opportunity has arisen from the new regulation (L. 431/98) on tourist accommodation, which allows private operators to enter the market. Since tourists demonstrate a far higher willingness to pay than do students, it is likely to 'crowd out' students from the market; at the same time, it forces landlords to formalise rental agreements. In fact, only a few landlords in the historical city have decided to use the new regulatory framework to rent to students; the situation is different in Mestre, a less attractive place for tourism accommodation, where 58 private flats have recently been put on the market for students (*La Nuova Venezia*, 15 June 2001).

Universities are among the main economic and institutional players operating in the historical city of Venice. In recent years, the debate has focused on the possibility that university reorganisation might be 'harnessed' to make it consistent with the general strategic plan for the historical city. The original plan aimed at redistributing university resources in the territory, developing an east-west polarity between the port area, where Ca' Foscari would resettle its faculties, and the old shipyard (*Arsenale*) areas, to be reconverted to civilian use, hosting the whole architecture campus. It was hoped, in this way, to counterbalance the progressive westward shift of the city's economic activities. The project was eventually abandoned because of emerging problems with the state office in charge of protecting monuments and historical assets. Moreover, it met with opposition from 'insiders', like conservative political and social groups, or university staff worried about the relative inaccessibility of the Arsenale area.

The city has seconded expansion plans when possible, conceding to pressure groups on the Arsenale option, which would have been an important achievement for two reasons. First, it would have been an opportunity for the regeneration of a socially declining area, with possibilities of fruitful synergies with the Biennale showcase located nearby. Secondly, concentrating the whole architecture campus in one enclosed area would have limited the opportunities for conflict between city and academy. However, the city had other plans. The underlying idea of the recently-approved master plan for the city of Venice (the first such plan in the last 40 years) is the functional 'reconnection' of the city and inland parts of Venice, with the economic weight of the city moving towards the two ends of the bridge over the lagoon, and the spreading of the cultural-tourist activities towards the periphery. In this context, higher education would be reorganised in a 'university cluster' around the two ends of the bridge that connects Venice to its region. The university would relocate in a 'triangle' (cf. Figure 10.3) with two vertices in the historical city, and one in the inland area. The latter campus would host all the science faculties and laboratories, plus student housing and facilities.

The only 'losing' actor in this process appears, again, to be the community, which bears the costs of an increased concentration of students in an area where conflicts are likely to emerge, when the opportunity existed to achieve a more balanced and sustainable settlement model in a peripheral area. Even though the southwestern student district offers plenty of variety and attractions to the resident population, the negative side seems to prevail at the present stage. The lack of communication between the university and the community has played a major role in nurturing the reluctance with which the community – well-represented by neighbourhood councils – sees the expansion process. Apart from this, there is generalised consensus on the present direction of the university settlements, clearly an outcome of a long-term strategic plan, in opposition to the hesitations of the past. *Fuori sede* students have an easier life, and the 'diffused campus' model promises to increase further the possibilities of integration with the local community. The shift of cumbersome (and hazardous) R&D activities to isolated areas in the mainland is also to be welcomed.

The city is also planning the expansion of student facilities and infrastructure, along with reorganisation of education poles, but these do not

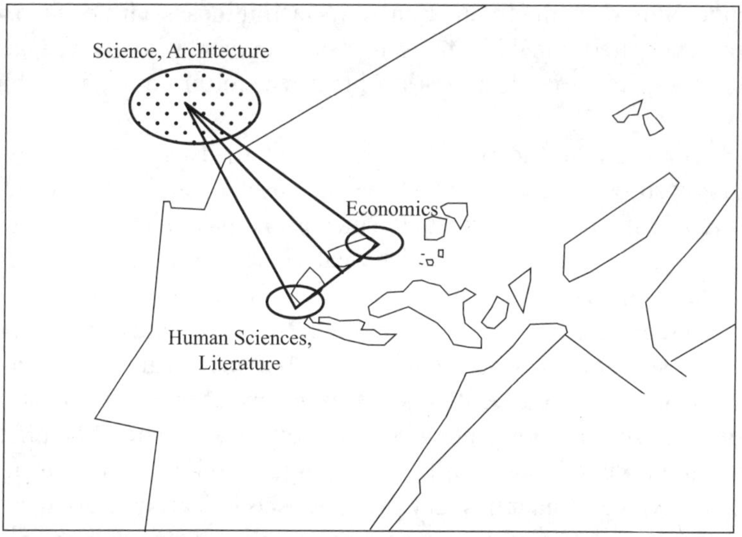

Figure 10.3 The plan for the reorganisation of the universities

always proceed at the same speed, nor can the city always get what it wants. Moreover, no real plan has been advanced so far to improve the mobility both within the 'triangle', and in the metropolitan regions, where students are supposed to circulate, participating in decentralised activities and programmes. In these conditions, the integration of educational activities in the territory is not guaranteed, and, moreover, students will not be free to exploit to the full the opportunities offered by the 'credits' system according to which they can build their curricula with a variety of activities.

10.2.3 The Student Community: Behaviour and Careers

Students come to Venice for two main reasons. Firstly, because of the prestige of its institutions. Until some decades ago, Venice was the only pole for education in social sciences in the Veneto Region, and IUAV is still the most prestigious architecture school in northern Italy and one of the most important in Europe. Secondly, many students like to spend part of their life in a city like Venice. Venice is clearly on a more human scale than other big education poles in northern Italy like Milan and Bologna, and has a reputation of 'cultural hub'. This applies especially to students from southern Italy and from abroad. Today, this attraction capacity of the city is on the wane – it is so difficult to find cheap housing in Venice that most 'regional' students prefer the new alternative faculties set up in recent years in Padua, Verona, and Udine. Moreover, the 'reputation' of its institutions is a little faded. They are seen as very traditional institutes in an environment where the capacity to adapt to new specialisations makes and breaks the fortune of long-established schools.

However, students do populate the urban landscape massively, and increasingly so, as Venetian residents become used to renting to students and see this as a profitable way to fill in the increasing mass of idle housing space in the old city. Student areas in the historical centre of Venice are immediately recognisable – lively and diverse even at night (something quite unusual for the city), and presenting a markedly different functional and commercial mix from the most tourist-crowded central areas. In recent years, sites such as S. Margherita Square and the 'Zattere' waterfront have been literally transformed by student activity, changing into favourite spots for sophisticated tourists and arts and culture practitioners, who often hang around in the dozens of bars, bookstores, typical restaurants and speciality shops. This has raised an issue with residents in these areas, who are not completely at ease with the noise and the vivacity.

A number of *student associations* operate within the two major Venetian HEIs. In Italy, organised student activity is a tool for the self-recognition and representation of students in university administration and government. Some are groups linked to national political and cultural organisations. Others are organised around special interests, student careers (e.g. AIESEC) or cultural themes, organising stages, student exchange programmes, special training courses, etc. At both Ca' Foscari and IUAV many groups are active on the cultural side, receiving grants and support for academic bodies to organise cultural programs, theatre activities, reading and performing groups, etc. Some of these groups participate in regular polls to elect their representatives on academic bodies such as the *Senate* (or Council) *of students*. This is a formal government body of the university, which serves as a discussion and advisory forum, with administrative and financial responsibility on some issues related to students' life and careers. The Senate of Students may acquire wider functions; for instance, at IUAV, the Senate runs a number of initiatives to improve student life. A number of *alumni networks*, constituted by former graduates of Ca' Foscari, connects graduates with the job market, providing information, links, support, and keeping databases to allow an effective matching between supply and demand. Among these, ALUC aims at promoting and valorising the human and scientific asset constituted by the graduates of Ca' Foscari, increasing their personal and professional relationships, and acting as a point of contact between the graduates, Ca' Foscari and the job market. ALEA organises all sorts of workshops, stages, expertise centres and newsletters to 'increase the dynamism of communication and knowledge exchange, overcoming cultural, territorial and sectoral barriers'.

There is little information regarding the residential choices of students after they complete their study period in the city, though the data on the job positions occupied by Ca' Foscari graduates indicates that the greatest part of these jobs are outside the city centre and even outside the municipality. The tightness of the housing market for new households, especially in the medium-high quality segment, discourages new graduates from settling in Venice. Moreover, the the most highly-skilled graduates from the university of Venice easily find better jobs in the hinterland, with economists and scientists working for the rich industry of northeast, architects and planners opening offices in growing urban areas like Padua or Trieste. Venice offers very little to them, at the present time. Those who are lucky enough to start a career in research or education in the local institutions, may profit from the city's incentives for setting-up new households.

10.2.4 Challenges for the City

Excluded from the broad network of international trade that made the fortune of the 'Serenissima' Republic, and increasingly disconnected from the booming backyard economy which is antithetical to its ancient splendour, Venice faced meagre growth in services economy of around 2 per cent in the decade 1981–91, compared to increases of 20 to 40 per cent in the surrounding cities (Rullani and Micelli, 2000), and scores badly in juvenile unemployment, having the highest rate in Veneto region at 13 per cent. Rather than being the result of deliberate policies, this situation is one in which no formal player has the authority and political strength to enforce a 'vision' of a modern, sustainable Venice. The dominant tourist vocation of the city is hardly a leverage to stimulate investments and synergies in other sectors.

The recent change in the political environment, with the direct election of a mayor and city government (who are granted in this way more responsibility and authority) has fuelled a wide-ranging debate on the strategies to reverse the existing trend and initiate a 'virtuous cycle' of development. The assumption is that the structural barriers that have kept old Venice disconnected from the economic boom of its hinterland can be removed with a modern and comprehensive infrastructure policy. Once this is achieved, the city can count on plenty of complementarities with its region. Moreover, in Venice there are underutilised locations and other 'immaterial' resources that can be put to good use. In this context, the possible new 'vocations' devised by scholars and practitioners are:

- a strong impetus to the cultural sector, no longer intended as a system of consumption of the existing assets (heritage tourism) or a 'window' on international art, but as a complete production cycle starting from intellectual thinking and education, and seeking cultural excellence in relation to a number of art forms, from fine arts to 'subcultural' movements;
- a focused promotion of the intangible location elements of Venice – its reputation and fascination, its international orientation – to make it the ideal seat for the representation functions of important local firms and companies;
- the development of the city as an R&D centre for new economic functions, overcoming the physical isolation through 'immaterial' linkages, and re-utilising the peripheral, dismantled production sites within and outside the centre as technology districts.

Whatever development path is chosen, there are common conditions to be fulfilled with urgency. First, even though the 'accessibility requirements' vary substantially between the three scenarios indicated above, Venice has to be faster and easier to reach, both physically and 'virtually'. Therefore, substantial investments in new and diverse transport modes and terminals, as well as in cables, routers, antennas and other ICT devices have to come fast enough to beat competing locations.

Secondly, all three scenarios require excellent, up-to-the-minute and market-oriented educational facilities and services. The higher education system must work *for* the city, providing the human capital, the knowledge, and the creative inputs required by development strategies. For practical purposes, it does not really matter whether this information input comes from the local universities; others would be interested to become 'providers' – for instance, the University of California and other international universities are opening offices, summer courses and representation facilities in the old city. Yet, the benefits for the city would be more 'localised' were the local institutions to play a central role.

The city has already begun, though with some difficulty, to take steps to enhance the impact of student communities and minimise the occasions of conflict, but it is not clear whether the private sector and the students themselves will comply with it: the first signs are not very promising. Under a recent law, the city pays part of the loan interests on selected housing. This measure has been particularly popular in the circle of academic workers, showing that good jobs always have to go hand-in-hand with accommodation policies. On a larger scale, there is still much room in Venice for the new 'knowledge workers', given the large number of empty dwellings that the city offers. But the constraint is, rather, on the availability of jobs. So far, entrepreneurial initiatives have been very unforthcoming, and it is not clear yet if Venice offers tangible advantages to those who wish to set up firms in the creative or high-tech industries. Clearly, both the entrepreneurial attitude of new graduates and the establishment of new activities have to be accompanied by strategic, concerted actions by City Hall, with its economic departments, the universities, and the private sector organisations. The information in the following sector can be utilised to find innovative solutions to this dead-end, building a general and integral strategy for the city development as a place for culture, education and research excellence.

10.3 Strategies and Integration. Venice as a 'Sustainable Student City'

10.3.1 Relations between the Students and the Community

The 'community of reference' of the city is not just that of the voters-residents, but extends to other 'invisible' – but highly relevant – categories such as businesspeople, commuting workers and students. A sensible administration has to interpret the needs expressed by these categories, increasing in this way the attractiveness of the city and the opportunities enjoyed by the resident population.

In this field, however, every player involved seems to go its own way. On one hand, the economic forces of the city are more interested in a short-term, predatory attitude towards students than in treating them like 'customers'. This attitude is not that different from the way in which they are used to dealing with tourists. During their years in Venice, students have to cope with eager landlords, unfriendly retailers and crowded study spaces. Commuting students hardly have time left for a drink after wandering from one place to the other to follow a number of classes before taking their train back home. As a consequence, *fuori sede* students create their own 'interaction spaces' – parties in private homes, the summer nights at S. Margherita, a few concerts and films around town. On these occasions, exchanges with the local community are limited. The occasions of conflict are more numerous, when citizens complain of the excessive noise or the increase in commodity prices driven by 'imported' demand. Students are hardly involved in decision-making; the only institution they have a dialogue with is the Region through its agency ESU. After the completion of their studies, they can take profit from one of the many private sector initiatives to look for a job that fulfils their aspirations, or to self-diagnose their capacity as entrepreneurs. But this is no guarantee that they will stay in Venice; it is most likely that would find better jobs in the hinterland or in their hometown.

The *Municipality* is an immediate point of reference for higher education policy because it makes available the areas and infrastructure that are needed for development. Formally, its role does not go much further. It has neither a competence on higher education management and orientation, nor has it sufficient financial resources to be directly involved the development process. However, the sensibility and strategic capacity of the city administration clearly influence higher education policies more than is written in its mandate. For instance, it negotiates the use of strategic areas, acting as a 'broker' between the owners of such areas and the universities; it puts together resources needed to form strategic partnerships such as consortiums of investors or science parks; and it defends the interests of categories and citizens when universities develop or relocate, for instance issuing permits for collateral activities. The assumption is that if the student housing situation is solved, then the balance of host-guest interaction will be again on the positive side. Therefore, the city started to prepare suitable housing units in a separate market from the residents, with the idea of tendering it to private operators, who would offer it at 'normal' market prices to students together with adequate facilities (e.g. laundry services, PC connections, study rooms, sport facilities, etc.).

A more empowered actor is the *Veneto Region*, which has been delegated by central government to deal with many matters concerning education, mobility and research. Moreover, most of the necessary funds for infrastructure development (like European grants) come from the regional government. However, the most 'visible' role of the Region is that of managing services for students such as housing, catering and curricula support, through the agency *ESU*. This agency targets 'disadvantaged' and 'meritorious' students especially though it has recently undergone a process of reorientation towards the whole student population. However, in this area there is competition with the City Council. ESU aims at expanding its activities to 'non-eligible' students, but has met resistance from the city which clearly prefers to go completely to the market.

A sign of ESU's 'business-orientation' is that they recently won the tender for managing a new student residence, beating competition from private operators. Moreover, they have devised creative arrangements to finance additional facilities in the expansion areas. Also, an important source of revenue comes from 'student tourism' – renting rooms to foreign students and young visitors during high season, when students are not away. This has created some problems with other players (the hoteliers have claimed unfair competition), but has generated a budget advance of €150,000 in 2001, making it possible to keep the normal rents unchanged for the next four years. Under the guidance of a student-president, ESU has started to address the issue of creating the conditions for the permanent residence of graduates in Venice. This will be achieved by increasing the occasions of 'shared use' of university resources with the residents (e.g. they have opened a ludothéque by one of the student residences, they organise open concerts and ethnic soirées). But, most noticeably, they support the professional delivery of services by organisations created by students, for instance a housing help desk and a project team, that may result in a number of 'life projects'.

On this side, they can count on the support of another local institution. The *Neighbourhood Councils* (elective bodies such as the City Council, responsible for one or more city districts) are important because they are the ideal interface between the host and guest communities. They have the pulse of the territory, and the capacity to communicate directly with citizens and with influential groups and associations in the community, such as the commerce retailers, the elderly, and committees of social housing tenants. However, the scarce resources with which they are endowed, as well as the lack of involvement by the city in the matter of planning, have limited their role; more often than not, these organs act as noisy but ineffective 'complaint rooms'. The most directly concerned of these councils, the neighbourhood no. 2 – West Venice, has recently tried a more comprehensive approach in negotiation, with some good results. Its president is active in negotiating with the city the concession of some buildings in which to locate such initiatives; and with the IUAV, that looks favourably on the 'outsourcing' of many services now provided internally by the university.

The local community could be right when they claim to bear the costs of the cumbersome presence of universities in the city, but they often neglect to consider the benefits, or they do not consider the long-term opportunities that are offered to the city by the students settling in the city. It is not infrequent to read on the same newspaper page of a new university facility having opened, or of the launching of a new research project, and of a deprived household chased away from its house under pressure from landlords who are going to speculate in the student market. Even if the two events are not necessarily related, this is what makes the news and this is how the locals perceive the university expansion. The city can change this state of things by improving planning procedures and communication techniques. Moreover, there should be a careful assessment of the consequences of physical expansion plans, so that the inevitable costs are always matched by comparable localised benefits.

In many instances, the community has not recognised the potential offered by student communities for social regeneration. This also needs to be communicated more clearly by competent bodies, but things can change only when a winning idea of a new Venice will start to become reality. If initiatives like the new cultural pole at Biennale or the regeneration of the shipyards for maritime economic activities were to break through, a generalised change of attitude in this respect is likely to follow. The recent administrations have taken action with the objective of facilitating the permanent settlement of students after the completion of their studies. The impression from insiders is that the beneficiaries of these measures, so far, have mainly been those postgraduate students who have had an opportunity to remain linked to the university with academic careers. There is little evidence, for instance, of private sector professionals moving to the HC of Venice. Obviously, the housing policy will be effective when the city has something more to offer to the skilled workers.

10.3.2 Relations between the Student Community and the Private Sector

The private sector is interested in the quality and quantity of graduates from the local HEI. Its organisations represent strong territorial interests that are affected by the university expansion strategies. Among the most active players, the *Chamber of Commerce* provides services for firms aimed at matching demand and supply of high skilled labour. Moreover, they run stage activities in professional schools and higher education institutions, and they organise a number of projects and services targeting specific groups or categories (female entrepreneurs, immigrants, craftsmen, etc.). The Treviso section of the Chamber had an important role in the launch of the education pole in that city, specialising in foreign commerce. The Union of the Industry Leaders (*Unindustria*) is also very active in this area. In addition, they organise the resources and actors for the development of areas of expansion (like the science pole in Mestre), and are concerned with the activation of new specialisations to serve the needs of associated firms, for instance in relation to e-commerce and the production of new materials. The TUTTOSTAGE project, a partnership between Unindustria and Chamber of Commerce, aims at building a web platform for matching demand and supply of in-company training, which has become an issue after a change in the academic requirements in terms of practical training, has led to a huge increase in demand for stages. A less active role is played by categories associations such as ASCOM, the city's union of retailers. This group is mainly concerned with negotiating with the city on occasions of the relocation of university settlements affecting the associated business. Moreover, they federate real estate agents, who are also clearly concerned about the university settlements.

Despite the opportunities offered by the recent reform for the diversification and the articulation of higher education curricula, and the efforts of the private sector associations in matching the firms' demand with the professional profiles created by the universities, there is still a long way to go. First of all, the universities are slow and bureaucratic in adapting to the new framework. The national ties of university administrations make any attention to the 'local' environment a break in routine. The rectors of the universities have to manage this change with care. This is a easier task for the Rector of IUAV, a smaller and more homogeneous university than Ca' Foscari. The latter has to mediate between the humanities orientation of a substantial part of its education, which is very reluctant to become 'market-oriented', and the pressure coming from the private sector. To identify opportunities and strategies in this sense, Ca' Foscari has created a 'brainstorm' body, the Committee of Social Partners, which brings together university, institutions, and the key figures in the private sector. In this way, they could connect with the *Province of Venice*, one of whose tasks is to keep the territorial balance of higher education activities, for which reason it participates in the university development plans that affect decentralised areas. In future, the Province will be responsible for the management of employment agencies, and so will play a crucial role in bringing together the demand and supply side of the job market, also influencing or orientating the curricula offered by higher education institutions.

However, the private sector still blames the university for the delays in adapting to the opportunities offered by the law, to the point that it has founded a consortium for the development of the university. This partnership between Unindustria, the *Generali* insurance company, the newspaper *Il Gazzettino*, and CARIVE, the main local bank, is in charge of negotiating the areas of development with the universities. This consortium had an active role in the planning of the Mestre expansion, and has accompanied its 'rooting' in the territory, helping the dialogue between universities and enterprises, with a focus on the 'immaterial' sectors. The consortium also funds the VIU's MBA, and has commissioned a study on the demand expressed by the companies in the metropolitan area for graduates that are skilled in e-commerce, with the idea of launching a new study programme at VIU. Substantial spin-off is expected from education on such issues, the more so since part of the training activities are planned at the Science Park VEGA, where the encounter between trainees, researchers and companies is likely to be intensive. It must be said that the strategy followed by Unindustria, coherently with its mission, is not just that of supporting the local universities whatever the cost. Instead, they favour competition between universities, so that they are ready to ally with non-local or even foreign universities (the University of California and MIT have already opened activities here) if they can provide skills and research of better quality. This should push Ca' Foscari into action; on the other hand, they are ready to help the universities when they make strategic choices, or to stimulate cooperation between regional institutions. An important dialogue regarding careers in business administration has been activated with the alumni association ALEA. It would be important that other disciplines express their own alumni network, a most effective instrument to provide a 'real world' interface between students and employers.

The private sector, which pays so much attention to job market conditions and physical relocation of university activities, is indifferent (or thinks it does not have a role) as far as the quality of life enjoyed by students in Venice is concerned. Unindustria is concerned with the entrepreneurs' location decisions, but they do not have a focus on keeping workers in the city. On one hand, this attitude confirms a common observation that investments in Venice are risky and entrepreneurs are not willing to engage in long battles with bureaucracies. On the other, it must be said that sometimes they are not ready to see the opportunities that are at hand. This attitude will have to change in the future, but it is clear that the city should support this change increasing the guarantees and speeding up the planning procedures.

10.3.3 Influence of Higher Education in the Relations between the Private Sector and the Local Community

Today, the universities of Venice face the challenge of being functional to the life of the city that hosts it. This necessitates a further evolution in the relationship with the city and territory, as an articulated and – socially and economically – complex entity. The international fame of IUAV and the specific local ties of its teaching class with Venetian society are guarantees that the physical fabric of the city is cared for and continually upgraded. The same kind of 'immaterial relationships' should be strengthened as regards other faculties. Despite the importance of economic studies, in fact, the output of the faculties, both in terms of research and human capital, tends to 'leak out' of the city. As a paradox, such integration has been achieved in Treviso, rather than in Venice, with a new short degree course in International Commerce, which responds both in orientation and articulation to the needs expressed by the local entrepreneurs. The same can be said as regards other courses of strategic importance, such as environmental science and information science. The decision of the two main universities to put together resources and cooperate in a wide-ranging programme of cultural studies is a sign that the universities are prepared to respond to stimuli from the local environment. In this respect, it is important that academies do not adopt a narrow conception of fine arts and are ready to embrace experimental or popular art forms. Arts and culture are one of the vocations on hand for a revitalisation of the local economy; the others, headquarter and 'new economy' functions, need comparable efforts. Some initiatives like the science and technology park and the VIU, promoted by the two universities, the local administrations and the business sector, go in the right direction, even if they have to be prepared for a 'scale jump' in their activity and necessitate of infrastructure improvements that at the moment are not on the agenda.

The Chamber of Commerce is also actively involved in the 'matching' effort in a number of national projects, some of them financed by the European Commission. The Region – through its agency ESU – is busy as well on this front and Province will follow suit, as it is now formally in charge of managing the employment help desks. The Province also had an important role in creating a strong and visible education and research pole at VIU, that provides a completely new model of small-scale facility but with high levels of interaction between research, education and environment, in an international orientation. From this point of view, VIU represents a case of 'best practice' for the city as a whole. To grow and become even more relevant, VIU needs to find ways to connect the human capital and skills formed with the city more systematically. The activity of research centres located there is highly strategic in this respect. The Province also learned the lesson, when accomplishing its mission of 'spreading' higher education on the territory, of focussing on local vocations and specificity. Hence, it promotes ancient classical studies in the archaeological area of Altino, or the Masters in Environmental Management at VIU in the middle of the Venice Lagoon. Clearly, this approach should be imitated by other institutional bodies. High profile skills in international law are highly sought after in the area, but the city's decision to give room to a prestigious Masters in International Human Rights organised by the University of Padua has not been greeted by many, who would have preferred a specialisation closer to the city's immediate aspirations.

10.3.4 Model Testing and Identification of Issues

The expansion of higher education functions in the city of Venice is not without problems, and if it carries on in the way it has begun, it might not respond to the demands posed by the necessity of building a new 'vocation' for the city, a precondition for a new stage of sustainable urban development. As with any other urban process of huge dimensions, this one has generated displacement and costs, causing resentment from sectors of the populations, before they can appreciate any of the expected benefits. Moreover, the cooperation between city and university has not always been smooth, not to mention the legitimate

– and rarely met – expectations of the private sector. Finally, students take decisions about their careers and residential habits that respond to a number of stimuli and intangible elements over which administrations have little control; but they are fundamental to the outcome of the regeneration process outlined above.

In this context, the ties between the HE sector and the community are not uncontroversial. Higher education is a fundamental function for the city, socially, culturally, and economically, and it pervades almost all aspects of city life. However, this importance has not been reflected in a strategic, proactive attitude of the community and its institutions towards it for a long time. The danger is that – disconnected, from the organisational point of view, from the context in which it operates – the Venetian HEIs will decide to follow the flow of many other organisations, leaving the historical city for a more favourable context in the hinterland, where the links with the community promise to be more rewarding, if not economically. The local ties between the two universities and the private sector are in fact on the weak side. HEIs are not seen as mattering as much as knowledge producers in the fields that could really bring Venice to a new stage of the development. Though the awareness of this distance is growing and there are initiatives to recoup it, the national regulations and ties are making this process of 'adaptation' slower than what is needed for a strong 'scale jump'. On the other hand, the private sector is today scarcely involved culturally or financially in higher education matters.

10.4 Summary and Conclusions

Building a Vision

In the conditions summarised in Figure 10.4, the city economy cannot really accomplish its definitive shift to the knowledge economy. A vision of a student-friendly city is needed to guide and coordinate many existing initiatives and programmes. This vision has to be generated and shared by all the categories of the society – simple inhabitants, commerce organisations, the business sector, universities and institutions. The city administration should take the leadership, complementing and 'harnessing' towards the public good the initiatives taken by the universities for the development and expansion of education and research activities. However, it must not be forgotten that many players in this process have interests and aspirations that the city has to cater for. It is especially so for the private sector, without whose support

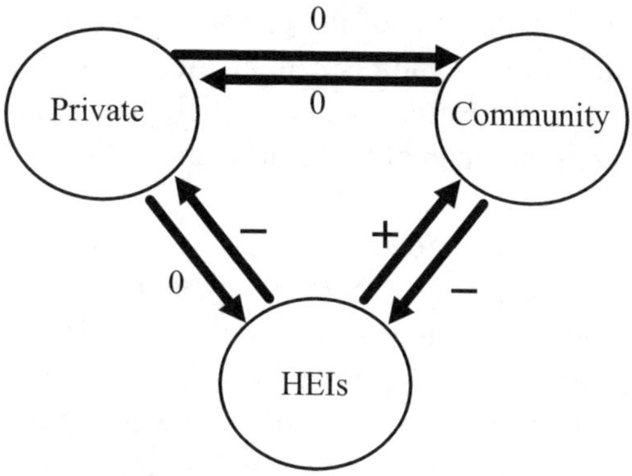

Figure 10.4 State-of-the-art: Venice as a student city

it is difficult to put the resources of the city to use and trigger a new cycle of economic development. The most critical issue to start with is student housing, according to the main parties involved. If effective solutions are not found, the community and the university will stay on opposite sides of the barricade, and any development plan will stay on paper. In this context, the degree of planning and the quality of negotiation of the city administration and the universities must certainly increase. A greater coordination between planning bodies, universities and those groups that have the 'pulse' of the housing market, like neighbourhood councils and the real estate agents' association, could help the city to identify the best possible solutions. One first step could be an agreement that foresees the realisation of well-equipped student housing units in specific areas of the city.

When this urgent problem is solved, administration, community and private sector can then unite to pursue the grand vision of Venice as the 'meeting place' of an economically thriving region, infrastructure node and education centre, capital of culture and artistic production, praised by well-educated and informed visitors. Culture is certainly a 'growth sector' for which the city has an advantage, inherited from history and preserved – though with some problems – in the age of mass tourism. Cultural tourism is clearly a way to put to value the city's resources, but one should not overlook the sectors connected with 'new' cultural production and consumption, knowledge generation and dissemination activities. That is, a whole cultural industry of which small but creative entrepreneurs, education and research facilities, and good communication infrastructure, are crucial elements (Rullani and Micelli, 2000).

The 'Venice problem' has so many facets that all the world looks at the Venetian academies to see how they deal with these issues, and look forward to establishing scientific links with them. A solution is feasible only to the extent that there is a dialogue between the involved players, the definition of a common vision, starting from the many advantages that Venice can boast as a knowledge excellence centre, and of the planning tools to achieve it. So far, the educational function of the universities has not been explicitly taken into account in this operation of vision building. The university is looked at as a source of physical and economic regeneration, but its role in forming the human capital that is needed by the city of the future has been overlooked (Mariotto, 1998). Now it is the time for the city to take an active lead also as regards such immaterial factors.

Higher Education Strategies

The situation of higher education in Venice is currently one of tension between two main objectives: 1) modernisation; 2) the accomplishment of the implicit mission of saving the city from socioeconomic decline.

For the first objective, it might be just a question of time before the bureaucracies involved learn to 'listen to the market', or at least become customer-oriented. A plethora of initiatives in this sense is supported both by private parties and institutions. However, time is pressing, as the representatives of the private sector suggest. If the local universities are not ready to take up the challenge of forming the skills and knowledge that are vital in a period of economic restructuring and opening borders, someone else should do it. The impending globalisation of the markets also extends to higher education: reputations easily fade, and local universities might become obsolete. This danger is especially relevant for Ca' Foscari, which faces strong competition from other national and international universities. At that point, the resource base of the city will be further eroded.

Therefore, *local universities need to take a more decisive step towards becoming competitive and market-oriented.* There is little danger that in this process the academic excellence of humanistic disciplines, which usually find it more difficult to 'sell' their output, would be undermined. On the contrary, this city offers the possibility of putting such skills to use in a cultural industry with an extraordinary market potential. The efforts made by IUAV and Ca' Foscari to join forces in order to provide the city with an up-to-date and integral programme of cultural studies and techniques proves that opportunities are there. Much more could be done. 'New economy' functions do not only require economists, who, anyway, have to be educated according to modern competencies: most economic tools taught today are obsolete or without any practical relevance. They also require the critical interpretation of recent social trends, communication strategies, modern marketing tools. *Education curricula of humanities faculties have to integrate such skills*, and this is indeed made possible by the credit system, allowing students to 'borrow' courses from other faculties or universities. On the other hand, the faculty heads have to acknowledge the demand from the private sector. Modern organisations require graduates with practical skills, otherwise they find it more profitable to recruit the workforce directly from high schools, and train it on the job according to their needs. In this way new entrants on the job market are paid less and lose flexibility. Therefore, an education curriculum that is heavily supported by on-the-job training, stages, foreign exchange, in-company final projects, etc., achieves the double goal of better preparing the students for an aggressive job market and improving their negotiation power with firms. In exchange, firms save on training costs and have more cultured, flexible and competitive human resources at their disposal.

Universities, however, face increasing competition from other institutions in the public and private sectors in delivering the resources – human capital and research – that is sought for by productive forces. Just because Venice is so attractive, it is very likely that many foreign institutions will begin to open offices here and compete with the local academies, for instance, as scientific consultants. *Therefore, it is necessary that universities start to promote themselves, to attract the best students, lecturers and scientific partners*. The issue of meritorious students is critical. For instance, universities should offer competitive scholarships to the best graduate students, who currently find it much more convenient to follow postgraduate careers abroad.

Finally, *communication and promotion are increasingly critical tasks for universities that are embedded in the local environment*. The university management should be concerned with negotiating (or at least auditing) development plans with all the concerned actors. The costs and benefits of such processes have to be assessed and communicated clearly. Resources and funds to minimise conflicts and impacts on development areas have to be offered to the community, for instance launching a programme of 'open spaces' of university facilities, targeting in particular deprived groups, young residents and the elderly.

As far as the second objective is concerned, many of the players involved recognise that a necessary condition to achieving a greater cohesion between host and guest community is to make the housing market for students more efficient and transparent. To this end, two things can be done more or less straight away. The first is to *pursue the implementation of non-subsidised student housing and facilities more decisively*, since this currently increases the 'separation' between the residents' and the students' market. Housing units for individual students with shared services in groups of three or four are the model in many European countries, and they are clearly inappropriate for families. These units should be located in attractive (but not 'secluded') areas like Giudecca and Zattere, whereas the creation of housing in the Mestre pole is clearly unattractive for students; a better solution would be to improve transport connections between the area and the historical centre. At the same time, more cooperation rather than competition with other institutions should be sought. ESU has shown itself able to diversify its tasks, becoming a useful instrument for the whole student community. This process of reorientation must be supported by the city.

Housing Market and Student Settlements

Since a consistent number of students will continue to stay on the 'residential' market, it is necessary to *monitor prices and availability* to the benefit of students *and* of the community. Neighbourhood Councils are the 'natural candidates' for this task, utilising the precious database on the consistence and quality of student settlements which will come from the 2001 population census, which, for the first time, will count also non-residents. It is advisable that the City Council organises the resources necessary so that Neighbourhood Councils can take over this task effectively. A special working group, possible coordinated by students themselves, could be put together involving ESU, Neighbourhood Councils and student associations for the identification of appropriate areas and facilities and the negotiation with their owners.

The 'open campus' solution – the organisation of research and educational activities in limited portions of the city – has prevailed on the 'closed campus' envisaged by IUAV with the relocation in the Arsenale shipyards. This is a positive outcome to the extent that it favours a closer integration of the university life with the community, providing the residents with an input of opportunities, cultural stimuli and social interaction on a daily basis. Also, the VIU – born as an 'education island' – will have to go 'into the city' – also physically – to meet the demand for its research output coming from the local environment. However, this model needs *adequate infrastructure*:

- *meeting places* where the population is not disturbed at night-time – the university facilities in the S. Marta area would an ideal location for clubs, dance floors and halls for cultural activities;

- *shared-use facilities* like theatres, cinemas, sport halls, libraries to be located in this area; keeping in mind that the more sophisticated tourists are a potential demand basin for such activities, their international orientation (multilingual facilities and programmes) is recommended;
- *fast water and road transport* connecting the various parts of the campus – to this end, it seems advisable to organise a shuttle for students connecting the various campus locations, travelling on water on the eastern side of the triangle, and on road to connect these with the Mestre pole (science park VEGA included) and the railway station. The idea is not to interfere with the normal traffic for commuters.

Initiatives for the Integration of Students into the Community

So far, the tools that the city has put forward for this huge but necessary operation of 'social engineering' have been blunt. Now, it has to readdress this issue from the beginning, considering students as a 'target group' that has to be attracted to the city, assisted during their period of residence, and convinced to stay after the completion of their studies. This issue is clearly related to that of increasing the range and quality of jobs available in the city centre, but, as seen earlier, other elements are also crucial. The quality of life enjoyed by students still determines to a large extent their postgraduate choices. First of all, students need to be welcomed and assisted in finding accommodation. *A reception office* (or '*student help desk*') *should be set up by the City*. After the payment of a small 'membership' that allows first year students to come back to this office in following years, this help desk should provide students with a database of addresses and information on housing. Obviously, this office has to be connected with the 'institutional' housing suppliers – ESU and the housing unit management – but also with the private market, for instance with the association of real estate agents within ASCOM. Moreover, the help desk could advise students (even without membership) in matters regarding health services, legal advice, cultural life, city services, and starters of small activities targeting the student community.

During their study years, *students have to be involved more closely in the decision-making process* concerning their presence in the city, but also the general life of the city. A first instrument is regular consultation, to be organised through surveys and meetings. Active democratic participation should also be incentivised. The students should be invited to participate in the activities of city and neighbourhood councils – for instance, taking part to commission activity, for which you do not need to be a resident. Moreover, the city should grant '*temporary citizenship*' to *fuori sede* students who have stayed in the city for more than one year, granting them the possibility of voting in local elections if they wish. This matter has to be negotiated with the central government, but it could be organised as a 'pilot project' associated with Italy's more general ongoing constitutional reforms. In addition, a certain number of seats on the City Council should be granted to representatives of 'non-resident' citizens (students but also immigrants and other temporary citizens), who organise their own lists. This could also be matter of discussion with regional and national powers.

The economic operators should be brought together in a *programme of 'customer orientation' recognising students as an important group of consumers*. The 'Cartateneo' experiment – which granted students a series of discounts in selected shops and public services – should be reproduced on a wider scale. The forthcoming issue of 'smart cards' which allow differential pricing of the city's facilities should be extended and personalised for student demand. According to students' satisfaction, the more student-friendly shops and facilities could be given an official label as a marketing strategy.

When *fuori sede* students are about to finish their study careers, the city should contact them through the appropriate organisations, such as the 'help desk', to *propose a 'life-project' in the city*. They should be given full information of job opportunities, postgraduate education and business start-ups, and also assisted in moving into 'resident housing' – the instruments are already there, with the city policy for new households.

Note

1 The island of S. Servolo is located in the Venetian lagoon, and can be reached with a ten-minute boat trip from S. Mark Square. The island covers an area of over 16,000m^2 which was once used for vineyards and gardens; an old monastery has been restored and adapted to academic use. Other buildings on the island have been transformed into guesthouses for students and staff, dedicated facilities and lecture halls.

Chapter 11

Comparative Analysis and Policy Guidelines

The objective of this final chapter is the *construction of a set of guidelines* that should assist decision-makers and all the other stakeholders involved in community development when planning for the development, relocation and organisation of higher education institutions. These guidelines are bound to be valid in all situations, but we also wish to indicate specific courses of action to be taken in the different cities studied.

We proceed by proposing three different profiles of student cities among those investigated, and by assessing in a comparative way the points of strength and weakness in city-university relations in each case.

11.1 City Profiles

The cities investigated in our research are quite different and find themselves at different stages of development. Highly successful cities like Lyon or Munich can hardly compare with cities that are still undergoing (sometimes painfully) a process of transition towards a post-industrial economy, like Rotterdam or Lille; nor can historical cities like Venice or Utrecht be analysed in the same way as modern science hubs like Eindhoven or Helsinki, and so forth. However, rather than looking at these general features, we deem it more interesting to classify the cities in our sample according to their 'student city' environment. Three groups can be devised, which present homogeneous features. We use boxes to illustrate our points directly relating to the case studies.

11.1.1 'Type 1' Student Cities: Rotterdam, Lille, Venice, Eindhoven

These cities are undergoing a process of economic transition in which higher education could play a key role in leading and strengthening the process. However, the existing potential is not utilised to the full. In this way, they risk missing a good opportunity to improve their position structurally.

The quality of higher education in these cities is not necessarily an issue, as it is generally good to excellent. Rather, problems arise with the capacity of these cities to put to value the human capital trained there. This is mainly due to two factors:

- the tightness of the local job market;
- the incapacity of the city to attract young graduates, offering ideal conditions to start a life project there.

The latter may be due to the unattractiveness of the city, at least as a location for the young. However, the *housing problem* is also a recurrent constraint on social dynamism: new groups like highly-skilled workers and foreigners have peculiar requirements as far as housing and quality of life are concerned. These cities, at present, are not able to match these demands. The housing stock may be of poor quality (Lille, Rotterdam) or excessively expensive (Venice). Eindhoven offers good quality of housing and environmental amenities in its region, but hardly any cultural or social attractions that may connect new graduates to the place.

Type 1 cities are therefore struggling to *become more attractive and competitive* on the job market, but a vision is not always clearly formulated around which to organise the action of the many players involved. In particular, the private sector is reluctant to invest in housing and infrastructure projects from which they may possibly never reap the benefits. A 'business model' is needed: a way to stimulate public-private partnerships so as to build a really student-friendly city out of self-interest.

In all the cities investigated, but in particular in the Type 1 cities, the housing market has proved to be utterly segmented. Only recently have city governments and other organisations started to realise that they should have *distinct policies to attract and accommodate different segments*: freshers, graduates, visiting fellows and young members of staff. Although student housing organisations and universities (like CROUS in France or ESU in Italy) may provide traditional, standardised solutions are seldom sufficient. These organisations are institutionally obliged to take care of the disadvantaged groups that require minimum levels of service, but nobody is there to take care of the top-end customers. Developers have this know-how and the means, but in most cases they have difficulties in participating in planning as empowered decision makers.

To be more attractive, Type 1 cities should not only be able to bring forward physical and cultural improvements, possibly making better use of the available

Rotterdam's Stadswonen Organisation, a semi-public company managing the student residence stock, has drawn up plans for a high-class, hospitable city centre that may appeal precisely to the target groups mentioned above: young skilled workers and 'creative talents'. Their failure to filter these plans through the local administration is an illustration of the incapacity of outsiders with good initiatives to matter for city policy. The same applies to Eindhoven, where a rigid planning rule has so far created a bottleneck in the accommodation capacity of the city. The new local government has promised to revise the planning regulation. It remains to be seen whether the plans of the local institutions and higher education institutions to make the city more culturally and socially exciting, will be backed up by important private partners as Philips.

Venice is undoubtedly an attractive city, but it still suffers from its image of a place that is only for tourists, with bars that close early at night and hardly any cinemas. This image may not correspond completely with the reality, but the side of the city unrelated with tourism, its cultural initiatives and its diverse social and ethnic composition, are not utilised by the city in its communication strategies. *Rotterdam* has the problem of becoming attractive for its residents. The cultural consumers of Rotterdam are more often not from the city, which is instead passive and reluctant to engage in world-class events like the European Cultural Capital 2001. *Lille* has bad weather and the reputation of being a grey, industrial city, but can boast a magnificent renaissance centre, good food and nice streetlife; however, most students live in campus and in unattractive suburbs, while many historical houses in the centre remain empty. Thanks to its optimal international accessibility, job opportunities in the region are not lacking as many companies have moved in, but top-level jobs cannot be found there; graduates would rather move to more congested but exciting cities like the neighbouring Paris and Brussels.

resources, but they should also be able to *communicate* their points of strength and their strategies.

In synthesis, *Type 1 student cities* have to improve their attraction capacity consistently to reap the full benefits of their educational and research facilities and in this way strengthen their economic development. Top-class human resources, social and cultural dynamism, and physical regeneration are badly needed as conditions for a full transition of these cities to a post-industrial, knowledge-intensive organisation of the economy and the society, where quality of life and flexibility in skills are appreciated.

11.1.2 Type 2 Student Cities: Munich, Birmingham, Lyon

These cities are economically strong, but the abundant higher education provision is not that relevant to the urban economy. Hence they find it hard to 'hook' knowledge to the regional economy, and to develop strategic functions in the global economic networks.

Type 2 cities are successfully positioned in the knowledge economy. For Birmingham, this has been a painful process and it is not yet complete, but today's Birmingham is a different city from that of 15 years ago, and not only from the physical point of view. All three cities are '*second cities*' in their national systems, but only Munich is using its position to attract cutting edge functions. Birmingham and Lyon still suffer from competition with their national capitals, and score worse than other smaller cities in their countries, like Manchester, Glasgow, Marseilles or Toulouse, as far as attractiveness and dynamism are concerned.

These cities have found their own recipe for urban growth, which is closely connected to the development of 'global' functions: financial services for Munich, congresses and legal services for Birmingham, healthcare and bio-research for Lyon. Their economy is diverse – industrial sectors like the automotive, textile and chemicals retain their importance, side by side with new growth industries like ICT, media and culture – but growth clusters have put these cities on the map of the global economy. *Today, the challenge for these cities is to maintain their competitive position and accelerate their development*. Birmingham and Lyon are seeking to overcome their 'second place' position, promoting complementary, unique specialisations, possibly linking up with international networks. Munich is just too good. Problems have arisen with managing its internal accessibility, which is making it difficult to retain growth within its city boundaries and is harming the environment, though projects to ease this pressure are under way. In any case, to confront emerging challenges in the global economy, these cities have to:

- be attractive for young workers in the new service sector, as the local supply is insufficient for the demand of skilled labour;
- develop high-class, internationally-oriented specialisations in education and research supply within their university institutes and technical schools.

In both respects, the score of Type 2 student cities is disappointing. All three are relatively dull and unanimated places when compared with other national cities. The offer of high culture and the calendar of events may be rich, but the point is that these places are not perceived as exciting. Families appreciate the suburban life and the environmental amenities to be found there, but the city centres are hardly 'cultural hubs' that may stimulate social interaction and bring creativity and dynamism to the local economy. In the future, they need to become much more attractive, offering up-to-date place amenities.

> *Munich* is a splendid city, very attractive to tourists, but living there is barely affordable for many students, who prefer to commute from their home towns in the region. Hence, there is very little 'student atmosphere' in the city. This city, very attached to its traditional image and way of life, is a consumption hub for wealthy people and foreigners, but has little social dynamism. Its point of strength, the human size, the aristocratic cityscape, are not necessarily attractive to students and skilled workers. The same holds true for *Lyon*, which is also perceived as rather dull. *Birmingham* is still a city centre with considerable problems in terms of appearance, safety, and lack of animation. The city is making plans for its East Side development to be a cultural hub, attractive to students as well as to citizens and visitors, but this vision has still to be finalised, and there is little certainty about its financial sustainability.

Possibly as a consequence, a Fordist, old-style organisation of the economy still dominates. Munich is still the city of BMW and Siemens, Birmingham the city of Cadbury and Jaguar, Lyon the city of Merieux and the other biomedical giants. This is not to say that small and medium-sized enterprises are not increasing in number and relevance, but rather that this is hardly reflected in the organisation of the society. They are ageing cities or cities where the more dynamic segments of the population are dominated by people in the low-skilled end of the job market. The 'entrepreneurial spirit' of the local communities is still inferior to that of other competitors. In these conditions, these strong cities face some threats in the future. *Their progresses could come to a halt if they do not make a more creative, proactive use of their academic community and knowledge.*

11.1.3 'Type 3' Student Cities: Utrecht, Helsinki

In *Type 3 student cities* higher education institutions have always been strong and beneficial stakeholders in the local economy; however, problems could arise in the future as further expansion may conflict with other urban functions. The danger is that higher education institutions create more problems than they contribute to solving.

In these cities, the recent history of economic development is closely-knit with that of universities and schools. These are prestigious, world-class institutes that occupy a central position as actors of urban change, both on the physical and the socioeconomic level. These have brought material and intangible benefits to otherwise somnolent places. The *networks* between higher education managers and city government, on one side, and the private sector, on the other, are well developed, even though in the case of Utrecht the scope is at the national rather than the local level. Both urban regions are wealthy. The two cities are hardly comparable in other respects, such as size and national relevance, but they share the cosy atmosphere and an open attitude towards students. Both host top-class universities of arts and design, which provide a relevant contribution to the *cultural landscape* of the cities, though the HKU in Utrecht has grown strategically disconnected from the city's cultural policy. In many respects, both from the point of view of the organisation and structure of student communities and from the point of view of the relations with the community, these cities may be considered examples of *successful student-community integration*.

Type 3 cities are very attractive to students. Graduates would build life careers in the places where they had such an enjoyable time as students, and where the job market offers many opportunities. Though there is not sufficient space to accommodate everybody, this natural fact is not perceived as a big problem, as their regions offer good alternatives to families attaching much importance to living space and quality. Their attractiveness is based on 'natural factors' that needed little promotion: the natural beauty and national centrality of Helsinki, and the cosy atmosphere and accessibility of Utrecht. Today, in both cases, the use of space for university-related functions seems to have been exhausted. Helsinki cannot accommodate the growth of its institute other than in the suburbs, while for Utrecht the problem is rather that of finding space for knowledge transfer activities that cannot be developed in the campus space. Both cities then have a problem with student settlements.

Helsinki may have to host an increasing number of students in suburban campuses, reducing the vitality of the city centre and making it necessary to move to a higher gear in mobility management in the metropolitan region, which may be a problem. *Utrecht* faces the risk of becoming a 'student-only' city, as a high number of residents will move to cheaper and more accessible dwelling in the new development area on the West Side. In both cases, the spatial balance of the city-university relation is in jeopardy.

The *metropolitan governance* of higher education expansion is a challenge as the decisional environment is problematic. This situation will have to change for these regions to remain strong and competitive in the future.

> The Province and Region of *Utrecht* are active in networks to improve job market matching, but they are hardly involved in planning the development of the knowledge sector. The neighbouring municipalities in *Helsinki* have so far been 'predatory' with respect to the attractiveness of Helsinki as a higher education centre.

11.2 Comparative Analysis

In the first chapter of this report an analytic framework was introduced that can be utilised to assess the points of strength and weakness in the dynamic relation between city and higher education for urban development. This relation is seen to stand on a *triangular link* between the higher education institutions and their communities (student and staff), the private sector and the local society. For a *virtuous cycle of development*, these relations have to score positively and to be in balance. The university has to be effectively oriented to the local economic sector in their study and research curricula, and the local business partners have to provide tangible and intangible resources to higher education expansion, like their own staff, research funds and internship placements. On the other side, the academic and student community has to enter into a symbiotic relationship with the local community, contributing their social and cultural dynamism, granting economic benefits and an enhanced image; in exchange, it is expected to receive good locations, services, and hospitality.

A good score on these two levels also has an indirect impact on the local economy: the 'rootedness' of the knowledge produced and the sense of integration in the community ensures that the process of economic development is sustainable: firms find the talents that they are looking for in the region, and top-class workers spend their salaries in the region, contributing to a strong, diverse and wealthy community. *If there is an imbalance in the relations between higher education institutions and private sector on one side, and the city on the other, this will be reflected in less than optimal job markets (excess demand or supply of labour), in leakages from the local economic circuit, and in constraints to development.*

Assessing and comparing how well each city in our study performs and checking whether there is a relation between city profiles as sketched out in the previous sector may help to deliver punctual suggestions for policy.

Our analysis proceeds in this way. In Table 11.1, we report on a grid the scores obtained by the cities in the three profile groups (Type 1, Type 2 and Type 3 student cities) with respect to the links between higher education institutions, community and private sector. We highlighted in bold type the areas in which the relation between city profile and source of problems for the sustainable city-university relationship seem the most significant.

11.2.1 University–Community relations

A balanced, positive relationship between universities and their communities is one in which top-quality higher education enhances the image and prestige of the city, and also contributes directly and indirectly to the generation of spin-offs. For instance, this happens when the academic community participates substantially ing the city's economy, as spenders and taxpayers, and when the universities use cities as 'real world laboratories', orienting their research programmes to the local contexts, as sources of intellectual stimulus. On the other hand, the local communities and their governing bodies contribute to the higher education institutions' development, providing good services and facilities, prestigious and accessible locations, and support in development programmes.

Only few cities in this research have scored positively on this relation: among them Helsinki and Utrecht, our *Type 3 student cities*. Even in these cases, there is room for improvement, especially in the way in which the governing bodies of Helsinki are dealing with the metropolitan expansion of the higher education institution infrastructure; in Utrecht the city is probably not paying enough attention to opportunities for a structural resolution of the student housing problems, and the university is still excessively geared to the national rather than the local level.

Type 1 student cities score moderately to negatively. In Rotterdam, there is reciprocal indifference between the higher education institution – and in particular the Erasmus University – and the community. Students (and especially foreign students, who are numerous in Rotterdam's higher education institutions) have the potential to bring forward change and regeneration to the city, but these opportunities are not given sufficient consideration by the local government. On the other hand, the university has so far planned its development without caring much about its interaction with the city. The city of

Table 11.1 Assessment grid for the nine partner cities

	Type 1 student cities				Type 2 student cities			Type 3 student cities	
	Eindhoven	**Lille**	**Rotterdam**	**Venice**	**Lyon**	**Munich**	**B'ham**	**Helsinki**	**Utrecht**
HEIs → community	0	0	–	+	–	–	–	++	0
Community → HEIs	+	0	–	–	**0**	**+**	–	++	+
HEIs → private sector	**0**	**0**	–	–	+	0	+	+	+
Private sector → HEIs	**+**	–	**0**	**0**	+	+	0	++	0
Private sector → community	+	+	0	0	++	++	0	**+**	**+**
Community → private sector	–	0	+	0	+	+	0	**+**	**0**

Venice can greatly improve its attitude with regard to university development being more strategic rather than opportunity-driven, and a more active role has to be taken also with regard to student housing and student integration into the local socioeconomic context. In the cases of Eindhoven and Lille, the student community is not visible enough, and as such, it contributes only marginally to the enhancement of the local quality of life and urban atmosphere.

In *Type 2 student cities*, these relations are decidedly on the bad side. Universities seem to matter very little to the city. This has to be compared with the relatively good quality of the institutions and the good state of the local economies. Higher education institutions are like islands, both physically, and in terms of network interaction. The student communities are dispersed in big cities and are not perceived as members of the community; more often than not their settlement is a source of friction with the local residents in congested and sensitive housing markets. While in Munich LMU and TU (and less so FHM) still receive support and good location from the local governments, in Birmingham and Lyon further university development is still being negotiated, with many problems. It can be foreseen that in the future, if the links between university and communities are not reinforced, this lack of cooperation will affect these cities' competitive capacity; for instance, the universities will not find the necessary support to expand their educational programmes and the city will be incapable of attracting the best talent, leading to a mismatch in the job market and to a downturn in the local economy.

11.2.2 University–Private Sector Relations

Type 3 cities perform well also as far as university-business relations are concerned. Helsinki's higher education institutions have developed an extensive network of formal and informal links with the local firms. The private sector contributes generously to research programmes, and Helsinki's R&D and knowledge transfer facilities are among the best in our sample of cities. Some problems exist regarding the way in which universities support the diversification of the local economy. Utrecht has some problems in the support of the private sector to the education facilities; moreover, room for the development of knowledge transfer initiatives is severely limited by physical space. All in all, though, these cities show that research and application to the business environment can go hand in hand. Helsinki has also proved that a study course that usually has difficulties in 'going to the market', like arts and culture, may fit perfectly the plans of large companies and important non-for-profit employers.

Type 2 cities score just as well in this respect. The linkages within the private sector are well articulated, and this has allowed the rapid development of local economies. The subcontractor networks of the big firms are extremely rich. In this process, proactive city governments and metropolitan authorities have played a key role, promoting growth clusters and investing in capacity and infrastructure. Especially in Lyon, positive and balanced links between higher education institutions and private sectors have been developed after a lengthy 'blockade' by a national law that in practice impeded this kind of connections. Room for improvement may come with time, as the cultural distance between the world of enterprise and that of the university will diminish. This very distance makes the universities in Munich not that active in establishing strong strategic links with the private sector, which are instead actively sought by private firms in innovative industries. The city takes a more active role in this respect, but it faces a centralised and bureaucratic university administration. The excellent quality of Munich's higher education institutions is therefore utilised to stimulate innovation to a lower degree than their potential would allow, though the general impression is that of a process of convergence. Birmingham has the opposite problem of a private sector still too 'young' to have developed strong links with the higher education institutions, and therefore not yet aware of the necessity to bear some of the costs of this connection, such as an active policy of funding research and student placements. In this way, they lose opportunities to be competitive and they do not succeed in changing the conditions that lead so many of the best graduates to leave Birmingham when they complete their studies.

Birmingham's and Lyon's experience with science parks was successful as far as knowledge transfer is concerned, less so in terms of starters. In all three cities, the strategic involvement of universities is somewhat limited, though they have been initiators of such activities. The successful development of Munich Bio-Region happened to a large extent *in spite of* the LMU, and hardly involves the technical schools. The knowledge transfer infrastructure of Lyon is decidedly nationally-oriented; the participation of university staff is limited due to regulatory and cultural barriers between the world of research and that of business, which are also present in the other two cases.

Type 1 cities present the most evident problems in the networking between private sector and education institutes. Leaving the case of Eindhoven aside, where Philips has been a fundamental actor in the development of higher education, in the other cities the higher education institutions do not always work in the interests of the local economy.

Rotterdam offers an exemplary illustration of such missing links: despite the importance of its main port, the relevance of port studies and business logistics in their faculties is limited, and the contribution of the port firms to the development of this knowledge is negligible. Rotterdam has looked towards the development of a growth sector in media and telecommunications, but there are only courses available at the level of HBO. In general, the local higher education institutions are not used to participating as active players in the promotion of the city's economic sectors, something that may change in the future. Lille has the same problem, though there is more attention from the local universities – a much more solid higher education apparatus is indeed present in the French city than in other cities in this group – for developments in the local markets. What is missing is the capacity of the universities to transfer their knowledge to the private sector. In Venice, the main University of Ca' Foscari is peculiarly 'disconnected' from the very local level, and instead links up with the regional economy. So far, only a few faculties have specialised in disciplines that could bring a change to the distressed economy of the city itself, caught between excessive tourism and strong de-industrialisation, unmatched by a firm growth in the service sector. The situation there is also in evolution, bringing more flexibility and capacity to adapting study courses to the requirements of the context, but constantly changing national university reform plans undermine the credibility and continuity of these efforts. Moreover, the local private sector is not generous or cohesive enough in supporting the university development plans.

11.2.3 Private Sector–Community Relations

There is a dynamic link between the higher education institutions and the city on one side, and higher education institutions and the private sector on the other. Weak links at these two levels are reflected in a decline in the local economy, which is generally captured by a mismatch in the job market or by an outflow of the revenues generated in region. However, this downturn may become evident later. Hence, situations that are now healthy, as in cities like Munich, Lyon, or Utrecht, may deteriorate in the future if such balance is not redressed. In the global age, very small differences in the performance of competing local economies are sufficient to cause considerable problems for disadvantaged cities. For example, Manchester and Nottingham, attractive cities for young people and new graduates, may take away human resources and investments from the economy of Birmingham, reaping the benefits generated in its excellent education centres and provoking a downturn in the rapidly-evolving regeneration process.

In the same vein, if we observe problems in the local economy or mismatch in the job market, it is likely that a positive development in the role of higher education institutions as engines of urban development will eventually bring development to its full potential. It is the case of *Type 3 student cities*, like Helsinki, where the recent problems in the ICT sector threaten to have huge impacts on the local (and national) economy, but the diverse range of professions and skills trained in the local higher education institutions is a guarantee that in the future Helsinki will be able to diversify its economy, making it more solid. In Utrecht, the lack of human resources produced locally will become less relevant to the extent that the city will be able to accommodate a larger pool of students, involving the region more closely in delivering education and research services. On the other hand, negative scores in the first two sets of links are signs that existing problems with local economic development are not being attacked consistently by making use of the potential of the universities and their student communities.

In *Type 1 cities* the relations between the city and the private sector are problematic (Venice, Rotterdam) or still in a process of reorientation (Lille, Eindhoven). These cities have problems in building an entrepreneurial class that may contribute to the strengthening of the local economy. This is the consequence of a higher education that is insufficiently geared to the economic environment. On account of this, the job market in the regions remains rather tight, and graduates can find good jobs elsewhere. The outflow of human resources ensues naturally, even if these cities should manage to improve their attraction capacity for students and young graduates.

In *Type 2 cities* students are not easily integrated into the community. They have a hard time all through their study career, be it in finding a room, going back and forth, with campuses and lecture halls scattered everywhere, or organising their cultural life. At the end of their study period, they might find good jobs in the region and remain in the area, but many of them will feel disaffected with the place and leave. In fact, these are the areas where local firms cannot find enough workforce and this determines a barrier to growth. The good performance of cities like Lyon and Munich could therefore be hampered in the future if these cities do not become more hospitable and friendly to incoming students.

11.2.4 Best Practices

Some initiatives, organisations, policy directions and projects, have been particularly successful and innovative in bringing forward the conditions for a sustainable city-university relationship. These cases might be utilised as 'best practice' in the policy for students. In any aspect of student management, every city in the set has something to learn from the others. It remains to be seen to what degree these best practices can be transferred as blueprints for action in other contexts.

- *Birmingham*: the organisation of a '*summer international university*' by the City Hall, in partnership with all the three universities of the city, is an important initiative to start a strategic collaboration between actors that have so far acted independently one from the others. With this project, the City shows concern about the international projection of the city as a top education centre, and the universities pool resources to achieve an important community objective.
- *Eindhoven*: the experience of the participation of students in local elections (the '*List 11*') is a demonstration that students may seek visibility and integration with the local community, even at the top level of decision-making. Participation in the campaign gave the students an opportunity to be heard by local stakeholders and to bring their ideas and projects into the debate.
- *Helsinki*: the *student union* of the University of Helsinki is an example of how empowered and 'rooted' student organisations may facilitate the integration of student communities, providing services and support of all kinds. They are also able to negotiate as peers with local politicians the decisions affecting the student community, and in this way, they bring more balance and inclusiveness to the city's higher education policy.
- *Lille*: the *École Superieur de Commerce* of Lille has a special contract with the city that allows students to be integrated into a company for two years without paying the school fees. This important initiative is a breakthrough in the city's training policy: the city gets directly involved in the way in which education orients itself to the job market, at the same time 'guiding' the knowledge transfer process in the desired direction, and is ready to take on itself the cost of this encounter.
- *Lyon*: the *Young Ambassadors of Lyon* is an initiative that allows the city to maintain a link with 'temporary citizens' after they leave. Moreover, they can use this project as a powerful promotional tool of the city's excellence in research and education. This initiative could be pursued by other cities suffering of a drain of graduates towards areas where they find better conditions, like Lille or Birmingham.

- *Munich*: *Studentenwerk Munich* is to the fore in solving the problem of the housing shortage for students in the city, thanks to an active acquisition policy and their capacity to negotiate the best solutions with the city and the HEIs, as well as with the private sector partners. The effectiveness of this organisation is in its good capacity to connect with students and involve them in creative solutions. They are trusted and helped by student associations.

  BioRegio is a non-profit organisation that has taken the lead in bringing together research and business in a period in which universities were not fully prepared to build these bridges and invest in innovative applied research. The success of Munich as an innovative cluster in biology and health sciences is owed to this organisation, which is now busy finding new areas and resources to purport its aims. The bottom-up approach of BioRegio can be exported as a business model to other areas where universities are reluctant to take the lead in technology transfer (Lyon, Lille, Birmingham, Venice).
- *Rotterdam*: the *Ichthus School* has been displaced by the city in a regeneration area, the Kop van Zuid, where they enjoy new purpose-built accommodation intended to facilitate a new concept of education based on teamwork. However, at the time of its instalment, they are still physically and functionally separated from the surrounding context. Ichthus made the best out of this unfavourable situation, starting a 'neighbourhood policy' that would show the benefits of the school's proximity in a distressed neighbourhood, through actions in the community and the setting-up of a permanent support centre.
- *Utrecht*: the *University College of Utrecht* is an illustration that in the field of higher education, excellence can be achieved by a city's initiative in pooling and making the best out of the existing resources. It also demonstrates that study curricula need not be completely geared to the city's specific economic structure or application-oriented to have an impact in terms of visibility and image.

  Vast Goed Belang is an organisation that illustrates how the private sector has a long-term interest in selling quality to familiarise new target segments with the local hosing market, at the same time adhering to more general community interests of maintaining the house stock value.
- *Venice*: *S. Margherita* square and the *S. Marta university complex* are examples of how planned or spontaneous student settlements may be functional to the regeneration strategies of the city. The university areas of the city offer plenty of cultural and leisure opportunities, which are

attractive to visitors and citizens alike. In this informal way, integration between the local context and the guest community is achieved. Students are now seen as a positive source of dynamism and the university a source of economic regeneration, overturning the prevailing perceptions of the past.

11.3 Guidelines for Policy

The basic assumption of this research is that an explicit strategy targeting the student community is necessary for the well-being and competitiveness of the city. The student community needs to be fully integrated in the local socioeconomic environment, culturally and socially active and motivated to stay after the completion of studies; the education facilities required by a competitive city are flexible, spatially balanced and well connected.

Among the aspects to be taken into consideration to this end are: a) the socioeconomic status of the city; b) the attractiveness of the city for students and graduates; c) the history and evolution of the student settlements in the city; d) the housing problem; and e) the governance structure of the higher education sector. Suggestions for improvement with respect to each of these factors can therefore be identified, as well as the most promising courses of action.

11.3.1 Socio-economic Positions

Universities and vocational schools have to build the intellectual capacity and the practical skills for a new class of entrepreneurs, social workers, educators and liberal professionals, who would operate in contexts radically different from 50 or even 20 years ago. In the new environment, key success factors are the capacity to work together and interchangeably, inclusion in social networks, orientation to the customer, innovativeness and high-tech aspects of the work. In this respect, the toughest challenge is for higher education institutes which were originally established to serve the industrialisation process of a region (as is the case in Lille, Birmingham, Eindhoven, Rotterdam), and are now obliged to accomplish a thorough change in orientation towards a more diverse range of disciplines which underscores the complexity of the 'knowledge society'. This change is not always without problems, as is shown in our research, but in cases where it is successful, universities may become real engines of economic regeneration. These cities should also use more explicitly the excellence of their higher education facilities to market the city as a convenient location for firms, which can find there knowledge provided by top research institutes and transfer facilities counting on extensive and well-developed networks; on their side, the academic staff and students may enjoy a vibrant intellectual environment and state-of-the-art research and teaching facilities. Cities can exploit network effects in promoting this potential, utilising the same international links activated by the universities, like the international research networks, the student exchange liaisons and the alumni networks. In targeting important emerging markets, like China or Eastern Europe, a first step would be to open student information points that also serve entrepreneurs to establish links with the local economic community.

Other cities are at a more advanced stage of their economic development processes, either because the transition to a service sector economy has been completed or because they maintain an economic mix that is the secret of their economic strength (Lyon, Munich, Helsinki). Cities are now pointing at the development of specific 'growth sector' strategies to maintain their competitiveness and reinforce their position in the global economic networks. In these cases, universities have to develop consistently, adapting their educational and research orientation to the requirements of the urban strategy, and stimulating innovation with their research activities. Local government can enhance these processes of reorientation bringing parties together and taking the leadership in the creation of 'excellence centres' (attracting funding and resources at EU and national levels) where the academic knowledge is networked with the local environment.

In the last set of cities, Utrecht and Venice, the local universities have a size and history that strongly characterises the local environment. These may be the only two examples in our set of 'student cities' in the traditional sense. However, this strong link has transformed itself into a sort of 'mortal embrace' (on different degrees in the two cases), and the functional links between city and higher education have been lost in the evolution of the two systems. The biggest challenge for these cities is to use the potential of their higher education sector in a better way while at the same time managing the challenges posed by a further development of the educational functions in a very sensitive urban environment.

Guidelines for local governments

- Use excellence in education to promote the city as a business location.
- Create 'excellence centres' for higher education.
- Act as a catalyst to fund locally-oriented research.

- Fund 'chairs' at local HEIs dealing with local issues.

For whom this is more compelling
- Rotterdam: not a knowledge-intensive economic development, lower-than-average skills in the region.
- Eindhoven: economy and region dominated by industrial production.
- Lille: not a top labour market.

Good examples to look at
- Birmingham European Summer University: http://www.birminghamesu.co.uk/aboutesu.html.
- Helsinki 'Culminatum Oy': http://www.culminatum.fi/english/index.cfm.
- Kenniscentrum Utrecht (Utrecht Centre for Knowledge): http://www.kennis-utrecht.nl/ (in Dutch only).
- Lyon's Young Ambassadors: http://www.jeune-ambassadeur-de-lyon.com/.
- University College of Utrecht: http://www.ucu.uu.nl/.
- Venice 'District for Innovation': http://www.venicedistrict.it/eng/home.asp.

11.3.2 University-driven Regeneration Policies and Models of Student Settlements

Another relevant factor in the city-university relationship concerns the settlement model of the student population and of the university itself. Two different models in this respect are foreseeable, one that foresees *'informal' downtown settlements*, and one that is organised in one or more *formal 'education citadels'*, that is a campus, often physically separated by the urban environment. Between these polar opposites, we observe a whole range of *intermediate situations*. Cities may host more than one higher education institution and distinct faculties within one university, with different location attitudes (Munich, Lille, Helsinki). Secondly, student settlements may be closely associated with campus locations (Lyon, Birmingham, Venice) or largely disconnected (Helsinki, Rotterdam, Utrecht).

Should the 'downtown' model be successful, fruitful opportunities for cultural cross-fertilisation are present, but there are also risks of conflict with the local population. The 'campus' model is clearly more conservative.

It is believed that the two models (as well as intermediate solutions) may be appropriate in specific contexts, provided the necessary conditions are present. Isolated and unattractive campuses, as well as unplanned downtown settlements, are both unfit.

The issue is also related to size. Large student settlements compared to the city mass, like those found in Utrecht, Venice, Lille and Helsinki, may present more risks but also more opportunities for integration of human resources into the community. In large cities, the relevant scale of observation is more often the neighbourhood or area, rather than the urban region as a whole. Therefore, cities like Birmingham and Lyon, where the student population is not that large, have a concentration of problems in specific neighbourhoods. They should instead look for a better spread of these settlements over the territory. It is one of the outcomes of our investigations that 'Latin quarters' with a high concentration of students and cultural producers, have a short lifecycle, and easily become a source of external costs for the community. A more balanced approach, combining the location of student facilities and meeting places in strategic locations and regenerating areas, might achieve better and longer-lasting results, containing the associated costs of gentrification. In this regard, organising accessibility among different campus locations and from these to the city centre is crucial. Lille's VAL metro system allowed the city to rejoin the campus to the city centre, making it possible for students to enjoy city life while living in cheap residences in the campus.

Different student populations use the city in different ways. The assumption that the student population is a homogeneous body with a typical behaviour and socio-cultural profile is simplistic. The attitudes, behaviour, study fields and lifestyles of students are many and varied. Some are more introverted and have a less symbiotic relationship with the city, others are more 'open' and influence the city's society and economy. A neat distinction has been underscored in our research. *Technical students* (students in faculties of engineering, maths and physical and natural sciences) attach much importance to campus life, as their progress depend peculiarly on the quality of research facilities. Their student life is a sort of 'full immersion' in a close community of peers, and typically the opportunities for interaction and integration with the local communities are limited. Career choices are made almost exclusively by looking at economic factors, over which cities may have little or no control. In cities like Eindhoven, Birmingham, Lille, it is therefore a task of the local policy leaders to find ways to reconnect student communities to the social contexts. The education policy of higher education institutions, with training periods organised in local firms, may partially offset this balance. However, a consistent policy of welcoming and attracting students into community life is also needed, for instance through regular events, in order to establish some empathy between local residents and the student population.

Other kinds of students (human and social sciences, art studies, liberal professions) are typically more used to 'going out' on the streets and having a social life. The city life represents for them not only a pleasant way to spend their free time, but also a source of intellectual stimulus, a laboratory for ideas. In these conditions, integration with the local community happens more easily. For instance, students may start to be interested in local politics and take part in associations and groups that have a community rather than university focus. Or they can be at the forefront of cultural innovation. In Utrecht, Venice and Rotterdam – cities that were rather 'sleepy' and far from dynamic before the big development of universities, this encounter has brought tangible improvements. However, in many respects *also this integration too needs to be aided*. For instance, the social pressure that student communities express can be felt as a threat by weak groups such as the elderly and unemployed. It is therefore necessary to *activate communication channels*, and to actively support and guide the activity of student groups. In this respect, it is important that city administrations are able to 'read' the cultural and social profile of student communities and 'use' them as an element of regeneration strategies. Regular surveys, information points and the organisation of 'clubs' and networks could achieve this result.

The last group of cities is utterly heterogeneous, both a result of their size and of the diversity within their higher education sector. In Helsinki, Munich and Lyon both suburban campuses and downtown settlements are present, though the first model seems to be on the wax as a concomitant to development in the future, due to the lack of opportunities in the city centres, with the associated threats that 'disconnected' campus developments may bring.

Guidelines for local governments

- Spread student settlements throughout the inner cities.
- Avoid peripheral/inaccessible campus development.
- Promote student meeting-places as 'cultural districts'.
- Provide facilities to support student community development in regenerating areas.
- Facilitate bureaucratic procedures for incoming international students.
- Organise/support events to familiarise students with the city culture and to bring together students and local residents.
- Organise/support student activities that are aimed at the community.

- Organise 'virtual communities' to stimulate cohesion and discussion between locals and students or between different student groups.
- Organise student audits/involve students in city politics.

For whom this is more compelling

- Munich: not a lively city with few opportunities for low-budget students, very traditional cultural life.
- Lyon: a dull city for being the second national city, with problematic surroundings.
- Rotterdam: a 'divided city' with a city centre in need of regeneration.
- Eindhoven: a 'tech-city' with student community as an external body.
- Birmingham: a polarised urban region with high unemployment areas and social exclusion.

Good examples to look at

- Helsinki Arabianranta: http://www.helsinki.fi/jarj/mao/urban/arabia.html.
- Rotterdam Ichthus: http://www.ichthus-rdam.nl/public/pages/algemeen/academie/co/master/hitech.cfm.
- Venice S. Margherita/S. Marta: http://www.pps.org/gps/one?public_place_id=478.
- Lille VAL metro system: http://www.metropla.net/eu/lil/lille.htm.
- Lille Montgolfier Event: http://www.ec-lille.fr/fr/assoces/montgolfiades.html.
- Eindhoven Kenniswijk (Knowledge Quarter): http://www.kenniswijk.nl/english/index.jsp?pageid=198695.

11.3.3 The Housing Problem

The 'downtown' model clearly implies that affordable housing opportunities for students have to be available and that this does not create unsustainable tensions in the real estate market. Negotiated solutions, public-private partnerships and close monitoring of market trends are necessary, as well as a rental policy that keeps the turnover in housing tenancy to reasonable levels. It is especially important that the city brings the private sector to recognise the importance of investing for student accommodation if they wish to reap the benefits from an increased pool of local labour.

Some *typical mistakes* will have to be avoided, like relying on rigid regulations to limit the inflow of students in residential areas (as in the case of the 10 per cent rule in Eindhoven), or allowing the subdivision of large flats to host numerous students, which ultimately lead to a decline of housing values and an irreversible decline in the overall housing quality of the city.

Guidelines for local governments

- Monitor housing market.
- Proactively stimulate housing development for students though the involvement of private parties.
- Organise solutions for short-stayers.
- Inform incoming students about housing through well-designed, interactive and multilingual webpages and well-distributed brochures.

For whom this is more compelling

- Venice: poor quality and high cost of housing, competition from tourists.
- Utrecht: expensive and overcrowded housing market.
- Eindhoven: rigid housing regulations.
- Birmingham: poor housing quality and increasing conflict with resident population in university neighbourhoods.

Good examples to look at

- Rotterdam Stadswonen:
  http://www.stadswonen.nl/ (only in Dutch).
- Eindhoven 'Fellowtel':
  http://www.tue.nl/stu/internationalisation/esk2.html.
- Munich Studentenwerk:
  http://www.studentenwerk.mhn.de/index-uk.html.

11.3.4 Attractiveness of the City for Students and Graduates

One peculiar piece of information that emerged from our research is that factors that are relevant in attracting starting students are not the same that matter when the graduates, the potential newly-skilled workers of the future, are concerned. Students may have a great time in a city that does not offer much to families, entrepreneurs and other social groups, provided student life is well organised and the university facilities are self-sufficient. It is the case of cities with a strong 'technical' orientation, like Eindhoven and Birmingham. *In these cases, though, students are hardly integrated into the local community*, and this affects their post-university decisions. Conversely, a city might be very attractive for visitors and a cosy, high-quality place to live for citizens, as it is the case with Munich, Lyon or Utrecht, but this does not mean that students have an easy life there; in fact, these places may be the hardest ones in which to find convenient accommodation.

'Formal' (e.g. the openness of the job market) as well as *less formal motivations* (the quality of life enjoyed in the period of study, the sense of integration in the community, the facilities and supports offered by the city, etc.) *count in this choice*. Cities who wish to retain human capital in the region, both as a workforce for local firms and as members of the community (and taxpayers), have to be active on both fronts.

To stimulate the attractiveness of these cities, the 'student resource' could be managed much better. *Technically-oriented student communities should be attracted to city centres*, and a good way to do this may be to displace library halls and other student facilities out of the campus into strategic central locations. In this way, they become familiarised with the city life and bring some animation to it. The issue of integration is tackled by creating occasions where students and residents meet. This could happen on campus, as in the case of Eindhoven, or in the city, as in Lille and Venice. Rotterdam does very little in this sense. In all cases, events must be coordinated and effectively promoted by a leading city government with a strategic attitude.

Cities like Lyon, Lille, Birmingham and Eindhoven have a problem of 'brain drain': being second (or secondary) cities in their national urban systems, they suffer from the competition from attractive areas like capital cities (Paris, London) or other large urban agglomerations (the Randstad). The challenge for them is to offer other advantages to new graduates, like higher quality of life and better conditions than those that can be found in congested and expensive national centres. The differentiation and quality in the housing market are factors that may convince graduates to stay on after completing their studies. It is important to understand that if graduates wish to start a life project in the city, they presumably require more quality and stability than normal students, and they are in the position to pay for these superior standards.

Other cities like Utrecht, Venice, Munich and Helsinki have the opposite problem of being 'too attractive' so that they cannot accommodate all the potential newcomers. Alternatives have to be found by involving a greater territory in their region in the education function. Finally, Rotterdam is a peculiar city that is attractive to foreign students but not enough to local students, with the result that many leave the city after the completion of their studies. In cities like Rotterdam, Eindhoven and Munich the size of the student community is negligible in relation to the size of the city, and hardly produces any relevant effects for the community. That may be the result of an excessive number of students commuting rather than staying in the city, itself a reflection of tightness in the housing market and of lack of attractiveness of the city, problems with which the city must come to terms.

Guidelines for local governments

- Make cities – and city centres in particular – more attractive for students through a targeted planning of leisure/cultural resources.
- Improve and differentiate housing standards for different categories of students and graduates.
- Have a sound welcoming policy that is negotiated and developed in partnership with HEIs.
- Involve neighbouring cities/province in a student welcoming policy when the university city is overcrowded.
- Upgrade housing stock and mini-flats in leisure districts of city centres.
- Market the city as a residential location for graduates providing special facilities and services.

For whom this is more compelling

- Venice: limited work opportunities in the historical city.
- Eindhoven: insufficient urban quality.
- Lille: a poor city image.
- Birmingham: a poor city image with accessibility problems.

Good examples to look at

- University College of Utrecht:
  http://www.ucu.uu.nl/.
- Utrecht *Vastgoed Belang*:
  http://www.vastgoedbelang.nl/ (in Dutch).
- Eindhoven Vestide:
  http://www.vestide.nl/wbvs/default.asp?wbv=vst.

11.3.5 Governance of Higher Education Development

Cities manage their student and academic communities as elements of the urban environment. Each city has a 'higher education policy', whether or not this is in their formal competence, and whether or not it is an explicit strategy.

Physical planning, housing regulations, international relations, economic development, are all fields of government which affect and connect with higher education and research and the way in which these functions develop in the urban environment. *However, this does not always happen in the best interests of the community and in the best possible ways.* Given its complex nature, the good management of higher education student communities, that we have come to term 'student-friendliness' throughout this study, is to be challenged with diverse, complex tools.

The role of *formal policy makers* is paramount in building a cohesive strategy to attract and manage students. However, the quality of the decisional networks, the cooperation and empathy established between the city and the higher education institutes, on one side, and the private sector, on the other, is fundamental to their effectiveness. *Informal stakeholders* (private sector/ students and community organisations), with their actions, can enhance or hinder impacts from student communities in the desired direction. This also requires strategic decisions on the education curricula to promote, the means to diffuse the knowledge generated, and the actions to facilitate the 'cultural empathy' between host and guest communities.

The case studies have highlighted a number of significant examples in which the interests and initiatives of the private sector and the resident community have been explicitly taken into account when planning for university development. Platforms of discussion have been established at all levels, but only in a few cases are they really effective in bringing forward more strategic cooperation between city and higher education institutions (examples are Culminatum in Helsinki, the Committee Grand Lille). In other cases they have not achieved their objectives because of lack of focus or because important actors were missing (examples are the Work Group Eindhoven Studentenstad, Rotterdam Morgen and Pole Universitaire Lyonnaise). In the cases of Utrecht, Venice and Birmingham discussions regarding specific problems like student housing and university locations are then enlarged to the wider area of development strategies.

Cooperation between different actors is easier if there are already cultural and social links bringing the key people together and stimulating their dialogue. However, the embeddedness of universities in the local economic context is something that should not be taken for granted. The origin and history of the universities is an important element in this respect, because it determines to a large extent their regulatory framework. In some cases, higher education institutions have been started as institutions focusing on the local level, often by the city or the region itself, or by the local economic elite, to serve local interests. This is the case of the Erasmus University in Rotterdam, the Technical University in Eindhoven, the 'Catho' of Lille and the Polytechnic Schools in Helsinki. In all these cases, a collaborative attitude and a networked structure of decision-making with a local perspective is integral to these institutions. Nevertheless, there could be a problem: local politics is subject to change and perturbations, and the very groups who started or gave strength to schools like the polytechnics and the universities of Birmingham, or Ca' Foscari University in Venice, may not be relevant any more. As a result, these institutions have had to restructure their links with the community in terms of strategic networking. Cities like Lyon, Utrecht and Munich instead host higher education institutions with a strong 'national' orientation, which until very recently had tenuous strategic links with the community (local institutions, firms, community organisations). The motives for strengthening these links are there, but the university bureaucracy may be reluctant or legally incapable.

It is important to recognise the role of *intermediary organisations* in the process of enhancing the communication between institutions of different nature. Associations of universities like the PULs in Lille and Lyon may be of help, but they need to acquire a real leadership status, dominating that of the single universities. Also, effective student associations can enhance the visibility of the student community, becoming an interlocutor of the institutions. *Student unions* in Helsinki, Eindhoven and Venice have acquired this importance. *Alumni networks* are also important organisations as they provide universities and host cities with a link with to the human resources that have been trained there, in spite of their pursuit of a work career elsewhere. These associations prove to be strong vectors for cities to establish network relations with the national or international business environment. City administrations can play a role in supporting these organisations through funding, locations and promotional backup.

Guidelines for local governments

- Bring all players together (HEIs and private sector) and stimulate dialogue of a vision of 'student-friendly', learning city.
- Support bottom-up initiatives of knowledge network creations.
- Establish intermediary organisations in situations in which formal bureaucracies cannot activate a dialogue; give them formal powers of dealing with joint university-city projects.
- Find ways to grant students a 'right of citizenship' (e.g. through democratic involvement).

- Support student associations and involve them as stakeholders in city policy.

For whom this is more compelling

- Lyon: conflicting policy making environment, insufficient cohesion between universities and *Grandes Ecoles*.
- Rotterdam: poor role of university in urban policy, poor networks with private sector.
- Birmingham: universities are 'political issues' in the community, no direct involvement in regeneration plans.
- Venice: conflict among different administrations with competence on higher education..

Good examples to look at

- BioRegio (partnership for development): http://www.bio-m.de/english/biotech/fs_biotech_index.html.
- Lille 'Committee Grand Lille': http://www.millenaire3.com/.
- Helsinki student union: http://www.helsinki.fi/hyy/en/.
- Eindhoven 'List 11': http://listen.to/lijst11 (only in Dutch).

Annex

List of Discussion Partners

Birmingham

J. Adams, Birmingham European Cultural Capital 2008
A. Coulson, Cabinet member, Birmingham City Council
R. Groves, Centre for Urban and Regional Studies, Birmingham University
U. Gudka, student representative, Aston University
T. Jordan, Excellence Challenge Coordinator, Birmingham City Council
N.B.R. Reeves, Pro-Vice Chancellor, Aston Univesrity
J. Robinson, Deputy Head European and International Affairs, City Council of Birmingham
S. Taylor, City Economic Advisor, Birmingham City Council
P. Walkling, Pro-Vice Chancellor, University of Central England

Eindhoven

F. Clair, student, SSRE
C. van Dijk, Director Strategic Planning, City of Eindhoven
T. van Hattum, Chamber of Commerce, Oost-Brabant
F.G.A. Huijink, Manager, Vestide
G.M. Mittendorf, Alderman for Young and Families, City of Eindhoven
G. Sikkema, ESC Eindhovens Studenten Corps
F.C. van Rijswijk, Fontys School.
M.M.M. Salimans, Fontys School
L. Scheepers, Fontys School
T. Schurgers, Syntens, Innovatie Centrum Oost-Brabant
A.J. Tindemans, Education Department, Municipality of Eindhoven
B. Vermeer, Cebra
R. van de Poel, ESC Eindhovens Studenten Corps
H. van der Wielen, beleidsmedewerker, Education Department, Gemeente Eindhoven
Ms Wishoupt, student, Demos
H.G.J. de Wilt, President of the Executive Board, Eindhoven University of Technology

Helsinki

I.-C. Björklund, Deputy Mayor, City of Helsinki
T. Cantell, Senior Researcher, City of Helsinki Urban Facts
P. Hellén, International Coordinator, Helsinki Polytechnic-Stadia

E. Holstila, Managing Director, Culminatum Oy
S. Lahti, Director of Planning and Development, UH
T. Laitio, Chairman, Helsinki University Students' Union Board
T. Luopajärvi, Rector, Helsinki Polytechnic-Stadia
M. Maartola, Planner, Business Development Unit, City of Helsinki
A. Manninen, Director, City of Helsinki Urban Facts
Y. Sotamaa, Rector, University of Art and Design
J. Sumuvuori, City Councillor

Lille

T. Baert, Agence d'urbanisme de Lille
C. Barre, Director, Lille Chamber of Commerce
M. Breuillard, CRAPS, University Lille 2
S. Caplier, Créative Allée
F. Clément, Director, Lille Chamber of Commerce
C. Delahoutre, Lille Chamber of Commerce
B. Delebecque, Lille Métropole Communauté Urbaine
A. Delpont, APIM Lille Métropole
C. Dubrulle, student representative, EGC Lille Métropole
M. Froissart, Director International Relations, Ecole Supérieure de Commerce de Lille
S. Gandin, student representative, Ecole Centrale de Lille
J.-C. Gentina, Director, Ecole Centrale de Lille
P. Gilles, student representative, EGC Lille Métropole
D. Paris, ENSAR, University Lille 1
D. Rosselle, University Lille 3
P. Souris, Director, Digiport – Declic.Net
N. Starkman, Director, Agence d'urbanisme de Lille
J.F. Stevens, ENVAR, Director, University Lille 1
J. Toulemonde, Director, Lille Chamber of Commerce
C. Vanhaecke, Chief of the department 'Vie de l'Etudiant', CROUS Lille
M. Vittu, Director, HEI, École Catholique de Lille

Lyon

D. Allaix, Student Services, University Lyon II
P. Bazin, Director of the Municipal Library
B. Bensoussan, Institute of Human Sciences (ISH)
G. Bertholon, Director, PUL
D. Beuneche, Framatome
B. Bigot, Director, ENS
E. Boissier, Mission Maeva, Lyon II
J.M. Daclin, Department for International Relations and Tourism, Municipality of Lyon
D. Debouzie, President, PUL

B. Falk, CCI, Education Department
L. Gandolfi, INSA LYON
E. Gros, International Information Centre (CII)
A. Guilhot, Young Ambassadors of Lyon (JAL)
G. Guyot, President of Lyon III
C. Hooge Chodkiewicz, Grand Lyon – Millenaire 3, International Relations
E. Janetzko, CROUS
T. Kalidou, student
P. Lareal, Alderman for Education and Research, Municipality of Lyon
M. Mahamood, student
S. Marcus, President, Student Association RFEI
J.J. Mathias, Real Estate Association CECIM
P. Michel, LyonLine
J.L. Molin, Millenaire, Adjoint, Grand Lyon
B. Mouton, Department of Education and Research, Municipality of Lyon
H. Ohanessian, Vice-Rector for International Affairs, INSA
C. Ortillez, International Relations, Lyon III
H. Plantier, International Relations, INSA
B. Redor, director, Tourist Office of Lyon
B. Regaldie, Municipality of Lyon
A. Storck, Director, INSA
P.Y. Tesse, Director, CCI
V. Vignal, Coordinator of JAL
M. Wilicki, student

Munich

Herr Bayer, Students' Association, Ludwig Maximilians University
H. Domdey, Director, BioM AG, coordinating agency of the BioTech-Region Munich
W. Kämmer, Senior Consultant, Siemens AG
S. Linack, Students' Association, Technical University München
D. Maßberg, Director, Studentenwerk München
M. Nebel, Studentenwerk München
M. Schick, President, Fachhochschule München
W. Schubö, Pro-Rektor, Ludwig Maximilians University
K. Schußmann, Director, Department of Labour and Economic Development
E. Schweigard, Department of Labour and Economic Development

Rotterdam

H.P. Beekman, Department of Social Econimic Affairs, City of Rotterdam
J. B. Benraad, Director, General Stadswonen
B.M. den Boogert, Student Dean, EUR
H. Brink, Chamber of Commerce of Rotterdam

J. Elbers, Director, Ichthus Hogeschool
S. Fleischeuer, Director, Ontwikkelingsberdijf Rotterdam
H. Kombrink, former Alderman for Spatial Development and Property and Land, Media Policy, and Culture Affairs, City of Rotterdam
E. Kuijper, former Alderwoman for Education, City of Rotterdam
P.J. Van der Maas, Dean of the Medical Faculty, EUR
I. Opstelten, Mayor of Rotterdam
A.D. van der Pijl, Student Affairs, EUR
H. Stam, Director, Ontwikkelingsberdijf Rotterdam
C. de Swart, Chairman of the Board, Stad Rotterdam Verzekeringsgroep (ASR)
S. van der Tak, Alderman for Social Integration, City of Rotterdam
J. van't Verlaat, Ontwikkelingsberdijf Rotterdam

Utrecht

H.P.M. Adriaansens, Dean, University College of Utrecht
S. Czerski, senior advisor, housing specialist, University of Utrecht
J. Goedhart, Cultural Center Parnassos
B. Groenemeijer, Chairman of the Central Board, Utrecht School of the Arts
A. Heijnen, Director of U-BLAD, university newspaper
J. Kessels, Director of Corporate Communications, University Strategic Programme, University of Utrecht
J.W.M. Kulik, Department for Urban Development, Municipality of Utrecht
W.A.W. Kuus, Director of Central Service Department, Hogeschool van Utrecht
J. Leppers, Director Social Affairs, Koninklijke Douwe Egberts
A. Loeve, Regional Advisor, Chamber of Commerce, Utrecht
M. Meijer, Department for Urban Development, Municipality of Utrecht
Ms. Strieder, USF Student Union
J. F. Sunderman, International Relations Manager, Hogeschool van Utrecht
J.L. Torenbeek, Vice-President for International Relations, University of Utrecht
R.M. Verdaasdonk, Head of Clinical Physics, University of Utrecht Medical Centre
J. de Viet, General Manager, Department for Urban Development, Municipality of Utrecht
P. Wolbers, Chamber of Commerce

Venice

F. Antonich, Young Entrepreneurs, Association of Industry of the Province of Venice (Unindustria)
L. Busatto, President of the Province of Venice and President of Venice International University
G. Capalbo, Chief Commissioner, Regional Agency for the Support to University Students
R. D'Agostino, Alderman for Strategic Planning, City of Venice
N. De Bei, President, Senate of Students IUAV
M. Folin, Rector of University Institute for Architecture of Venice (IUAV)

M. Grimani, President, Responsible Relations with Categories, Chamber of Commerce, Industry and Crafts, Venice
R. Magliocco, President, Real Estate Agents, Confcommercio-ASCOM Venice
P. Mainardi, Economic Development, Young Entrepreneurs, Association of Industry of the province of Venice (Unindustria)
M. Miotton, Responsible Relations with Categories, Chamber of Commerce, Industry and Crafts, Venice
P. Politeo, responsible for formation, Young Entrepreneurs, Association of Industry of the Province of Venice (Unindustria)
F. Reberschegg, President, Neighbourhood Council No. 2, Venice
M. Rispoli, Rector University Ca' Foscari of Venice
L. de Salvo, Communication Office, Province of Venice
S. Scavone, Real Estate Agents, Confcommercio-ASCOM Venice
G. Zacchello, President, Young Entrepreneurs, Association of Industry of the Province of Venice (Unindustria)

References

General Bibliography

Berg, L. van den, Braun, E. and van der Meer, J. (1997), *Metropolitan Organising Capacity: Experiences with Organising Major Projects in European Cities*, Ashgate, Aldershot.

Berg, L. van den, van der Borg, J. and van der Meer, J. (1995), *Urban Tourism: Performance and Strategies in Eight European Cities*, Avebury, Aldershot.

Berg, L. van den and Braun, E. (1999), 'Urban Competitiveness, Marketing and the Need for Organising Capacity', *Urban Studies*, Vol. 36, No. 5–6, pp. 987–99.

Berg, L. van den, Braun, E. and Otgaar, A.H.J. (2003), *City and Enterprise. Corporate Community Involvement in European and US Cities*. Ashgate, Aldershot.

Berg, L. van den, Braun, E. and van Winden, W. (2001), *Growth Clusters in European Metropolitan Cities*, Ashgate, Aldershot.

Bianchini, F. (1993), 'Culture, Conflicts and Cities: Issues and Prospects for the 1990s', in F. Bianchini and M. Parkinson (eds), *Cultural Policy and Urban Regeneration: the West European Experience*, Manchester University Press, Manchester.

Bonifacio, M., Vassallo, M. and Zanus Fortes, L. (1995), *Indagine sulla Condizione Abitativa degli Studenti Fuori Sede, Anno Accademico 1993/1994*, working paper, Ufficio Sudi e Programmazione, University Ca' Foscari, Venice.

Borg, J. van der, Rumi, G. and Russo, A.P. (1999), *The Visitor-Friendly Metropolis: A Comparative Analysis of Visitor-Friendliness in Lyon, Nantes, Rotterdam, Lisbon, Turin and Birmingham*, EURICUR report, mimeo Rotterdam.

Brunello, G. (2001), 'On the Complementarity between Education and Training in Europe', *IZA Discussion Paper*, No. 309.

Chatterton P. (2000), 'The Cultural Role of Universities in the Community: Revisiting the Cultural-Community Debate', *Environment and Planning A*, Vol. 32, pp. 165–81.

Crosby, B.L. (1996), 'Policy Implementation: The Organisational Challenge', *World Development*, Vol. 24, No. 9, pp. 1403–15.

Dubet, F. and Sembel, N. (1994), 'Les étudiants, le campus et la ville', *Les annales de la recherche urbaine*, No. 62–63, pp. 224–34.

Dziembowska-Kowalska, J. and Funck, R.H. (2000), 'Cultural Activities as a Location Factor in European Competition Between Regions: Concepts and Some Evidence', *Annals of Regional Science*, Vol. 34, pp. 1–12.

Elliott, J., Francis, H., Humphreys, R. and Istance, D. (1996), *Communities and Their Universities. The Challenge of Lifelong Learning*, Lawrence and Wishart, London.

Felsenstein, D. (1995), 'The University in the Metropolitan Arena: Impacts and Public Policy Implications', *Urban Studies*, Vol. 33, No. 9, pp. 1565–80.

Florax, R. (1992), *The University: A Regional Booster?*, Avebury, Aldershot.

Hall, P. (1997), 'The University and the City', *GeoJournal*, Vol. 41, No. 4, pp. 301–9.

Indovina, F. (1998), 'Sinergia tra città e università', *Archivio di Studi Urbani e Regionali*, Vol. XXVIII–XXIX, Nos 60–61.

Jansen-Verbeke, M. (1988), *Leisure, Recreation and Tourism in Inner Cities*, KU Nijmegen, Nederlandse Geografische Studies, 58, Nijmegen.

Karlsson, C. and Wei-Bin, Z. (2001), 'The Role of Universities in Regional Development: Endogenous Human Capital and Growth in a Two-region Model', *Annals of Regional Science*, Vol. 35, pp. 179–97.

Kooiman, J. (1993), *Modern Governance*, SAGE, London.

Kotler, P., Haider, D.H. and Rein, I. (1993), *Marketing Places*, The Free Press, New York.

Martin, B.S. and Uysal, M. (1990), 'An Examination of the Relationship Between Carrying Capacity and the Tourism Life-cycle: Management and Policy Implications', *Journal of Environmental Management*, Vol. 31, pp. 327–33.

Martinotti, G. (1997), *Perceiving, Conceiving, Achieving the Sustainable City*. European Foundation for the Improvement of Living and Working Conditions, Dublin.

Noyelle, T. and Stanback, T. (1983), *The Economic Transformation of Cities*, Rowman and Allanheld, Totowa, NJ.

Nyström, L. (ed.), in cooperation with Colin Fudge (1999), *City and Culture: Cultural Processes and Urban Sustainability*, The Swedish Urban Environment Council, Stockholm.

Pallares, J. and Feixa, C. (2000), 'Studenti e città. Spazi e tempi dei giovani universitari a Lleida', *Archivio di studi urbani e regionali*, No. 69, pp. 69–93.

Parsons, K.C. and Davis, G.K. (1971), 'The Urban University and its Urban Environment', *Minerva*, Vol. 9, No. 3, pp. 361–85.

Phelps, R.P. (1998), 'The Effect of University Host Community Size on State Growth', *Economics of Education Review*, Vol. 17, No. 2, pp. 149–58.

Quadrado, L., Loman, S. and Folmer, H. (2001), 'Multi-dimensional Analysis of Regional Inequality: The Case of Higher Educational Facilities in Spain', *Papers in Regional Science*, Vol. 80, pp. 189–209.

Russo, A.P. (2001), 'The "Vicious Circle" of Tourism Development in Heritage Destinations', *Annals of Tourism Research*, Vol. 29, No. 1.

Savino, M. (1998), 'Università, città, studenti: aspetti complessi di interdipendenze non sempre note', *Archivio di Studi Urbani e Regionali*, Vol. XXVIII–XXIX, Nos 60–61.

Vassal, S. (1987), 'L'Europe des Universités', *Cahiers de sociologie economique et culturelle. Ethnopsychologie*, Vol. 8, pp. 149–57.

VVAA (2001), 'Universite: droite de cité', *Urbanisme*, No. 317, 3/4.

Williams, C.C. (1997), *Consumer Services and Economic Development*, Routledge, London and New York.

Birmingham Case Study

Birmingham Economic Information Centre (BEIC) (various years), *Economic Reviews*, Birmingham.

Groves, R., Revell, K. and Leather, P. (2002), *The Changing Nature of the Housing Market in Selly Oak. Final Report*, Centre for Urban and Regional Studies, University of Birmingham.

Loftman, P. and Nevin, B. (1994), 'Prestige Project Developments: Economic Renaissance or Economic Myth? A Case Study on Birmingham', *Local Economy*, Vol. 8, pp. 307–14.

Newman, P. and Thornley, A. (1996), *Urban Planning in Europe*, Routledge, New York and London.

Eindhoven Case Study

Adang, A.J.V.M. and van Oorschot, J.M.P. (1996), *Regio in bedrijf, Hoofdlijnen van industriële ontwikkeling en zakelijke dienstverlening in Zuidoost-Brabant.*

Berg, L. van den, Braun, E. and van Winden, W. (2001), *Growth Clusters in European Metropolitan Cities*, Ashgate, Aldershot.

Berg, L. van den and van Winden, W. (2002), *ICT as Potential Catalyst for Sustainable Urban Development: Experiences in the Cities of Eindhoven, Helsinki, Manchester, Marseilles and The Hague*, EURICUR Series, Ashgate, Aldershot.

City of Eindhoven (2000), *Stadsvisie 2000*, Eindhoven.

Commissie Regionale Kansen (2002), *Programma Horizon Zuid-Oost Nederland. Kansen voor de technologieregio 'Eindhoven'*, on-line document.

ETIN Adviseurs (2002a), *Statistisch Zakboek Noord-Brabant*, ETIN, Tilburg.

ETIN Adviseurs (2002b), *Werkgelegenheid in de Noord-Brabant Gementen 1997–2001*, ETIN, Tilburg.

Kusters, A. and Minne, B. (1992), *Technologie, marktstructuur en internationalisatie: de ontwikkeling van de industrie*, Den Haag, Centraal Planbureau.

NEI (1996), *Economische betekenis van industriële kennisclusters in de regio Eindhoven/Venlo*, NEI Rotterdam.

Stimulus (1998), *Nieuwsbrief Januari 1998.*

Helsinki Case Study

Berg, L. van den and van Winden, W. (2000), 'ICT as Potential Catalyst for Sustainable Urban Development; Experiences in the Cities of Eindhoven, Helsinki, Manchester, Marseilles and The Hague', EURICUR Report.

Berg, L. van den, Braun, E. and van Winden, W. (2001), *Growth Clusters in European Metropolitan Cities*, Ashgate, Aldershot.

City of Helsinki Urban Facts (2001), *Education in Helsinki*, Helsinki.
City of Helsinki Urban Facts (2002a), *Helsinki Region Statistical Comparisons*, Helsinki.
City of Helsinki Urban Facts (2002b), *Helsinki – An Educating, Training and Learning City*, Helsinki.
City of Helsinki Urban Facts, City of Helsinki Education Department (2002), *Education in Helsinki.*
Helsinki Club (1997), *Success Strategies and Partnership Projects for the Helsinki Region*, Helsinki.
Helsinki Polytechnic Stadia (2000), *Annual Report.*
Helsinki Polytechnic Stadia, *Study Guide for International Students*, ECTS Guide 2001–2002.

Lille Case Study

Agence de développement et d'Urbanisme Lille Métropole (1998), *Lille Métropole en 2015? Le schema directeur de développement et d'urbanisme de Lille Métropole*, Lille.
Agence de développement et d'Urbanisme Lille Métropole (2001), *Lille Métropole dans tous ses etats 1990–2000*, Lille.
CUDL/CCILM (2002), Chiffres Clé de Lille Métropole, Lille.
Face (2001), *Region Transmanche Top 2001. Les enterprises leaders*, Chambre de Commerce de Lille Métropole, October.
Newman, P. and Thornley, A. (1996), *Urban Planning in Europe*, Routledge, London and New York.
VVAA (2001), 'Université: droit de cite', *Urbanisme*, Vol. 317, Nos 3/4.

Lyon Case Study

Aderly, Chiffres clé, on-line at www.aderly.com.
Berg, L. van den, Braun, E. and van Winden, W. (2001), *Growth Clusters in European Metropolitan Cities*, Ashgate, Aldershot.
CCI Rhone-Alpes, Chiffres clé, on line at http://www.rhone-alpes.cci.fr/economie/chiffres/index.php.
Grand Lyon (2002), 'Schéma de development Economiuqe: etat d'advancement – 18 Octobre 2002', on line-document.
Michel, P. (2002), *The Student City: Lyon*, PUL, Lyon.
Opale (2002), 'Emploi and chomage repères dans l'aire urbaine de Lyon', on-line document.

Munich Case Study

Berg, L. van den, Braun, E. and van Winden, W. (2001), *Growth Clusters in European Metropolitan Cities*, Ashgate, Aldershot.

Biotech Region München (2000), *Annual Report 2000*, Munich.

De Welt (2001), 'TUM schafft zwölf neue Studiengänge', 28 June 2001.

Domdey, H. (ed.) (2002), *Company Profiles – Research, Finance and Infrastructure, Part 2, Biotech-Region*, Munich.

Domdey H. (ed.) (2002), *Report: Company profiles – Biotech and Pharma Companies, Part 1, Biotech-Region.* Munich.

IHK-Magazin für Munchen und Oberbayern (2001), 'Direkt am Campus', Serie Wirtschaft, 5/2001.

Landeshauptstadt München, referat für Arbeit und Wirtschaft (2000), *München – führendes europäisches Technologie-Zentrum.*

Landeshauptstadt München, referat für Arbeit und Wirtschaft (2000), *München – die Biotechnologie-Region Deutschlands.*

Landeshauptstadt München, referat für Arbeit und Wirtschaft (2002), *Munich, the Economic Location. Facts and Figures*, Munich.

LMU München (2001), *Annual Report 1998–2000*, Munich.

Ministry for Education and Research (2000) *Die wirtschaftliche und soziale lage der Studierenden in der Bundesrepublik Deutschland, 2000.*

Münchner Unimagazine (2001), 'Zimmer verzweifelt gesucht, vor allem auslandische studenten gehen leer aus', No. 4/2001.

Siemens (2000), *Initiatives in the Community. Corporate Citizenship Report*, Munich.

Siemens (2000), *Lösungen für Städte von Morgen*, Munich.

Siemens (2001), *Annual Report 2000*, Munich

Studentenwerk München (1999), *75 Jahre studentenwerk München. 'Wo geht's hier zum Studenthaus?'*, Munich.

Studentenwerk München (2001), *Annual Report 2000*, Munich.

Sud-deutsche Zeitung (2001) 'Medizintechnic-Neubau Ende 2001 bezugsfertig', 5 April 2001.

Sud-deutsche Zeitung (2001), 'Lieber einen Europäer als einen Portugiesen', 30 October 2001.

Sud-deutsche Zeitung (2001) 'Die Universitäten zieht's ins Umland', 2 November 2001.

VVAA (1999), 'Das habilitationswesen an den Universitäten In Bayern', praxis und perspectiven, IHF (Faculty of Applied Science).

VVAA (2000), *Der student als Nomade*, Perspective TUM Workshop.

VVAA (2000), *Die Universiät – das kulturelle Herz der Metropolen*, Proceedings of workshop 'Students as Nomads' Munich.

Rotterdam Case Study

Berg, L. van den, Braun, E. and Otgaar, A.H.J. (2003), *City and Enterprise. Corporate Community Involvement in European and US Cities*, Ashgate, Aldershot.

Berg, L. van den, Braun, E. and van Winden, W. (2001), *Growth Clusters in European Metropolitan Cities*, Ashgate, Aldershot.

Berg, L. van den, van Klink, H.A. and Langen, P. de (1997), *Logistieke denkkracht als springplank naar het achterland*, EURICUR, Rotterdam (R1997–9).

Berg, L. van den, van den Borg, J. and Russo, A.P. (2003), 'The Infrastructure for Urban Tourism: a European Model? A Comparative Analysis of Mega-Projects in Four Eurocities', in D. Judd (ed.), *The Infrastructure of Play*, M.E. Sharpe, Chicago.

COS (2000), *Statistiche jaarboek Rotterdam en Regio 2000*, on-line document.

De Langen, P.W. (1998), *The EUR and the Region, An Overview of the Relations between the EUR and its Regional Partners*, EURICUR Report, Rotterdam.

Eramsus Magazine (2002), Vol. 6, year 6, pp. 20–22.

EUA (European University Association) (2001), 'Internationalisaton Quality Review of Erasmus University Rotterdam', Report by Peer Review team, mimeo.

Hitters, E. and Richards, G. (2002), 'The Creation and Management of Cultural Clusters', *Creativity and Innovation Management*, Vol. 11.

Rotterdams Dagblad (2002), 'En nu verder: Speciale uitgave van Rotterdams dagblad', November.

VVAA (2001), *Studenten monitor 2000*, OcenW, RISBO, SCO Kohnstam Instituut, SEO Report.

Utrecht Case Study

Berg, L. van den and Pol, P.M.J. (1999), *The High-Speed-Train Station and Urban Revitalisation; A Comparative Analysis of Station Areas in the Cities of Amsterdam, Antwerp, Berlin, Cologne, Dortmund, Lille, Munich, Rotterdam and Utrecht*, EURICUR Report.

Berg, L. van den, van der Meer, J. and Pol, P.M.J. (2001), *Social Challenges and Organising Capacity in Cities*, EURICUR Report.

Broere, A., Czerski, S., Jochems, T., Mul, G. and van der Werf M. (2002), *Studeren op Kamers in Utrecht – Knelpunten en oplossingen voor de kamernood in studentensteden.*

Gemeente Utrecht (2001), *Vastgoed Monitor: Utrecht 2001*, Utrecht.

Gemeente Utrecht (2002), *Utrecht Werkt: Trendrapportage economie 2002*, Utrecht.

Venice Case Study

Bonifacio, M., Vassallo, M. and Zanus Fortes, L. (1995), *Indagine sulla Condizione Abitativa degli Studenti Fuori Sede, Anno Accademico 1993/1994*, Working Paper, Ufficio Sudi e Programmazione, University Ca' Foscari, Venice.

Costa, P. (1993), *Venezia. Economia ed analisi urbana*, Etas Libri, Milano.

Costa, P. (1996), 'Università. Ca' Foscari "a" e "per" Venezia', in COSES (ed.), *Alfabeto Veneziano*, Il Mulino, Bologna.

Dente, B. et al. (2000), 'Governing the Sustainable Development of Venice: Elements of the Institutional Planning Procedure', in I. Musu (ed.), *Sustainable Venice: Suggestions for the Future*, Kluwer, Dordrecht.

Klaassen, L.H., van den Berg, L. and van der Meer, J. (eds) (1989), *The City: Engine Behind Economic Recovery*, EURICUR Series, Avebury, Aldershot.

La Nuova Venezia (2001), 'Affitto studenti, Venezia non risponde. Reperiti 68 alloggi: 63 sono a Mestre', www.nuovavenezia.it/, 15 June.

La Nuova Venezia (2001), 'Notti Roventi a S. Margherita', www.nuovavenezia.it/, 29 July.

Mariotto, A. (1998), 'L'Università a Venezia: Scelte Localizzative e Funzione Formativa nelle Politiche Territoriali', *Archivio di Studi Urbani e Regionali*, Vol. XXVIII–XXIX, Nos 60–61.

Paladini, G. (1996), *Profilo Storico dell'Ateneo*, ed. Università Ca' Foscari. www.unive.it/!wda/ateneo/storia.

Rullani, E. and Micelli, S. (2000), 'Immaterial Production in Venice: Towards a post-Fordist Economy', in I. Musu (ed.), *Sustainable Venice: Suggestions for the Future*, Kluwer, Dordrecht.

Russo, A.P. (2001a), 'The "Vicious Circle" of Tourism Development in Heritage Destinations', *Annals of Tourism Research*, Vol. 29, No. 1.

Russo, A.P. (2001b), 'Cultural Clusters and Tourism Development', paper presented at the international seminar *Culture: A Driving Force for Urban Tourism: Application of Experiences to Countries in Transition*, 18–19 May 2001, Dubrovnik, Croatia.